INTRODUCTION TO GRAVITATION

INTRODUCTION TO GRAVITATION

Venzo de Sabbata
Dipartimento di Fisica dell'Università, Bologna
and Dipartimento di Fisica dell'Università Ferrara

Maurizio Gasperini
Istituto di Fisica Teorica dell'Università, Torino
and Istituto Nazionale di Fisica Nucleare, Torino

World Scientific

Published by

World Scientific Publishing Co Pte Ltd.
P. O. Box 128, Farrer Road, Singapore 9128

INTRODUCTION TO GRAVITATION

ISBN 9971-50-049-3
9971-50-050-7 (pbk)

Printed in Singapore by Singapore National Printers (Pte) Ltd.

PREFACE

This book is primarily intended for undergraduate students in physics who wish to begin the study of the gravitational interaction. A fairly good knowkedge of special relativity and of the electromagnetic theory are the only basic requirements needed for the reading of this book. In fact, for the reader's convenience, we have tried to make the book self-contained as much as possible, recalling and introducing briefly each time all the notions necessary for the develoment of the theory, such as the tensor calculus, the variational formalism, elementary notions of astronomy, the basic principles of the gauge theories, the calculus of the exterior forms, and so on.

Born of an elaboration of the lectures that one of the authors (V.D.S.) gives and has given for over twenty years to the students in physics, astronomy and mathematics of the universities of Bologna and Ferrara, this book is conceived as a test-book, and all the computations are displayed in great detail, in order that the student may get some practice with the formalism used in the gravitational theories, in particular with the manipulation of the indices.

Starting from the principle of equivalence and the connection between geometry and gravitation, the first nine chapters of this book are devoted to the theory of general relativity and its classical macroscopic applications, with particular attention to cosmology and to the problem of the gravitational waves. In the subsequent chapters the geometrical notion of torsion is introduced, the various aspects and the main consequences of a theory of gravity formulated using a manifold with a nonsymmetric affine connection are discussed, and in particular it is stressed the fundamental role played by torsion in the framework of a gauge theory of gravity and of the supersymmetric gravitational theories.

Besides introducing torsion, this book differs from many other text-books on general relativity as it contains a fairly detailed discussion of the spinor equations in a curved space-time, and of the possible approach to a gauge theory of gravity. Moreover, we have tried to emphasize the use of the anholonomic language, that is the "tangent space formalism" which nowadays is more and more used, mainly in the context of the unified and multidimensional theories.

Of course we don't conceal the fact that many interesting gravitational topics have been skipped, but, as noted previously, the content of this book has been choosen following a certain didactic trend. Moreover it must be noted that we have decided to omit a detailed bibliography, providing, at the end of each chapter, only a limited number of essential references.

Finally, we will be very grateful to anyone who will provide us with criticism and suggestions aiming at the correction of the possible errors and at the improvement of the present form of this book.

The authors
Venzo De Sabbata
Maurizio Gasperini

CONTENTS

Preface v

CHAPTER I: GEOMETRY AND GRAVITATION 1

CHAPTER II: THE FORMALISM OF GENERAL RELATIVITY

1. Introduction 13
2. Covariant and controvariant tensors 15
3. Covariant derivation 20
4. The geodesic equation 30

CHAPTER III: GRAVITATIONAL FIELD EQUATIONS

1. The curvature tensor 34
2. Einstein's field equations 42
3. Field equations from a variational principle 44
4. Maxwell's equations in a Riemann space-time 51
5. The initial values problem in general relativity 55

Appendix A — The totally antisymmetric tensor in curvilinear coordinates 58

Appendix B — Four-dimensional variational formalism: the canonical energy-momentum tensor and the spin density tensor 61

CHAPTER IV: THE THREE CLASSICAL TESTS OF EINSTEIN'S THEORY

1. The Schwarzschild solution 68
2. Planetary motions 73
3. Deflection of light rays 79
4. The shift of spectral lines 85

CHAPTER V: ELEMENTS OF COSMOLOGY

1. The cosmological problem 88
2. Newtonian cosmology 92
3. Relativistic cosmology: the cosmic time 98

CHAPTER VI: RELATIVISTIC COSMOLOGICAL MODELS

1. The Einstein static model 104
2. The De Sitter universe 113

CHAPTER VII: NON-STATIC MODELS OF UNIVERSE

1. The Robertson-Walker metric 120
2. The magnitude-redshift relation as a test of the non-static models 135
3. Other observational tests 139
4. Concluding remarks 149

Appendix A — Visual, photographic and bolometric magnitude 156

CHAPTER VIII: GRAVITATIONAL WAVES

1. The weak field approximation 159
2. Plane gravitational waves 163
3. Emission of gravitational radiation 167
4. Possible sources of radiation 170
5. Detection of gravitational waves 178
6. Interaction between gravitational waves and electromagnetic fields 183

Appendix A — Infinitesimal coordinate transformations 190

Appendix B — The energy-momentum pseudotensor 194

Appendix C — Quadrupole radiation 200

Appendix D — Geodesic deviation 204

Appendix E — On the polarization of a plane wave 209

CHAPTER IX: DENSE AND COLLAPSED MATTER

1. White dwarfs 215
2. Neutron stars 219
3. Black-holes 222

CHAPTER X: THE EINSTEIN-CARTAN THEORY

1. Physical motivations supporting a generalization of the Einstein theory 228
2. The geometry of a Riemann-Cartan manifold U_4 230
3. Field equations 239
4. Bianchi identities and equations of motion 243
5. The Dirac equation in the U_4 theory 249
6. Propagating torsion 264

CHAPTER XI: THE STRONG GRAVITY THEORY

1. The tensor meson dominance hypothesis 271
2. Field equations 273
3. Linear approximation 276
4. Strong gravity with torsion 278

Appendix A — Strong gravity and the large numbers hypothesis 286

CHAPTER XII: GAUGE THEORY OF GRAVITY

1. Gauge invariance and compensating fields 293
2. Electromagnetism as a gauge theory 299
3. Local Lorentz invariance and the gravitational interaction 302

CHAPTER XIII: SUPERGRAVITY

1. Introduction 315
2. Global supersymmetry 317
3. Supergravity and local supersymmetry 321

CHAPTER XIV: GRAVITATIONAL THEORY IN THE LANGUAGE OF THE EXTERIOR FORMS

1. Forms, exterior product and exterior derivative 327
2. Gauge theory 330
3. Torsion, curvature, and the algebra of the Poincaré group 331

CHAPTER I

GEOMETRY AND GRAVITATION

General relativity, as formulated by Einstein, is the first purely geometric theory of gravity, in which matter dynamics is prescribed by the geometry of space and, conversely, geometry is determined by matter, so that the notion of absolute space of classical mechanics is definitively dropped.

In the special theory of relativity the notion of velocity is a relative concept, because all inertial frames are physically equivalent; in general relativity also the accelerations loose an absolute meaning, inertial frames play no longer a preferred role, as all the observers are physically equivalent, including those on accelerated systems of reference.

In this sense general relativity is a generalization of special relativity. From a mathematical point of view, this generalized equivalence of coordinate systems implies, as we shall see, the replacement of the "flat" Minkowski space-time with a more general Riemann manifold, which has, in general, a non-vanishing curvature directly related to the presence of a gravitational field: in this sense we have a geometrical description, or a "geometrization", of the gravitational interaction.

The starting point toward the formulation of general relativity is then the abolition of the absolute space and of all preferred frames.

As is well known, the Newton law of motion $\vec{F} = m\,\vec{a}$ is valid in this form only in an inertial frame, while in an accelerated system one must introduce the so-called inertial or "fictitious" forces. Newton believed to the existence of an absolute space and then to the possibility of defining an absolute acceleration.

It is famous his pail experiment, which seems to give an experimental support to the hypothesis of absolute space. Newton put in a state of rotation a pail half-full of water, suspended by a rope: in the beginning the pail rotate alone, water is not dragged along the pail and the water surface is flat. Gradually, also the water starts rotating, and its surface assumes an hollow shape, due to centrifugal forces. If the pail is stopped, the water continues to rotate and retains its parabolic surface. In the first and last case we have the relative rotation of water and pail, but in one case the water surface is plain, in the other parabolic (see Fig. 1.1). Why this difference? Newton claimed that this difference is a demonstration of the existence of absolute space. In fact centrifugal forces cannot be explained as due to a relative rotation, because such a relative motion between the pail and the water exists at the beginning as well as the end of the experiment, while the centrifugal forces appear only at the end. According to Newton, we have then an absolute rotation: when the pail rotates and the water is at rest there are no centrifugal forces, which appear however when the water rotates, and the pail is at rest. If the rotation would be a relative concept, no differences should appear; this result proves then the existence of the absolute space.

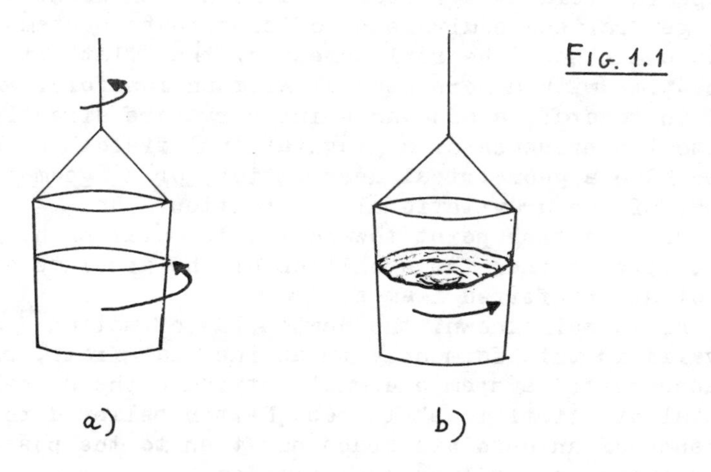

Fig. 1.1 . In both cases there is the same relative motion between the water and the pail. However in the first case a) the water surface is flat, while in the second case b) the surface is parabolic.

The interpretation of this experiment given by Newton was criticized by Berkeley[1], who observed that pail and water rotate not in vacuum, but in the presence of all the matter of the universe. Berkeley says in his "De Motu": "58.........Therefore if we suppose that everything is annihilated except one globe, it would be impossible to imagine any movement of that globe: it is in fact necessary the presence of another body respect to position of which the movement must be determined.........
59. Let us imagine two globes, and that besides them nothing else material exists, then the motion in a circle of these two globes round their common centre cannot be imagined. But suppose that the heaven of fixed stars was suddenly created and we shall be in a position to imagine the motion of the globes by their relative position to the different parts of the heaven.........".

According to Berkeley, therefore, we have not an absolute motion, but only a motion relative to fixed stars. It is true that his considerations regarding the possible existence of only one body in the universe cannot be verified experimentally, but his criticism of the absolute space of Newton is substantially correct, as from the pail experiment one can deduce only that centrifugal forces appear when water rotates relatively to fixed stars: it is just the existence of these other bodies which gives a meaning to the motion.

The ideas of Berkely were resumed and deepened by Mach, about one hundred and fifty years later. Mach says[2]: "......Obviously it does not matter if we think of the earth as turning round on its axis, or at rest while the fixed stars revolve round it. Geometrically these are exactly the same case of a relative rotation of the earth and the fixed stars respect to one another. But if we think of the earth at rest and the fixed stars revolving round it, there is no flattening of the earth, no Foucault's experiment, and so on - at least according to our usual conception of the law of inertia. Now one can solve the difficulty in two ways.

Either all motion is absolute, or our law of inertia is wrongly expressed. I prefer the second way. The law of inertia must be so conceived that exactly the same thing results from the second supposition as from the first. By this it will be evident that in its expression, regard must be paid to the masses of universe".

According to Mach, therefore, the motion of a body, and in particular the rotation of the Earth, are relative to the fixed stars, and not to an absolute space. The well known result of the Foucault's pendulum experiment is due to the existence of the fixed stars: in fact no movement would be possible if the earth would be the only body in the universe, and in particular no rotation would be possible. It is then the presence of the matter of the universe which produces those effects interpreted by Newtón as inertial forces (in particular centrifugal forces due to an absolute rotation, in the pail experiment).

This in agreement with Berkeley. Bat Mach makes a further step, suggesting that not only inertial forces are depending on the presence and the behaviour of surrounding matter, but also the inertia itself of a body is not an intrinsical property of that body, but is due to the existence of all the other matter of the universe.

These ideas as a whole were called, by Einstein, "Mach's principle", and this principle even if not yet experimentally proved, neither theoretically clarified, played an important role for the development of the general relativistic theory.

In fact, the idea that inertial forces may be interpreted as a dynamical gravitational effect (remember: following Mach the centrifugal force does not indicate a rotation relative to the absolute space, but only a rotation relative to the masses of the universe; if we consider the water with its hollow surface as being at rest, the earth and the fixed stars rotate around the water; in this conception the centrifugal force is a dynamic gravitational effect of rotating masses), leads to the concept of relativity of gravitational forces: according to this idea of Mach, the gravitational field is deprived then of its absolute character and recognized as a quantity which depends on the state of motion of the observers, or on the system of coordinate chosen. This represents one of the most significant aspects of Mach's view point, and leads formally to the principle of general covariance.

Another fundamental result of classical physics, which strongly influenced the formulation of the Einstein's relativistic theory of gravity, was the experiment of Galileo which proved that bodies with different masses fall with the same acceleration in the gravitational field of the Earth. This result may be stated as "gravitational forces are proportional to the masses of the bodies on which they are acting" or "gravitational masses (m_g) are proportional to inertial masses (m_i).

Consider in fact two bodies with masses m(1) and m(2) falling in the earth gravitational field. We have

$$m_i(1)\,a(1) = -\frac{G M m_g(1)}{r^2} \qquad (1.1)$$

$$m_i(2)\,a(2) = -\frac{G M m_g(2)}{r^2} \qquad (1.2)$$

where the indices "i" and "g" denote respectively inertial and gravitational masses. M is the earth mass and G the Newton coupling constant

$$G = 6.67 \times 10^{-8}\ cm^3 g^{-1} sec^{-2}$$

By taking the ratio of these two equations one obtains

$$\frac{m_i(1)}{m_g(1)}\,\frac{a(1)}{a(2)} = \frac{m_i(2)}{m_g(2)} \qquad (1.3)$$

and because one finds experimentally a(1) = a(2) (all bodies fall with the same acceleration) we have

$$\frac{m_i(1)}{m_g(1)} = \frac{m_i(2)}{m_g(2)} \qquad (1.4)$$

The ratio between gravitational and inertial mass is a constant, or, in other words the inertial mass is proportional to the gravitational mass.

High-precision experiments confirm this proportionality. The first experiment was performed by Newton, using two pendulums with the same length, but different composition (their periods are proportional to $(m_i / m_g)^{1/2}$). No difference in their periods was found, with a precision of one part in 10^3.

Eotvos (Budapest 1889 and 1922) used a torsion-balance apparatus: he compared the vertical defined by two weights of different composition. Cutting the wires bearing the bodies, the two masses would fall with the same acceleration. However, Eotvos performed a static experiment, trying to measure the directions of the accelerations of the two masses. In fact, these directions show how much the two masses leave the line of free fall, because of the centrifugal forces due to the earth rotation.

In his pratical apparatus, two weights hang from the ends of a bar suspended on a fine wire at its center, and any inequality in the ratios m_i / m_g for the two weights would twist the suspension wire. Obviosly in order to reveal this effect one must repeat the experiment exchanging the positions of the two masses, (just this fact of exchanging the masses is very harmful for the precision of the experiment); no twist was detected, so that m_i would differ then from m_g by at most 1 part in 10^9, in the case of wood and platinum bodies (however this precision was surely an optimistic estimation).

An even better precision has been recently obtained by Dicke[3] and Braginski[4], who considered the relative accelerations of two bodies free-falling in the solar gravitational field, avoiding to perform explicitly the exchange of the masses, because of their interchange (every twelwe hours) due to the diurnal rotation of the earth. It was found that the accelerations of aluminum and platinum test-bodies do not differ by more than 1 part in 10^{11} (Dicke) or 10^{12} (Braginski).

The proportionality between inertial and gravitational masses is the starting point for the Einstein theory, and gives an experimental ground to the so-called "principle of equivalence".

In fact, Einstein noticed that there are other forces, in nature, which are proportional to the mass of the bodies on which they act: the inertial (or fictitious) forces, arising in accelerated frames. He was led then to the conclusion that inertial and gravitational forces cannot be locally distinguished.

This local equivalence between gravity and inertial fields can be well described by the famous elevator example.

Consider an observer toghether with some test bodies closed in a box, sufficiently far away from any gravitational source, and traveling with a rectilinear, uniform motion. The observer and the bodies will float freely inside the box. But if the box is suddenly accelerated by a rocket, then the observer and the test-bodies will fall towards the wall of the box opposite to the direction of the acceleration, as if a gravitational field were applied along that direction. That wall of the box appears then to the observer like the "floor" of the box, and he cannot say if its box is being accelerated by a rocket or if it lies on the surface of a planets, subject to its gravitational attraction.

In other words, gravitational and inertial forces are locally undistinguishable, and then non-inertial frames and gravitational fields are locally equivalent.

This equivalence principle is closely related to the ideas of Mach. In fact, if gravitational and inertial forces are physically equivalent, then one may interpret inertia as a gravitational effect; and since the source of gravity is matter, then inertial forces should be produced by the matter present in the universe. In particular then centrifugal forces are generated by a rotation relative to the matter of the universe, and they are not a consequence of the existence of an absolute space.

Therefore, starting from the Galileo experiment, and using the equivalence principle, one is led back to Mach's ideas.

Inspired by these principles, Einstein formulated the theory of general relativity, in which all reference systems, non-inertial frames included, are physically equivalent for the description of natural phenomena. Mathematically this implies the introduction of a non-euclidean geometry, replacing the usual euclidean geometry of the flat space.

It is easy to provide an intuitive justification for the necessity of introducing non-euclidean geometries, if we consider a non-inertial coordinate system (which is equivalent to a system subject to the action of a gravitational field).

Consider for example a frame K' rigidly fixed to the surface of a disk, rotating relatively to an inertial frame K, and suppose to measure the circumference and the radius of the disk.

Imagine that we have a large number of measuring rods all equal to each other. Suppose to put them in series along the periphery and the diameter of the circle corresponding to the disk, in a frame at rest with respect to K'. If U is the number of the rods along the periphery and D the number along the diameter, if K' does not rotate with respect to K one must find $U/D = \pi$.

But if K' rotates, then all the rods aligned along the disk periphery will suffer, with respect to K, a Lorentz-contraction, while the rods along the diameter do not experience it,being their velocity, relative to K, perpendicular to the length of the rods. It follows that $U/D > \pi$.

We find then that in a non-inertial system the euclidean geometry is no longer valid. Using the equivalence principle, we can argue then that in the presence of a gravitational field (physically and locally equivalent to a non-inertial frame), a non-euclidean geometry must be used, or, in others words, in the presence of matter, space is no longer euclidean.

Therefore, starting from the Galileo experiment on the proportionality between inertial and gravitational masses, and assuming this experimental result as the basis of the equivalence principle, we are led to say that in the presence of a gravitational field the geometry of space is, in general, non-euclidean.

Non-euclidean geometries were first studied by Gauss, Lobachevski and Bólyai. Since it is possible to construct different geometries, one has to decide which is the geometry of physical space. Gauss was the first who tried to answer experimentally this question, measuring the angles of a triangle, whose vertices were represented by the tops of three mountains. The result gave 180° for the sum of the three angles, in agreement with euclidean geometry, within experimental errors.

However, the experimental determination of the geometry of physical space cannot be reduced simply to a problem of empirical observations: a discussion of the measuring method itself is needed, before drawing a conclusion about the kind of geometry.

Suppose in fact that the sum of the angles measured by Gauss were different from 180°. Could we conclude that the world geometry is non-euclidean? Let us discuss the method of performing angular measurements. If we use a sextant, looking at the tops of the mountains through the lenses of the optical instrument, then the triangle sides are represented by light rays.

Are we certain that light rays travel along stright lines, i. e. that the tops of the mountains cannot be connected by shorter paths? (If not, the measurement is not conclusive). We could repeat the measure, for example, using rods aligned along the path of light-rays. Suppose that we obtain the same numerical result as before. May we conclude now that geometry is non-euclidean?

The situation has not changed, since now we can dispute about the rod's behaviour. We are not sure that, during the measurement, rods do not contract or dilate under the action of some unknown force, acking in the same way on all objects.

In order to get a deeper insight into this problem, we may consider the example made by Reichenbach[5].

Let us imagine (see Fig.2.1.) a surface G consisting of a big hemisphere made of glass which merges gradually in a glass plane, and an opaque plane E located below G.

The observers of G (G-people)determine the shape of their world by geometrical measurements, and find that they are living in a plane with a hemispherical hump in the middle. Suppose now that on E a mysterious force acts on the length of all measuring rods, so that their length is alway equal to the corresponding shadows of the rods projected from the surface G. For example, if the distances A'B' and B'C' are found to be equal by G-people, the corresponding distances AB and BC of their shadows will be inequal. Not only the measuring rods, but all instruments and the body of the people themselves are affected in the same way, so that E-people cannot perceive this change.

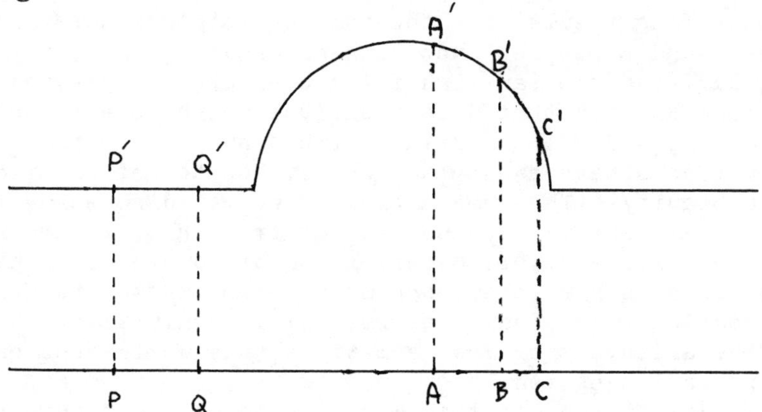

FIG. 2.1

It follows that the geometry determined by E-people in the same of G, even if their surface is different, since the deformation of the rods is not noticed. Why? Let us suppose, for a moment, that under the E-plane there is a source of heat, concentrated under the hump, which deformates the measuring rods so that their length is alway equal to the shadow of the G-rods. In this case, however, E-people would discover the mysterious force which expands or contracts their measuring rods: in fact heat affects different materials in different ways. Thus E-people would obtain different geometries wether they use copper or wooden rods. In this way they would notice the existence of a force. The fundamental point to stress is that E-people is able to discover the existence of a force because different materials are affected in different ways.

But if the force affects all materials in the same way, then E-people will not be able to reveal its existence. These forces are called by Reichenbach "universal forces", while all other forces are called "differential forces". This means that if the E-people does not know a priori the geometry of their surface, they will not be able to discover it through measuring rods because they cannot put in evidence the existence of forces if these forces are "universal".

In other words we started by saying that G is a surface with a hump and E is a plane which appears to have a hump. But what right do we have to make this assertion? The measuring results are the same on both surfaces and as the G-and E-peoples know nothing about the geometry of their surfaces, we can rightfully overturn the previous assertion and say that G is a surface which appears to have a hump and E is a surface with a hump. Or perhaps both surfaces have a hump and so on. How to get out from this ambiguity? (The same ambiguity we met discussing the Gauss experiment). We cannot go out from this dilemma if we do not give a priori a definition of congruence. That is to solve this problem, one must establish the behaviour of measuring rods when trasported along. Only after this step the discovery of the geometry of the world becomes an empirical question.

We are facing, in this way, the so-called "conventionalism" problem about the choice of the geometry of a space. Following Reichenbach, this question may be solved noticing that we should establish a criterion in order to single out the "natural" geometry of a given space. In fact we cannot obtain a statement about its real geometry if, in addition, we do not specify its field of universal forces. But the more natural assumption is to put all universal forces equal to zero (since they cannot be revealed, all rods being affected in the same way). With this convention the determination of the geometry of a space may obtain an empirical solution; from the Gauss experiment, in particular, we may deduce that the natural geometry of our world, on a little geographical scale, is euclidean.

On the other hand we have seen, using the equivalence principle, that the geometry of the physical space, in the presence of matter, should be non-euclidean, on a sufficiently large scale.

This statement is no longer ambiguous if, following the Reichenbach convention, the gravitational forces are put equal to zero (in fact they are universal forces, all bodies fall with the same acceleration). In this way the gravitational field disappears from the theory as a classical field of force, but now we have a space whose geometry is non-euclidean: gravitational forces and, more generally, all gravitational effects appear in the theory as a consequence of a deviation of the geometry of the world from the standard euclidean geometry.

In this sense, as already stated at the beginning of this chapter, general relativity is a geometrical theory of gravity.

REFERENCES

1) G.Berkeley: "De Motu" (Oxford-Clarendon Press, 1901) Vol. I, 58 and 59, p.522

2) E.Mach: "History and Root of the Principle of the Conservation of Energy" (1872), p.76

3) P.G.Roll, R.Krotkov and R.H.Dicke: Ann. Phys. 26, 242 (1967)

4) V.B.Braginski and V.I.Panov: Zh. Eksp. Teor. Fiz. 61, 873 (1971)

5) P.Reichenbach: "The phylosophy of space and time" (Doves ed., 1958) p.11

CHAPTER II

THE FORMALISM OF GENERAL RELATIVITY

1.- Introduction

One of the fundamental properties of a gravitational field, as we have seen in the previous chapter, is that different test bodies, subject to the same initial conditions, fall with the same acceleration independendly of their masses, and this leads to identify locally the physical effects of gravitation with those of a noninertial frame of reference (principle of equivalence).

In an inertial frame, the infinitesimal proper time interval ds (also called line-element) is given, using cartesian coordinates (x,y,z,ct), by

$$ds^2 = c^2 dt^2 - dx^2 - dy^2 - dz^2 \qquad (2.1)$$

and, as is well known, it is invariant under Lorentz transformations, keeping the same form in all inertial frames.

But if we consider a noninertial system of reference, then ds^2 is no longer given, in general, by the sum of the squares of the four differentials of the coordinates. For example if (x',y',z',ct') are the coordinates of a frame rotating uniformly with angular velocity ω along the $\hat{z}$ axis, the transformation to this frame is given by

$$\begin{aligned} x &= x' \cos \omega t - y' \sin \omega t \\ y &= x' \sin \omega t + y' \cos \omega t \\ z' &= z \end{aligned} \qquad (2.2)$$

and the line-element becomes

$$ds^2 = [c^2 - \omega^2(x'^2 + y'^2)]\,dt^2 - dx'^2 - dy'^2 - dz^2$$

$$+ 2\omega y' dx' dt - 2\omega x' dy' dt \qquad (2.3)$$

Therefore, in a noninertial system of reference, the line-element is, in general, a quadratic form in the differentials of the coordinates $x^1, x^2, x^3, x^4 = ct$, and we may write, in general, (see also eq.(2.20) below)

$$ds^2 = g_{\mu\nu}\, dx^\mu dx^\nu \qquad (2.4)$$

where Greek indices run from 1 to 4, and $g_{\mu\nu} = g_{\nu\mu}$ are ten functions of the coordinates $\vec{x}$ and t. The system of coordinates described by eq.(2.4) is called curvilinear, and it corresponds, physically, to an accelerated frame of reference. The functions $g_{\mu\nu}$ describe the geometrical properties of a curvilinear coordinate system, and are called the metric of the space-time manifold. In the case of an inertial frame they reduce to $g_{11} = g_{22} = g_{33} = -1$, $g_{44} = 1$, and $g_{\mu\nu} = 0$ for $\mu \neq \nu$; this frame is called galilean, and the corresponding metric is the Minkonski metric $\eta_{\mu\nu}$

$$\eta_{\mu\nu} = \mathrm{diag}(-1, -1, -1, +1) \qquad (2.5)$$

Since accelerated frames are equivalent to gravitational fields, gravitational effects are to be described by the metric $g_{\mu\nu}$. In this framework, gravitation is to be understood as a deviation of the metric of the space-time manifold from the flat Minkonski metric. Therefore the metric $g_{\mu\nu}$ is not fixed arbitrarily on the whole space-time, as in the case of special relativity, but, as we shall see, it depends on the local distribution of matter.

We wish to stress, however, the existence of a fundamental difference between "real" gravitational fields and accelerated frames, in spite of their local equivalence. In fact, given the fundamental quadratic form (2.4), in the case of a noninertial system of reference we can always reduce it to the galileian form (2.1) globally on the whole space-time by means of a transformation of coordinates (see for example eq. (2.1), (2.3)).

On the contrary, a "real" gravitational field cannot be globally eliminated by any coordinate transformation, and the metric may be reduced to the Minkowski form only locally.

This corresponds, as we shall see in the following chapters, to a space-time manifold described by a nonvanishing curvature tensor $R_{\mu\nu\alpha}{}^{\beta} \neq 0$, while in the case of a non-inertial frame this tensor is vanishing, that is the space is intrinsically flat, even if the line-element is expressed in curvilinear coordinates.

2.- Covariant and controvariant tensors

We start considering a general coordinate transformation from a sistem of coordinates x^μ to $x'^\mu = f^\mu(\vec{x},t)$. It is well known that the differentials dx^μ transform according to the rule

$$dx^\mu \longrightarrow dx'^\mu = \frac{\partial x'^\mu}{\partial x^\nu} dx^\nu \tag{2.6}$$

We call "controvariant vector" a four components geometrical object $A^\mu = (A^1, A^2, A^3, A^4)$ which, under a coordinate transformation, follows the transformation rule of the differentials:

$$A^\mu \longrightarrow A'^\mu = \frac{\partial x'^\mu}{\partial x^\nu} A^\nu \tag{2.7}$$

If φ is a scalar function, its gradient $\partial_\mu \varphi$ transforms as

$$\partial_\mu \varphi \equiv \frac{\partial \varphi}{\partial x^\mu} \longrightarrow \frac{\partial \varphi}{\partial x'^\mu} = \frac{\partial x^\nu}{\partial x'^\mu} \frac{\partial \varphi}{\partial x^\nu} \tag{2.8}$$

A "covariant vector" is a four dimensional object A_μ, $A_\mu = (A_1, A_2, A_3, A_4)$ following the same transformation rule, i.e.

$$A_\mu \longrightarrow A'_\mu = \frac{\partial x^\nu}{\partial x'^\mu} A_\nu \tag{2.9}$$

Therefore lower (upper) indices are covariant (controvariant). The generalization to tensor of higher rank is straightforward: for example $A^{\mu\nu}$ is a controvariant tensor if

$$A'^{\mu\nu} = \frac{\partial x'^{\mu}}{\partial x^{\alpha}} \frac{\partial x'^{\nu}}{\partial x^{\beta}} A^{\alpha\beta} \tag{2.10}$$

while $A_{\mu\nu}$ is covariant if

$$A'_{\mu\nu} = \frac{\partial x^{\alpha}}{\partial x'^{\mu}} \frac{\partial x^{\beta}}{\partial x'^{\nu}} A_{\alpha\beta} \tag{2.11}$$

and $A_{\mu}{}^{\nu}$ is called a mixed tensor if

$$A'_{\mu}{}^{\nu} = \frac{\partial x^{\alpha}}{\partial x'^{\mu}} \frac{\partial x'^{\nu}}{\partial x^{\beta}} A_{\alpha}{}^{\beta} \tag{2.12}$$

In order to see the geometrical difference between covariant and controvariant components, consider two non-orthogonal unity vectors $\{ \hat{e}_1, \hat{e}_2, |\hat{e}_1| = |\hat{e}_2| = 1 \}$ defining two axis x^1 and x^2 in a plane. The controvariant components (A^1, A^2) of a vector $\vec{A}$ are then given by the projections of $\vec{A}$ parallel to the axis x^1, x^2, while the orthogonal projections of $\vec{A}$ give the covariant components (A_1, A_2) (see Fig. 2.1) (obviously $A_1 = A^1$ and $A_2 = A^2$ if $\hat{e}_1 \perp \hat{e}_2$).

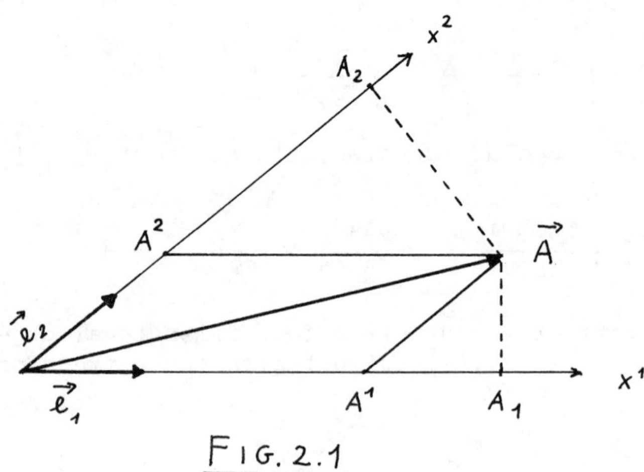

FIG. 2.1

In four dimensions, given a vector $\vec{A}$ and four basis vectors $\vec{e}_\mu$, we have then

$$\vec{A} = A^\mu \vec{e}_\mu \quad , \quad A_\mu = \vec{A} \cdot \vec{e}_\mu \tag{2.13}$$

(a dot denotes scalar product). Defining

$$g_{\mu\nu} = \vec{e}_\mu \cdot \vec{e}_\nu \tag{2.14}$$

scalar products become

$$\vec{A} \cdot \vec{A} = A^\mu \vec{e}_\mu \cdot \vec{e}_\nu A^\nu = g_{\mu\nu} A^\mu A^\nu$$

and we have

$$A_\mu = \vec{A} \cdot \vec{e}_\mu = A^\nu \vec{e}_\nu \cdot \vec{e}_\mu = g_{\mu\nu} A^\nu \tag{2.15}$$

Moreover, defining the inverse matrix $g^{\mu\nu}$ such that

$$g_{\mu\alpha} g^{\nu\alpha} = \delta_\mu^{\ \nu} = \begin{cases} 1 , & \mu = \nu \\ 0 , & \mu \neq \nu \end{cases} \tag{2.16}$$

we obtain also

$$g^{\mu\nu} A_\nu = g^{\mu\alpha} g_{\alpha\beta} A^\beta = \delta^\mu_{\ \beta} A^\beta = A^\mu \tag{2.17}$$

Equations (2.15) and (2.17) show that $g^{\mu\nu}$ and $g_{\mu\nu}$ may be used respectively to raise or lower indices, in other words to relate the covariant components to the controvariant ones, for any tensor, including the metric tensor itself.

In flat space, in particular using the Minkowski metric (2.5), we obtain, from eq. (2.15), that

$$\begin{aligned} A_4 &= \eta_{44} A^4 = A^4 \ , \quad A_1 = \eta_{11} A^1 = - A^1 \\ A_2 &= \eta_{22} A^2 = - A^2 \ , \quad A_3 = \eta_{33} A^3 = - A^3 \end{aligned} \tag{2.18}$$

Therefore, with our metric convention, transforming controvariant into covariant components, we obtain that in the Minkowski space the time-like components are unchanged, while the space-like ones change their sign.

(If one uses, on the contrary an euclidean space-time metric (that is $g_{\mu\nu} = \delta_{\mu\nu}$ = diag (1,1,1,1)) one has simply $A_\mu \equiv A^\mu$.

With the definition (2.14), the scalar product of two vectors $\vec{A}$, $\vec{B}$ is given by

$$\vec{A} \cdot \vec{B} = A^\mu \vec{e}_\mu \cdot B^\nu \vec{e}_\nu = A^\mu B^\nu \vec{e}_\mu \cdot \vec{e}_\nu =$$

$$= A^\mu B^\nu g_{\mu\nu} = A_\mu B_\nu g^{\mu\nu} = A^\mu B_\mu = A_\mu B^\mu \tag{2.19}$$

and then, in particular, the squared modulus of an infinitesimal displacement dx^μ may be written

$$ds^2 = |dx|^2 = g_{\mu\nu}\, dx^\mu dx^\nu \tag{2.20}$$

To conclude this section we give explicitly the infinitesimal change of a tensor under an infinitesimal coordinate transformation

$$x'^\mu = x^\mu + \xi^\mu \tag{2.21}$$

that is

$$\delta x^\mu \equiv x'^\mu - x^\mu = \xi^\mu \tag{2.22}$$

with $|\xi| << 1$.

A scalar field φ is affected only by the translation, and since $\varphi'(x') = \varphi(x)$ we have

$$\varphi'(x') = \varphi(x) = \varphi(x' - \xi) \simeq \varphi(x') - \xi^\alpha \partial_\alpha \varphi \tag{2.23}$$

neglecting terms of order $(\xi^\mu)^2$ and higher.

Therefore

$$\delta\varphi = \varphi'(x') - \varphi(x') = -\xi^{\alpha}\partial_{\alpha}\varphi \tag{2.24}$$

In the case of a covariant vector A_{μ}, we notice that

$$\frac{\partial x^{\nu}}{\partial x'^{\mu}} = \frac{\partial}{\partial x'^{\mu}}(x'^{\nu} - \xi^{\nu}) = \delta^{\nu}_{\mu} - \partial_{\mu}\xi^{\nu} \tag{2.25}$$

Therefore, to first order in ξ,

$$A'_{\mu}(x') = \frac{\partial x^{\nu}}{\partial x'^{\mu}} A_{\nu}(x) = (\delta^{\nu}_{\mu} - \partial_{\mu}\xi^{\nu}) A_{\nu}(x' - \xi)$$

$$\simeq (\delta^{\nu}_{\mu} - \partial_{\mu}\xi^{\nu})(A_{\nu} - \xi^{\alpha}\partial_{\alpha}A_{\nu}) \simeq$$

$$\simeq A_{\mu}(x') - \xi^{\alpha}\partial_{\alpha}A_{\mu} - A_{\nu}\partial_{\mu}\xi^{\nu} \tag{2.26}$$

from which

$$\delta A_{\mu} = A'_{\mu}(x') - A_{\mu}(x') = -\xi^{\alpha}\partial_{\alpha}A_{\mu} - A_{\alpha}\partial_{\mu}\xi^{\alpha} \tag{2.27}$$

For a controvariant vector we have

$$\frac{\partial x'^{\mu}}{\partial x^{\nu}} = \delta^{\mu}_{\nu} + \partial_{\nu}\xi^{\mu} \tag{2.28}$$

and then

$$A'^{\mu}(x') = \frac{\partial x'^{\mu}}{\partial x^{\nu}} A^{\nu}(x) = (\delta^{\mu}_{\nu} + \partial_{\nu}\xi^{\mu})(A^{\nu}(x') - \xi^{\alpha}\partial_{\alpha}A^{\nu})$$

$$\simeq A^{\mu}(x') - \xi^{\alpha}\partial_{\alpha}A^{\mu} + A^{\alpha}\partial_{\alpha}\xi^{\mu} \tag{2.29}$$

$$\delta A^{\mu} = -\xi^{\alpha}\partial_{\alpha}A^{\mu} + A^{\alpha}\partial_{\alpha}\xi^{\mu} \tag{2.30}$$

Notice that the sign of the second term changes from the case of a covariant to the case of a controvariant vector, while the first term (corresponding to a translation) is the same.

Following the same procedure we find, for a covariant tensor $A_{\mu\nu}$,

$$A'_{\mu\nu}(x') = \frac{\partial x^\alpha}{\partial x'^\mu} \frac{\partial x^\beta}{\partial x'^\nu} A_{\alpha\beta}(x' - \xi) =$$

$$= (\delta^\alpha_\mu - \partial_\mu \xi^\alpha)(\delta^\beta_\nu - \partial_\nu \xi^\beta)(A_{\alpha\beta} - \xi^\mu \partial_\mu A_{\alpha\beta})$$

$$\simeq A_{\mu\nu} - \xi^\alpha \partial_\alpha A_{\mu\nu} - A_{\alpha\nu} \partial_\mu \xi^\alpha - A_{\mu\alpha} \partial_\nu \xi^\alpha \qquad (2.31)$$

and then

$$\delta A_{\mu\nu} = - \xi^\alpha \partial_\alpha A_{\mu\nu} - A_{\alpha\nu} \partial_\mu \xi^\alpha - A_{\mu\alpha} \partial_\nu \xi^\alpha \qquad (2.32)$$

while, for a controvariant tensor $A^{\mu\nu}$,

$$\delta A^{\mu\nu} = - \xi^\alpha \partial_\alpha A^{\mu\nu} + A^{\mu\alpha} \partial_\alpha \xi^\nu + A^{\alpha\nu} \partial_\alpha \xi^\mu \qquad (2.33)$$

These formulas, applied to the metric tensor $g^{\mu\nu}$, will be used in particular to investigate the invariance of the matter Lagrangian under local infinitesimal translation (in Chapter 3, Section 3).

3.- Covariant derivation

Unlike the case of cartesian coordinates, in a curvilinear system of coordinates the differential of a vector, dA^μ, is not a vector, and the partial derivative, $\partial_\mu A^\nu$, is not a tensor under general coordinate transformations. Essentially this is due to the fact that dA^μ is computed by taking the difference between two vectors lying at two different points (infinitesimally separated) of the space-time, and the transformation laws of vectors are position dependent.

We can see explicitly that the differential dA^μ does not transforms homogeneously as a controvariant vector. In fact, putting

$$A^\mu = \frac{\partial x^\mu}{\partial x'^\nu} A'^\nu \tag{2.34}$$

(notice that we have used the inverse of the transformation law (2.7)) we have

$$dA^\mu = \frac{\partial x^\mu}{\partial x'^\nu} dA'^\nu + A'^\nu d\left(\frac{\partial x^\mu}{\partial x'^\nu}\right) =$$

$$= \frac{\partial x^\mu}{\partial x'^\nu} dA'^\nu + A'^\nu \frac{\partial^2 x^\mu}{\partial x'^\alpha \partial x'^\nu} dx'^\alpha \tag{2.35}$$

which reduces to the correct transformation for a vector (see (2.34)) only if $\partial^2 x^\nu / \partial x'^\mu \partial x'^\alpha = 0$, namely only if the considered transformations $x^\mu = f^\mu(x')$ are linear functions of x^μ (this is the case, for example, of the Lorentz transformations in flat space-time).

In order to define a suitable generalization of the differential operator in a curvilinear frame, the difference between two vectors must be performed at the same point of the space-time. In other words, it is necessary to transport one of the two vectors from its position to the (infinitesimally close) position of the other one. This transport operation is to be performed so that, in cartesian coordinates, this difference coincides with the usual differential dA^μ. As dA^μ is the difference between the components of two infinitely close vectors, it follows that during the displacement (in cartesian coordinates) the components of A^μ are unchanged: this transport is then the displacement of a vector parallelly to itself.

In a curvilinear system, the components of a vector undergoing parallel transport are however in general changed, unlike the case of a cartesian system. Therefore, if A^μ are the components of a vector in x^μ, and $A^\mu + dA^\mu$ the components in $x^\mu + dx^\mu$, the parallel transport of A^μ from x^μ to $x^\mu + dx^\mu$ produces a variation of its components, δA^μ. It follows that, after the displacement, the difference DA^μ between the two vectors is given by

$$D A^{\mu} = d A^{\mu} - \delta A^{\mu} \qquad (2.36)$$

Assuming that δA^{μ} depends on the infinitesimal displacement dx^{μ}, and also linearly on the components of A^{μ} itself (the linearity must be required in order that the sum of the two vectors transforms like each vector) we may write

$$\delta A^{\mu} = - \Gamma_{\beta\alpha}{}^{\mu} A^{\alpha} d x^{\beta} \qquad (2.37)$$

where the functions $\Gamma_{\alpha\beta}{}^{\mu}$ are the coefficients of the so-called affine connection, which are in general nonsymmetric in theis lower indices, $\Gamma_{\alpha\beta}{}^{\mu} \neq \Gamma_{\beta\alpha}{}^{\mu}$. In general relativity, however, one consides only the symmetric part of the connection, assuming that $\Gamma_{\alpha\beta}{}^{\mu} = \Gamma_{\beta\alpha}{}^{\mu}$, and in this case the functions $\Gamma_{\alpha\beta}{}^{\mu}$ coincide with the so-called Christoffel's symbols (see the discussion at the end of this Section). Their explicit form depends on the choice of the system of coordinates, and since in a cartesian frame $\Gamma_{\alpha\beta}{}^{\mu} = 0$, it follows that $\Gamma_{\alpha\beta}{}^{\mu}$ are not the components of a tensor, because if a tensor is vanishing in some coordinate system, it is then identically zero in all frames.

To visualize intuitively parallel transport along a given curve, consider a bidimensional surface. Suppose that the surface may be developed on a plane: transport the vector parallelly to itself and then let the surface assume its initial shape. If the surface cannot be developed, we must choose the path of parallel transport and then define a tangent plane at each point of the path. The surface enveloped by these tangent plane can now be developed, and then we can follow the same procedure as before (see Fig.2.2).

In particular, if the vector is displaced along a geodesic curve (defined in the following Section), which, developed in a plane, becomes a straight line, then the angle between the vector and the geodesic is unchanged under parallel displacement. In general, the operation of parallel displacement depends on the choice of the path.

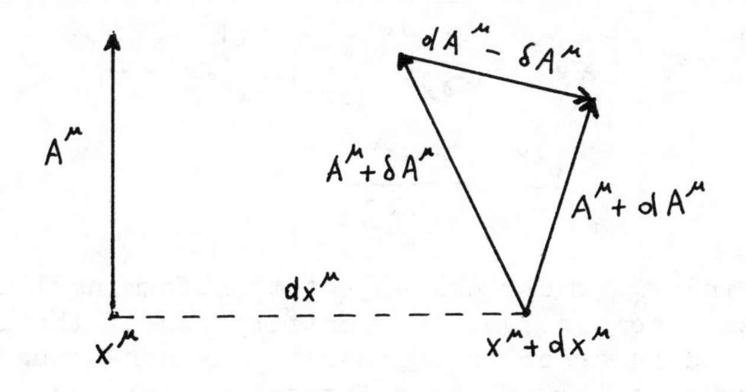

FIG. 2.2.

In a cartesian frame we have $\delta A^\mu = 0$ (Fig. 2.3).

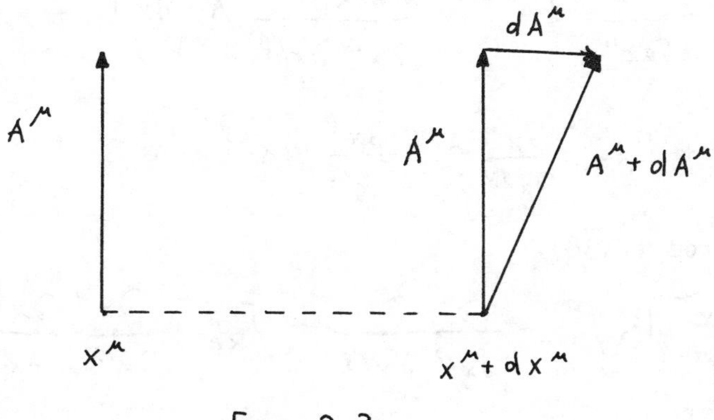

FIG. 2.3

Notice that δA^μ, like dA^μ, is not a vector, being the difference between two vector lying at two different places. The difference $DA^\mu = dA^\mu - \delta A^\mu$ transforms however like a vector; it follows that, under a general coordinate transformation $x'^\mu = f^\mu(x)$, the transformation rule for $\Gamma_{\alpha\beta}{}^\mu$ is

$$\Gamma_{\lambda\rho}{}^{\nu} \longrightarrow \Gamma'_{\lambda\rho}{}^{\nu} = \frac{\partial x'^{\nu}}{\partial x^{\mu}} \frac{\partial x^{\alpha}}{\partial x'^{\lambda}} \frac{\partial x^{\beta}}{\partial x'^{\rho}} \Gamma_{\alpha\beta}{}^{\mu} + $$

$$+ \frac{\partial x'^{\nu}}{\partial x^{\mu}} \frac{\partial^2 x^{\mu}}{\partial x'^{\lambda} \partial x'^{\rho}} \qquad (2.38)$$

A three index geometrical object transforming like this is called a connection; the contribution of the last term is required in order to compensate the non-homogeneous transformation law of the differential (see eq. (2.35)).

Using eq. (2.34-2.38), we can verify explicitly that DA^{μ} transforms like A^{μ}, i.e. it is a true vector.

In fact

$$DA^{\mu} = dA^{\mu} + \Gamma_{\beta\alpha}{}^{\mu} A_{\alpha} dx^{\beta} =$$

$$= \frac{\partial x^{\mu}}{\partial x'^{\nu}} dA'^{\nu} + \frac{\partial^2 x^{\mu}}{\partial x'^{\alpha} \partial x'^{\beta}} A'^{\alpha} dx'^{\beta} +$$

$$+ \Gamma_{\beta\alpha}{}^{\mu} \frac{\partial x^{\alpha}}{\partial x'^{\nu}} A'^{\nu} \frac{\partial x^{\beta}}{\partial x'^{\rho}} dx'^{\rho} \qquad (2.39)$$

But, from (2.38),

$$\frac{\partial x^{\sigma}}{\partial x'^{\nu}} \Gamma_{\lambda\rho}{}^{\sigma} - \frac{\partial^2 x^{\sigma}}{\partial x'^{\lambda} \partial x'^{\rho}} = \Gamma_{\alpha\beta}{}^{\sigma} \frac{\partial x^{\alpha}}{\partial x'^{\lambda}} \frac{\partial x^{\beta}}{\partial x'^{\rho}} \qquad (2.40)$$

therefore

$$DA^{\mu} = \frac{\partial x^{\mu}}{\partial x'^{\nu}} dA'^{\nu} + \frac{\partial^2 x^{\mu}}{\partial x'^{\alpha} \partial x'^{\beta}} A'^{\alpha} dx'^{\beta} +$$

$$+ A'^{\nu} dx'^{\rho} \left(\frac{\partial x^{\mu}}{\partial x'^{\alpha}} \Gamma_{\nu\rho}{}^{\alpha} - \frac{\partial^2 x^{\mu}}{\partial x'^{\nu} \partial x'^{\rho}} \right) =$$

$$= \frac{\partial x^{\mu}}{\partial x'^{\nu}} \left(dA'^{\nu} + \Gamma'_{\alpha\beta}{}^{\nu} A'^{\alpha} dx'^{\beta} \right) =$$

$$= \frac{\partial x^{\mu}}{\partial x'^{\nu}} \left(D A^{\nu} \right)' \tag{2.41}$$

It is important to notice that the connection can be locally eliminated by a suitable transformation of coordinates, but this does not imply that the connection is vanishing everywhere, because it is not a tensor. Consider in fact the following transformation around the origin

$$x^{\mu} = x'^{\mu} + \frac{1}{2} a_{\alpha\beta}{}^{\mu} x'^{\alpha} x'^{\beta} \tag{2.42}$$

where $a_{\alpha\beta}{}^{\mu}$ are arbitrary coefficients. From eq. (2.38, 2.42) the coefficients of the transformed connection, $\Gamma'_{\lambda\rho}{}^{\nu}$, at the point $x'^{\mu} = x^{\mu} = 0$, are

$$\Gamma'_{\lambda\rho}{}^{\nu}(0) = \Gamma_{\lambda\rho}{}^{\nu}(0) + a_{\lambda\rho}{}^{\nu} \tag{2.43}$$

Therefore, choosing $a_{\lambda\rho}{}^{\nu} = -\Gamma_{\lambda\rho}{}^{\nu}(0)$, we can obtain, in the origin, $\Gamma'_{\lambda\rho}{}^{\nu} = 0$. Notice that only the symmetric part of the connection can be eliminated with the previous transformation. Notice also that the derivatives of the connection, $\partial_{\alpha} \Gamma'_{\lambda\rho}{}^{\nu}$ are nonvanishing at $x^{\mu} = x'^{\mu} = 0$, and, in general, the connection and its derivatives cannot be eliminated at the same time (unless, obviously, the space is flat, *i.e.* $g_{\mu\nu} = \eta_{\mu\nu}$). From the differential

$$D A^{\mu} = d A^{\mu} - \delta A^{\mu} = \left(\frac{\partial A^{\mu}}{\partial x^{\beta}} + \Gamma_{\beta\alpha}{}^{\mu} A^{\alpha} \right) dx^{\beta} \tag{2.44}$$

we can define the so-called covariant derivative (usually denoted by a semi-colon)

$$\frac{D A^{\mu}}{D x^{\beta}} \equiv A^{\mu}{}_{;\beta} = \partial_{\beta} A^{\mu} + \Gamma_{\beta\alpha}{}^{\mu} A^{\alpha} \tag{2.45}$$

It is called covariant derivative because the index β, of the tensor $A^{\mu}{}_{;\beta}$, transforms like the index of a covariant vector. Therefore

$$A^{\mu;\beta} = g^{\beta\nu} A^{\mu}{}_{;\nu} \tag{2.46}$$

Notice that, in a cartesian system, $\Gamma = 0$, and $A^{\mu}{}_{;\beta}$ reduces then to the usual partial derivative $\partial_\beta A^{\mu}$.

The law of parallel displacement in the case of a covariant vector A_μ can be obtained simply observing that the scalar $A^2 = g_{\mu\nu} A^\mu A^\nu = A^\mu A_\mu$ must be left invariant during the displacement, i.e.

$$\delta(A^\mu A_\mu) = A^\mu \delta A_\mu + A_\mu \delta A^\mu = 0 \tag{2.47}$$

(this implies that the coefficients of the affine connections appearing in the trasformation laws of covariant and controvariant tensors must be the same).
Using eq. (2.37) we obtain

$$A^\mu \delta A_\mu = A_\mu \Gamma_{\beta\alpha}{}^{\mu} A^\alpha dx^\beta = A_\alpha \Gamma_{\beta\mu}{}^{\alpha} A^\mu dx^\beta$$

$$\delta A_\mu = \Gamma_{\beta\mu}{}^{\alpha} A_\alpha dx^\beta \tag{2.48}$$

The differential is then

$$DA_\mu = dA_\mu - \delta A_\mu = dA_\mu - \Gamma_{\beta\mu}{}^{\alpha} A_\alpha dx^\beta \tag{2.49}$$

and the covariant derivative

$$A_{\mu;\beta} = \partial_\beta A_\mu - \Gamma_{\beta\mu}{}^{\alpha} A_\alpha \tag{2.50}$$

The equations (2.45) and (2.50) can be easily generalized to the case of higher rank tensors. For example, to obtain the covariant derivative of a two index controvariant tensor, we consider the product of two vectors, putting $A^{\mu\nu} = A^\mu B^\nu$. We have then

$$\delta A^{\mu\nu} = \delta(A^\mu B^\nu) = A^\mu \delta B^\nu + B^\nu \delta A^\mu =$$

$$= - A^\mu \Gamma_{\beta\alpha}{}^{\nu} B^\alpha dx^\beta - B^\nu \Gamma_{\beta\alpha}{}^{\mu} A^\alpha dx^\beta$$

$$= - (\Gamma_{\beta\alpha}{}^{\nu} A^{\mu\alpha} + \Gamma_{\beta\alpha}{}^{\mu} A^{\alpha\nu}) dx^\beta \tag{2.51}$$

From the definition $DA^{\mu\nu} = dA^{\mu\nu} - \delta A^{\mu\nu}$ we obtain

$$A^{\mu\nu}{}_{;\beta} = \partial_\beta A^{\mu\nu} + \Gamma_{\beta\alpha}{}^{\nu} A^{\mu\alpha} + \Gamma_{\beta\alpha}{}^{\mu} A^{\alpha\nu} \qquad (2.52)$$

In an analogous way we have also

$$A^{\mu}{}_{\nu;\beta} = \partial_\beta A^{\mu}{}_{\nu} - \Gamma_{\beta\nu}{}^{\alpha} A^{\mu}{}_{\alpha} + \Gamma_{\beta\alpha}{}^{\mu} A^{\alpha}{}_{\nu} \qquad (2.53)$$

$$A_{\mu\nu;\beta} = \partial_\beta A_{\mu\nu} - \Gamma_{\beta\nu}{}^{\alpha} A_{\mu\alpha} - \Gamma_{\beta\mu}{}^{\alpha} A_{\alpha\nu} \qquad (2.54)$$

In general, for each index of the tensor, we must introduce a term containing the Christoffel symbol, acting on that indices.

In general relativity, the connection is not an independent field variable, because its coefficients can be expressed as a function of the metric tensor and of its first derivatives. The relation between $g_{\mu\nu}$ and $\Gamma_{\mu\nu}{}^{\alpha}$ can be obtained noticing that the covariant derivative operator commutes with the metric. In fact, as DA^{μ} is a controvariant vector, the corresponding covariant one is given by

$$DA_\mu = g_{\mu\nu} DA^{\nu} = D(g_{\mu\nu} A^{\nu}) \qquad (2.55)$$

so that $D g_{\mu\nu} = 0$, or

$$g_{\mu\nu;\alpha} = 0 = g^{\mu\nu}{}_{;\alpha} \qquad (2.56)$$

This is the so-called "metricity-postulate", and the connection Γ, when eq. (2.56) is satisfied, is said to be "metric-compatible". Writing explicitly eq. (2.56), and performing a cyclic permutation of the index, we have

$$\begin{aligned} \partial_\alpha g_{\mu\nu} - \Gamma_{\alpha\nu}{}^{\beta} g_{\mu\beta} - \Gamma_{\alpha\mu}{}^{\beta} g_{\beta\nu} &= 0 \\ \partial_\mu g_{\nu\alpha} - \Gamma_{\mu\alpha}{}^{\beta} g_{\nu\beta} - \Gamma_{\mu\nu}{}^{\beta} g_{\beta\alpha} &= 0 \\ \partial_\nu g_{\alpha\mu} - \Gamma_{\nu\mu}{}^{\beta} g_{\alpha\beta} - \Gamma_{\nu\alpha}{}^{\beta} g_{\beta\mu} &= 0 \end{aligned} \qquad (2.57)$$

By multiplying the first of these equations by 1/2, the second and the third by -1/2, and summing the three equations, we obtain

$$\frac{1}{2}\left(\partial_\alpha g_{\mu\nu} - \partial_\mu g_{\nu\alpha} - \partial_\nu g_{\alpha\mu}\right) + \Gamma_{(\mu\nu)}{}^{\beta} g_{\beta\alpha}$$

$$- \Gamma_{[\alpha\nu]}{}^{\beta} g_{\mu\beta} + \Gamma_{[\mu\alpha]}{}^{\beta} g_{\beta\nu} = 0 \qquad (2.58)$$

(square brackets denote antisymmetrization, $A_{[\mu\nu]} = \frac{1}{2}(A_{\mu\nu} - A_{\nu\mu})$, round brackets symmetrization, $A_{(\mu\nu)} = \frac{1}{2}(A_{\mu\nu} + A_{\nu\mu})$).

Imposing on the connection coefficients to be symmetric in the first two indices, $\Gamma_{[\mu\nu]}{}^{\alpha} = 0$, we obtain

$$\Gamma_{\mu\nu\alpha} = \frac{1}{2}\left(\partial_\mu g_{\nu\alpha} + \partial_\nu g_{\alpha\mu} - \partial_\alpha g_{\mu\nu}\right) = \Gamma_{\nu\mu\alpha} \qquad (2.59)$$

The Christoffel symbols are then

$$\Gamma_{\mu\nu}{}^{\beta} \equiv \left\{ {\beta \atop \mu\nu} \right\} = g^{\beta\alpha}\Gamma_{\mu\nu\alpha} = \Gamma_{\nu\mu}{}^{\beta} =$$

$$= \frac{1}{2} g^{\beta\alpha}\left(\partial_\mu g_{\nu\alpha} + \partial_\nu g_{\alpha\mu} - \partial_\alpha g_{\mu\nu}\right) \qquad (2.60)$$

and in this case the connection is completely determined as a function of the metric tensor.

It is important to stress that the fact that the connection can be expressed completely in terms of the Christoffel symbols, is related to the invariance of the scalar product under parallel displacement, see eq. (2.47) (which is equivalent to the metricity condition (2.56)) and to the requirement that the connection be torsion-free, $\Gamma_{[\mu\nu]}{}^{\alpha} = 0$ (the antisymmetric part of the connection transforms as a tensor, as can be seen from eq. (2.38), and it is called the torsion tensor).

These two requirements are satisfied in general relativity, which turns out to be then the simplest geometrical theory of the gravitational field. The reader, however, should remember that, in principle, one can also construct theories of gravity in which the connection is not metric-compatible, $g_{\mu\nu;\alpha} \neq 0$, (like the Weyl theory), or the torsion is nonzero, $\Gamma_{[\mu\nu]}{}^{\alpha} \neq 0$ (like the Einstein-Cartan theory, see Chapter X) or both $g_{\mu\nu;\alpha}$ and $\Gamma_{[\mu\nu]}{}^{\alpha}$ are nonzero (in this case the theory is called "metric-affine"[9]).

In the most general case, to obtain a relation between the connection and the metric, one must start considering, instead of eq. (2.57), the following equations:

$$\partial_\alpha g_{\mu\nu} - \Gamma_{\alpha\nu}{}^{\beta} g_{\mu\beta} - \Gamma_{\alpha\mu}{}^{\beta} g_{\beta\nu} = g_{\mu\nu;\alpha}$$

$$-\partial_\mu g_{\nu\alpha} + \Gamma_{\mu\alpha}{}^{\beta} g_{\nu\beta} + \Gamma_{\mu\nu}{}^{\beta} g_{\beta\alpha} = -g_{\nu\alpha;\mu} \qquad (2.61)$$

$$-\partial_\nu g_{\alpha\mu} + \Gamma_{\nu\mu}{}^{\beta} g_{\alpha\beta} + \Gamma_{\nu\alpha}{}^{\beta} g_{\beta\mu} = -g_{\alpha\mu;\nu}$$

Putting

$$\Gamma_{[\mu\nu]}{}^{\alpha} = Q_{\mu\nu}{}^{\alpha} = -Q_{\nu\mu}{}^{\alpha} \qquad (2.62)$$

$$g_{\mu\nu;\alpha} = N_{\mu\nu\alpha} = N_{\nu\mu\alpha} \qquad (2.63)$$

where $Q_{\mu\nu}{}^{\alpha}$ is the torsion tensor, and $N_{\mu\nu\alpha}$ is called the non-metricity tensor, and performing the same procedure as before, we obtain, from eq. (2.61)

$$\frac{1}{2}\left(\partial_\alpha g_{\mu\nu} - \partial_\mu g_{\nu\alpha} - \partial_\nu g_{\alpha\mu}\right) + \Gamma_{(\mu\nu)\alpha} -$$

$$- Q_{\alpha\nu\mu} + Q_{\mu\alpha\nu} = \frac{1}{2}\left(N_{\mu\nu\alpha} - N_{\nu\alpha\mu} - N_{\alpha\mu\nu}\right) \qquad (2.64)$$

By adding to the two members of eq. (2.64) the torsion tensor $Q_{\mu\nu\alpha}$, and raising the index α by contraction with $g^{\beta\alpha}$, as before, we obtain the most general expression for the connection coefficients:

$$\Gamma_{\mu\nu}{}^{\beta} = g^{\beta\alpha}\left(\Gamma_{(\mu\nu)\alpha} + Q_{\mu\nu\alpha}\right) =$$

$$= \left\{ {\beta \atop \mu\nu} \right\} - K_{\mu\nu}{}^{\beta} - V_{\mu\nu}{}^{\beta} \qquad (2.65)$$

where the Christoffel symbols $\left\{ {\beta \atop \mu\nu} \right\}$ are given in terms of the metric as in eq. (2.60), the tensor $K_{\mu\nu}{}^{\beta}$ is called the contorsion-tensor, and is given by

$$K_{\mu\nu}{}^{\beta} = - Q_{\mu\nu}{}^{\beta} - Q^{\beta}{}_{\mu\nu} + Q_{\nu}{}^{\beta}{}_{\mu} \qquad (2.66)$$

and finally the tensor $V_{\mu\alpha}{}^{\beta}$ is given in terms of the non-metricity as follows

$$V_{\mu\nu}{}^{\beta} = \frac{1}{2}\left(N_{\mu\nu}{}^{\beta} - N^{\beta}{}_{\mu\nu} - N_{\nu}{}^{\beta}{}_{\mu} \right) \qquad (2.67)$$

It should be noticed that the symmetric part of the connection, $\Gamma_{(\mu\nu)}{}^{\beta} = \left\{ {\beta \atop \mu\nu} \right\} - K_{(\mu\nu)}{}^{\beta} - V_{(\mu\nu)}{}^{\beta}$, coincides with the Christoffel symbols only in general relativity, where $K = V = 0$.

4.- The geodesic equation

As an application of the notion of covariant derivative discussed in the previous Section, we can obtain the equation of motion of a point-like test particle in general relativity (the general eq. of motion for a macroscopical body will be discussed in Chapter X). The geodesic equation is a consequence of the following variational principle

$$\delta \int ds = 0 \qquad (2.68)$$

where ds is the square root of the line-element (2.20), and the variation is vanishing at the limits of the integration interval.

In special relativity, as is well known, in the case of a free particle one gets the equation[2)]

$$\frac{d^2 x^\mu}{ds^2} = 0 \qquad (2.69)$$

If the particle is subject to the action of a gravitional field, the motion equations follows from the same variational principle (2.68), provided that the Minkowski metric $\eta^{\mu\nu}$ is replaced by the metric of the curved space, $g_{\mu\nu}(\vec{x},t)$. The geodesic equation obtained in this way is no longer a straightline, in general, because space-time is no longer flat, and then the particle motion is not rectilinear and uniform, in general. However, the particle will follow a trajectory corresponding to the shortest path between two given points, because its motion is described by a minimal action principle.

Instead of computing explicitly the variation in eq. (2.68), the geodesic equation can be obtained simply as the covariant generalization of eq. (2.69). The special relativistic law of motion $du^\mu = 0$, where $u^\mu = dx^\mu / ds$ is the 4-velocity vector, must become, in a curved space, $D u^\mu = 0$, i.e.

$$du^\mu + \Gamma_{\beta\alpha} u^\alpha dx^\beta = 0 \qquad (2.70)$$

Therefore

$$\frac{du^\mu}{ds} + \Gamma_{\beta\alpha}{}^\mu u^\alpha u^\beta = \frac{d^2 x^\mu}{ds^2} + \Gamma_{\beta\alpha}{}^\mu \frac{dx^\alpha}{ds}\frac{dx^\beta}{ds} = 0 \qquad (2.71)$$

is the covariant law of motion. Notice that the terms $-m\Gamma_{\beta\alpha}{}^\mu u^\alpha u^\beta$ may be interpreted as the gravitational forces, and then the components of the metric tensor $g_{\mu\nu}$ play the role of the classical gravitational potential (as $\Gamma \propto \partial g$).

Choosing a local inertial frame in which $g_{\mu\nu} = \eta_{\mu\nu} =$ const, we have $\Gamma = \partial\eta = 0$, and $du^\mu / ds = 0$. Therefore the gravitational forces can be locally eliminated, and the geodesic equation can be locally reduced to the special relativistic law of motion, in agreement with the equivalence principle.

Finally we present explicitly the deduction of the geodesic equation (2.71) from the variational principle

(2.68). To simplify the computation, it is convenient to start with the following remark. Putting $\dot{x}^\mu = dx^\mu / ds$, and

$$y = (g_{\mu\nu} \dot{x}^\mu \dot{x}^\nu)^{1/2} \qquad (2.72)$$

then eq. (2.68) can be written

$$\delta \int ds = \delta \int (g_{\mu\nu}\, dx^\mu dx^\nu)^{1/2} = \delta \int y\, ds = 0 \qquad (2.73)$$

Considering a general function of y, $F(y)$, with $\partial F/\partial y \neq 0$, one can show that the variational problem

$$\delta \int F(y)\, ds = 0 \qquad (2.74)$$

leads to the same geodesic equations as (2.73). In fact, eq. (2.74) gives the following Euler - Lagrange equations

$$\frac{d}{ds} \frac{\partial F}{\partial \dot{x}^\mu} = \frac{\partial F}{\partial x^\mu} \qquad (2.75)$$

that is

$$\frac{d}{ds}\left(\frac{\partial F}{\partial y} \frac{\partial y}{\partial \dot{x}^\mu} \right) = \frac{\partial F}{\partial y} \frac{\partial y}{\partial x^\mu} \qquad (2.76)$$

from which

$$\frac{\partial F}{\partial y}\left(\frac{d}{ds} \frac{\partial y}{\partial \dot{x}^\mu} - \frac{\partial y}{\partial x^\mu} \right) + \frac{dy}{ds} \frac{\partial^2 F}{\partial y^2} \frac{\partial y}{\partial \dot{x}^\mu} = 0$$

But $dy / ds = 0$ (as $\ddot{x}^\mu \dot{x}_\mu = 0$) and then we obtain, from eq. (2.74), the Lagrange equations

$$\frac{d}{ds} \frac{\partial y}{\partial \dot{x}^\mu} = \frac{\partial y}{\partial x^\mu} \qquad (2.77)$$

which coincide exactly with the ones following from eq. (2.73). In this case it is convenient to choose, as the lagrangian function, $F(y) = y^2 = g_{\mu\nu} \dot{x}^\mu \dot{x}^\nu$. From eq. (2.77) we get then

$$\frac{d}{ds}\left(2 g_{\mu\alpha}\dot{x}^{\alpha}\right) = \partial_{\mu} g_{\alpha\beta}\, \dot{x}^{\alpha}\dot{x}^{\beta} \qquad (2.78)$$

$$2 g_{\mu\alpha}\ddot{x}^{\alpha} + 2\dot{x}^{\alpha}\partial_{\beta} g_{\mu\alpha}\dot{x}^{\beta} - \partial_{\mu} g_{\alpha\beta}\dot{x}^{\alpha}\dot{x}^{\beta} = 0 \qquad (2.79)$$

$$g_{\mu\nu}\ddot{x}^{\nu} + \frac{1}{2}\left(\partial_{\beta} g_{\mu\alpha} + \partial_{\alpha} g_{\mu\beta} - \partial_{\mu} g_{\alpha\beta}\right)\dot{x}^{\alpha}\dot{x}^{\beta} = 0 \qquad (2.80)$$

which can also be written

$$\ddot{x}^{\nu} + \frac{1}{2} g^{\nu\mu}\left(\partial_{\beta} g_{\mu\alpha} + \partial_{\alpha} g_{\mu\beta} - \partial_{\mu} g_{\alpha\beta}\right)\dot{x}^{\alpha}\dot{x}^{\beta} = 0 \qquad (2.81)$$

and which coincides with the geodesic equation (2.71), provided that the connection is given in terms of the Christoffel symbols, as in equation (2.60).

This method of obtaining the geodesical equation provides also, incidentally, an alternative variational procedure to derive the explicit expression of the Christoffel symbols.

Finally we notice that the geodesic equations can also be obtained as the equation of motion, for a spin - less test particle, following from the energy-momentum conservation law, as shown explicitly in Chapter X, Section 4.

REFERENCES

1) as regards the discussion of metric-affine theories of gravity see for example F.W.Hehl, E.A.Lord and L.Smalley: Gen. Rel. Grav. 13, 1037 (1981)

2) see for example L.Landau and E.Lifchitz: "Theorie du Champs" (ed. MIR, Moscow 1966) § 9, p.38

CAPTER III

GRAVITATIONAL FIELD EQUATIONS

1.- The curvature tensor.

In order to formulate the basic equations of Einstein' gravitational theory, the introduction of the four-indices Riemann tensor, also called the curvature tensor, is needed.

Consider the parametric equation of a curve, $x^\mu = x^\mu(s)$, and the tangent vector $u^\mu = dx^\mu / ds$. If the curve is a geodesic, $Du^\mu = 0$, i.e. $du^\mu = \delta u^\mu$, (see the previous Chapter). Therefore the parallelly displaced tangent vector $u^\mu + \delta u^\mu$ coincides with the tangent $u^\mu + du^\mu$ to the geodesic at the point $x^\mu + dx^\mu$. More generally, transporting a vector along a geodesic, the angle between the vector and the tangent to the curve is constant.

The parallel displacement of a vector, performed in a non-euclidean space-time, is generally path-dependent. Therefore, if we displace a vector along a closed curve consisting, for example, of three geodesic arcs, like in Fig. 3.1, the final vector does not coincide, in general, with the initial one.

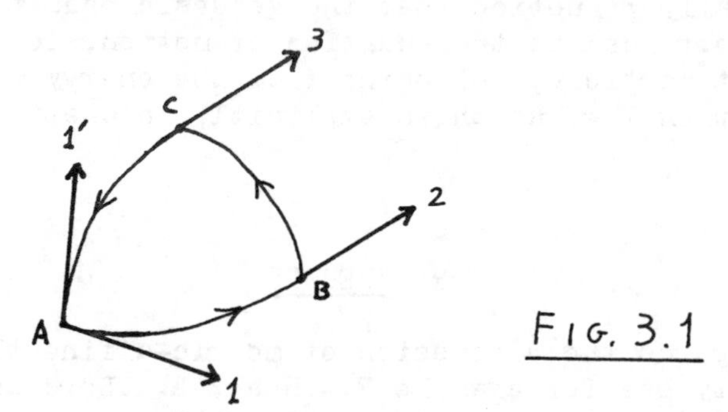

FIG. 3.1

The Riemann tensor, as we shall see, determines the variations of a vector A_μ during its parallel displacement along an infinitesimal close contour γ, defined as the following line integral

$$\Delta A_\mu = \oint_\gamma \delta A_\mu = \oint \Gamma_{\alpha\mu}{}^\beta A_\beta \, dx^\alpha \tag{3.1}$$

By using the four-dimensional version of the Stokes theorem[1] for a vector V_μ , we have

$$\oint_\gamma V_\mu dx^\mu = \int_F df^{\mu\nu} \partial_\mu V_\nu = \frac{1}{2} \int_F df^{\mu\nu} (\partial_\mu V_\nu - \partial_\nu V_\mu) \tag{3.2}$$

where $df^{\mu\nu} = -df^{\nu\mu}$ is the infinitesimal element of a bidimensional surface, and the surface integral is performed over a surface F bounded by γ. From eq. (3.1) and (3.2), one obtains then

$$\Delta A_\mu = \frac{1}{2} \int df^{\alpha\nu} [\partial_\alpha (\Gamma_{\nu\mu}{}^\beta A_\beta) - \partial_\nu (\Gamma_{\alpha\mu}{}^\beta A_\beta)] =$$

$$= \frac{1}{2} \int [\partial_\alpha \Gamma_{\nu\mu}{}^\beta A_\beta - \partial_\nu \Gamma_{\alpha\mu}{}^\beta A_\beta + \Gamma_{\nu\mu}{}^\beta \partial_\alpha A_\beta$$

$$- \Gamma_{\alpha\mu}{}^\beta \partial_\nu A_\beta] \, df^{\alpha\nu} \tag{3.3}$$

As the change in the vector A_μ is due to a parallel displacement along a geodesic, we have $DA_\mu = 0$, and then

$$\partial_\mu A_\nu = \Gamma_{\mu\nu}{}^\alpha A_\alpha \tag{3.4}$$

Therefore

$$\Delta A_\mu = \frac{1}{2} \int df^{\alpha\nu} [\partial_\alpha \Gamma_{\nu\mu}{}^\beta - \partial_\nu \Gamma_{\alpha\mu}{}^\beta +$$

$$+ \Gamma_{\nu\mu}{}^\rho \Gamma_{\alpha\rho}{}^\beta - \Gamma_{\alpha\mu}{}^\rho \Gamma_{\nu\rho}{}^\beta] A_\beta \tag{3.5}$$

We are considering an infinitesimal closed contour; therefore we can evaluate this integral, replacing the integrand function by its value at some point inside the closed area, and taking it out from the integral sign. The integration provides then simply the area $\Delta f^{\alpha\nu}$ of the infinitesimal surface bounded by the contour, and we obtain finally

$$\Delta A_\mu = \frac{1}{2} R_{\alpha\nu\mu}{}^{\beta} A_\beta \, \Delta f^{\alpha\nu} \tag{3.6}$$

where

$$R_{\alpha\nu\mu}{}^{\beta} = \partial_\alpha \Gamma_{\nu\mu}{}^{\beta} - \partial_\nu \Gamma_{\alpha\mu}{}^{\beta} + \Gamma_{\alpha\rho}{}^{\beta} \Gamma_{\nu\mu}{}^{\rho} - \Gamma_{\nu\rho}{}^{\beta} \Gamma_{\alpha\mu}{}^{\rho} \tag{3.7}$$

is the Riemann curvature tensor. The fact that it transforms correctly as a tensor can be also directly verified using the transformation law (2.38) of the affine connection . For a controvariant vector A^μ we obtain, with the same procedure,

$$\Delta A^\mu = -\frac{1}{2} R_{\alpha\nu\beta}{}^{\mu} A^\beta \, \Delta f^{\alpha\nu} \tag{3.8}$$

An alternative method to introduce the curvature tensor is to evaluate the commutator of two covariant derivatives, because $A^\mu{}_{;\alpha;\beta} \neq A^\mu{}_{;\beta;\alpha}$.

In fact, starting from eq. (2.45), and noticing that $A^\mu{}_{;\alpha}$ is a mixed tensor of rank two, one has

$$(A^\mu{}_{;\alpha})_{;\beta} = \partial_\beta (\partial_\alpha A^\mu + \Gamma_{\alpha\nu}{}^{\mu} A^\nu) + \Gamma_{\beta\rho}{}^{\mu} (\partial_\alpha A^\rho + \Gamma_{\alpha\nu}{}^{\rho} A^\nu) - \Gamma_{\beta\alpha}{}^{\rho} (\partial_\rho A^\mu + \Gamma_{\rho\nu}{}^{\mu} A^\nu) \tag{3.9}$$

Using the symmetry of the connection, $\Gamma_{[\beta\alpha]}{}^{\mu} = 0$, one obtains then

$$A^\mu{}_{;\alpha;\beta} - A^\mu{}_{;\beta;\alpha} = 2 A^\mu{}_{;[\alpha;\beta]} = - R_{\alpha\beta\nu}{}^{\mu} A^\nu \tag{3.10}$$

where $R_{\alpha\beta\nu}{}^{\mu}$ is the same tensor of eq. (3.7). For a covariant vector, we obtain

$$A_{\mu;\alpha;\beta} - A_{\mu;\beta;\alpha} = R_{\alpha\beta\mu}{}^{\nu} A_{\nu} \qquad (3.11)$$

(if the connection is not symmetric, a term proportional to the torsion must be added to the commutator of two covariant derivatives, see Chapter X).

In a euclidean space, the curvature tensor is vanishing (as we can find a coordinate transformation eliminating the Christoffel connection on the whole space-time and reducing the metric to the Minkowski form). Conversely, if $R_{\mu\nu\alpha}{}^{\beta} = 0$, the space-time is flat.

Consider in fact four orthonormal vectors $A_{\mu}^{(\alpha)}$, $(\alpha) = 1,2,3,4$, at a point P, with $|A^{(4)}|^2 = 1$, $|A^{(1)}|^2 = |A^{(2)}|^2 = |A^{(3)}|^2 = -1$. Since $R_{\mu\nu\alpha\beta} = 0$, parallel transport around a close curve does not change a vector, so we can define $A_{\mu}^{(\alpha)}$ away from P by parallel transport. For a symmetric connection we have then

$$\partial_{\nu} A_{\mu}^{(\alpha)} = \Gamma_{\nu\mu}{}^{\rho} A_{\rho}^{(\alpha)} = \partial_{\mu} A_{\nu}^{(\alpha)} \qquad (3.12)$$

or $\partial_{\mu} A_{\nu}^{(\alpha)} - \partial_{\nu} A_{\mu}^{(\alpha)} = 0$, and therefore we can introduce four functions $\varphi^{(\alpha)}$ such that

$$A_{\mu}^{(\alpha)} = \partial_{\mu} \varphi^{(\alpha)} \qquad (3.13)$$

Since scalar products are preserved by parallel transport, we have (because of the orthonormality of $A_{\mu}^{(\alpha)}$)

$$g^{\mu\nu} A_{\mu}^{(\alpha)} A_{\nu}^{(\beta)} = \eta^{\alpha\beta} = g^{\mu\nu} \partial_{\mu} \varphi^{(\alpha)} \partial_{\nu} \varphi^{(\beta)} \qquad (3.14)$$

Therefore, performing the coordinate transformation $x'^{\mu} = \varphi^{(\mu)}(x)$, the transformed metric tensor

$$g'^{\alpha\beta} = \frac{\partial x'^{\alpha}}{\partial x^{\mu}} \frac{\partial x'^{\beta}}{\partial x^{\nu}} g^{\mu\nu} = \eta^{\alpha\beta} \qquad (3.15)$$

coincides with the Minkowski metric. The space-time is then flat, because the metric can be reduced everywhere to the Minkowski form, by a suitable transformation of coordinates.

It is important to notice that, in the case of flat space-time, the curvature tensor cannot be eliminated because, even if we can obtain, locally, $\Gamma_{\mu\nu}{}^{\alpha} = 0$ by a suitable coordinate transformation, the derivatives $\partial_{\beta}\Gamma_{\mu\nu}{}^{\alpha}$ are generally nonzero, and then $R_{\mu\nu\alpha\rho} = 0$.

The Riemann curvature tensor satisfies the following properties, which can be easily verified using its explicit definition, eq. (3.7):

$$R_{\mu\nu\alpha\beta} = R_{\mu\nu[\alpha\beta]} \tag{3.16}$$

$$R_{\mu\nu\alpha\beta} = R_{[\mu\nu]\alpha\beta} \tag{3.17}$$

$$R_{\mu\nu\alpha\beta} = R_{\alpha\beta\mu\nu} \tag{3.18}$$

$$R_{\nu\alpha\beta}{}^{\mu} + R_{\alpha\beta\nu}{}^{\mu} + R_{\beta\nu\alpha}{}^{\mu} = 0 \tag{3.19}$$

Another important relation is the so-called Bianchi identity:

$$R_{\mu\nu}{}^{\alpha\beta}{}_{;\rho} + R_{\nu\rho}{}^{\alpha\beta}{}_{;\mu} + R_{\rho\mu}{}^{\alpha\beta}{}_{;\nu} = 0 \tag{3.20}$$

This relation can be easily derived in a locally inertial frame, i.e. using a coordinate system in which $\Gamma = 0$: in fact, in this case we have

$$R_{\mu\nu\alpha}{}^{\beta} = \partial_{\mu}\Gamma_{\nu\alpha}{}^{\beta} - \partial_{\nu}\Gamma_{\mu\alpha}{}^{\beta} \tag{3.21}$$

$$R_{\mu\nu\alpha}{}^{\beta}{}_{;\rho} = \partial_{\rho}\partial_{\mu}\Gamma_{\nu\alpha}{}^{\beta} - \partial_{\rho}\partial_{\nu}\Gamma_{\mu\alpha}{}^{\beta} \tag{3.22}$$

and, cyclically permuting the indices μ, ν, ρ and summing, eq. (3.20) is obtained. Because of its tensorial character, this relation holds in the form (3.20) in every coordinate system. Notice that, in the presence of torsion, the properties (3.19, 3.20) of the curvature tensor are no longer satisfied, and must be suitably generalized (see Chapter X).

Moreoverthe curvature is no longer antisymmetric in the last two indices, eq. (3.16), if the connection is not metric-compatible, i.e. $g_{\mu\nu;\alpha} \neq 0$.

From the curvature tensor $R_{\mu\nu\alpha\beta}$ one can define the following contractions:

$$R_{\mu\nu} = R_{\alpha\mu\nu}{}^{\alpha} = R_{\nu\mu} \qquad (3.23)$$

called the Ricci tensor (other authors follow however a different sign convention), and

$$R = g^{\mu\nu} R_{\mu\nu} = R_{\mu}^{\cdot\,\mu} \qquad (3.24)$$

called the curvature scalar (notice that, because of the relations (3.16), (3.17), one has $g^{\mu\nu} R_{\mu\nu\alpha\beta} = 0 = g^{\alpha\beta} R_{\mu\nu\alpha\beta}$.

The symmetry of the Ricci tensor can be verified by its explicit expression

$$R_{\mu\nu} = R_{\alpha\mu\nu}{}^{\alpha} = \partial_{\alpha} \Gamma_{\mu\nu}{}^{\alpha} - \partial_{\mu} \Gamma_{\alpha\nu}{}^{\alpha} + \Gamma_{\alpha\rho}{}^{\alpha} \Gamma_{\mu\nu}{}^{\rho} - \Gamma_{\mu\rho}{}^{\alpha} \Gamma_{\alpha\nu}{}^{\rho} \qquad (3.25)$$

In fact the first and third terms are symmetric because $\Gamma_{\mu\nu}{}^{\alpha} = \Gamma_{\nu\mu}{}^{\alpha}$; the last term is symmetric, because

$$\Gamma_{\mu\rho}{}^{\alpha} \Gamma_{\alpha\nu}{}^{\rho} = \Gamma_{\rho\mu}{}^{\alpha} \Gamma_{\nu\alpha}{}^{\rho} = \Gamma_{\nu\alpha}{}^{\rho} \Gamma_{\rho\mu}{}^{\alpha} = \Gamma_{\nu\rho}{}^{\alpha} \Gamma_{\alpha\mu}{}^{\rho} \qquad (3.26)$$

The symmetry of the second term can be shown using the following useful formula (see the Appendix A at the end of this Chapter)

$$d\sqrt{-g} = \frac{1}{2} \sqrt{-g}\, g^{\mu\nu} d g_{\mu\nu} = -\frac{1}{2} \sqrt{-g}\, g_{\mu\nu}\, d g^{\mu\nu} \qquad (3.27)$$

where $g = \det g_{\mu\nu}$. From the definition the Christoffel symbols we have

$$\Gamma_{\mu\alpha}{}^{\alpha} = \frac{1}{2} g^{\alpha\nu} \partial_{\mu} g_{\alpha\nu} \qquad (3.28)$$

and then, combining eq. (3.27), (3.28),

$$\Gamma_{\mu\alpha}{}^{\alpha} = \frac{1}{\sqrt{-g}}\,\partial_\mu \sqrt{-g} = \partial_\mu \ln(\sqrt{-g}) \qquad (3.29)$$

Therefore

$$\partial_\nu \Gamma_{\mu\alpha}{}^{\alpha} = \partial_\nu \partial_\mu \ln(\sqrt{-g}) = \partial_\mu \partial_\nu \ln(\sqrt{-g}) = \partial_\mu \Gamma_{\nu\alpha}{}^{\alpha} \qquad (3.30)$$

It should be noticed that the Ricci tensor, being a symmetric second rank tensor, has, in four dimensions, only 10 independent components. The Riemann tensor, however, has, in four dimensions, 20 independent components. In fact, its $4^4 = 256$ components are reduced first of all by the antisymmetry properties (3.16) and (3.17), so that $R_{\mu\nu\alpha\beta}$ behaves like a combination of two second rank antisimmetric tensor, $R'_{[\mu\nu]}$ and $R''_{[\alpha\beta]}$, each with 6 components; therefore the components are first reduced to $6 \times 6 = 36$. Moreover, from eq. (3.18), among the 30 components with $\mu,\nu = \alpha,\beta$ only 15 are independent, and we are left with $15 + 6 = 21$ components. The last condition (3.19) can be used to eliminate one additional component, and then we obtain the following 20 independent components (for simplicity, $R_{\mu\nu\alpha\beta}$ is symbolically written as $\mu\nu\alpha\beta$)

1212	1223	1313	1324	1423	2323	2424
1213	1224	1314	1334	1424	2324	2434
1214	1234	1323	1414	1434	2334	3434

where $1234 - 1324 + 1423 = 0 \qquad (3.31)$

In a three-dimensional space (omitting the components containing the index 4) we are left only with 6 independent components, while in two dimensions we have only one component, R_{1212}. In this case the curvature scalar, $R = g^{\mu\beta} g^{\alpha\nu} R_{\mu\alpha\nu\beta}$, coincides with the Gauss curvature, $R = 2K$, where $K = 1/R_1 R_2$, and R_1, R_2 are the two principal radii of curvature of the surface.

Before discussing the equations for the gravitational field, it is useful to derive another identity.

Starting with the Bianchi identity (3.20), and contracting μ with β, we obtain

$$R_\nu{}^{\alpha}{}_{;\rho} - R_\rho{}^{\alpha}{}_{;\nu} + R_{\nu\rho}{}^{\alpha\beta}{}_{;\mu} = 0 \qquad (3.32)$$

The contraction of ν and α gives

$$R_{;\rho} - R_{\rho}{}^{\alpha}{}_{;\alpha} - R_{\rho}{}^{\beta}{}_{;\beta} = 0 \qquad (3.33)$$

which can be written

$$G_{\mu}{}^{\nu}{}_{;\nu} = 0 \qquad (3.34)$$

where

$$G_{\mu}{}^{\nu} = R_{\mu}{}^{\nu} - \frac{1}{2}\delta_{\mu}{}^{\nu} R \qquad (3.35)$$

is called the Einstein tensor, and eq.(3.34) is called the contracted Bianchi identity. Note that $G_{\mu\nu} = g_{\mu\alpha} G^{\alpha}{}_{\nu} = G_{\nu\mu}$. In the presence of torsion, on the contrary, $G_{[\mu\nu]} \neq 0$, and also the contracted Bianchi identity is modified (see Chapter X).
Concluding this Section, we wish to define the Weyl tensor, $C_{\mu\nu\alpha\beta}$, which is related to the curvature tensor by

$$C_{\mu\nu\alpha\beta} = R_{\mu\nu\alpha\beta} - \frac{1}{2}(g_{\mu\beta}B_{\nu\alpha} + g_{\nu\alpha}B_{\mu\beta} - g_{\mu\alpha}B_{\nu\beta} - g_{\nu\beta}B_{\mu\alpha}) - \frac{1}{12}R(g_{\mu\beta}g_{\nu\alpha} - g_{\mu\alpha}g_{\nu\beta}) \qquad (3.36)$$

where

$$B_{\mu\nu} = R_{\mu\nu} - \frac{1}{4}g_{\mu\nu}R \qquad (3.37)$$

is the traceless part of the Ricci tensor ($g^{\mu\nu}B_{\mu\nu} = 0$). Contracting (3.36) by $g^{\mu\beta}$ we have

$$g^{\mu\beta}C_{\mu\nu\alpha\beta} = 0 \qquad (3.38)$$

$C_{\mu\nu\alpha\beta}$ is also called the conformal curvature tensor, because it is left invariant by a conformal transformation of the metric, $g'_{\mu\nu} = \varphi\, g_{\mu\nu}$, where φ is a scalar function. In fact $C'_{\mu\nu\alpha\beta} = C_{\mu\nu\alpha\beta}(g'_{\mu\nu}) = C_{\mu\nu\alpha\beta}(g_{\mu\nu})$ (unlike the Riemann tensor).

The Weyl tensor satisfies the same identities as the Riemann tensor, eqs.(3.16-3.19), and in addition the relation (3.38), which is symmetric in ν and α (representing then 10 additional conditions). The independent components of the Weyl tensor are then 20 - 10 = 10 .

2.- Einstein's field equations

In the nonrelativistic gravitational theory of Newton, the gravitational scalar potential ϕ satisfies the well known Poisson equation

$$\nabla^2 \phi = 4\pi G \rho \tag{3.39}$$

(ρ is the matter density, and G the newton constant), whose general solution is

$$\phi(x) = -G \int d^3x' \frac{\rho(x')}{|\vec{x}-\vec{x}'|} \tag{3.40}$$

The relativistic generalization of the matter energy density is, as we know from special relativity, the energy momentum tensor $T_{\mu\nu}$ (see the appendix B at the end of this Chapter); on the other hand we have seen that, in a Riemann space, the metric tensor $g_{\mu\nu}$ plays the role of the gravitational potential. A relativistic generalization of eq.(3.39) should contain therefore, on the right-hand side, the energy momentum tensor of matter. In order to mantain the analogy with the Newton theory, the other side of the generalized equations must satisfy the following conditions: a) it must contain no derivatives of $g_{\mu\nu}$ of order higher than two, and b) it must be linear in the second derivatives of $g_{\mu\nu}$ (these requirements follows immediately from the analogy with the Poisson equation (3.39).

Moreover the covariant divergence of the left-hand side of the generalized equations must be vanishing. This follows from the energy-momentum conservation law, which in special relativity is written $\partial_\nu T_\mu{}^\nu = 0$ (see the Appendix B), and it becomes $T_\mu{}^\nu{}_{;\nu} = 0$ in a general curvilinear system (see the next Section), in agreement with the principle of equivalence (which requires that locally, in a frame in which $\Gamma = 0$, the special relativistic law, $\partial_\nu T_\mu{}^\nu = 0$, should be recovered).

All the three previous conditions are satisfied by the Einstein tensor (3.35). We are led to postulate then the following equations

$$G_{\mu\nu} = \chi T_{\mu\nu} \tag{3.40}$$

Tracing the Einstein tensor we have

$$G_{\mu}{}^{\mu} = g^{\mu\nu} G_{\mu\nu} = R - 2R = -R = \chi T \tag{3.41}$$

where $T = T_{\mu}{}^{\mu}$. Therefore these field equations can also be rewritten

$$R_{\mu\nu} = \chi \left(T_{\mu\nu} - \frac{1}{2} g_{\mu\nu} T \right) \tag{3.42}$$

In order to determine the value of the constant χ we require that, in the nonrelativistic limit, the equations (3.42) are reduced to the Poisson equation (3.39). To this aim, consider the motion of a particle described by the geodesic equation

$$\frac{d^2x^{\mu}}{ds^2} + \Gamma_{\alpha\beta}{}^{\mu} \frac{dx^{\alpha}}{ds} \frac{dx^{\beta}}{ds} = 0 \tag{3.43}$$

In the nonrelativistic limit, neglecting terms of second order in the velocity, we have $ds^2 \simeq c^2 dt^2 = (dx^4)^2$, and eq.(3.43) becomes

$$\frac{d^2x^{\mu}}{dt^2} = - \Gamma_{44}{}^{\mu} \left(\frac{dx^4}{dt} \right)^2 = - c^2 \Gamma_{44}{}^{\mu} \tag{3.44}$$

In the limit of weak gravitational field (see also Chapter VIII, Sect.1), we can put $g_{\mu\nu} = \eta_{\mu\nu} + h_{\mu\nu}$, with $|h_{\mu\nu}| \ll 1$, and we can neglect terms of order h^2 and higher. We obtain then, from the definition (2.60) of the Christoffel symbols, supposing that the field is static (i.e. $\partial_4 g_{\mu\nu} = 0$),

$$\Gamma_{44}{}^{k} = \frac{1}{2} g^{km} (- \partial_m g_{44}) = \frac{1}{2} \partial_k h_{44} \tag{3.45}$$

where Latin indices takes only spatial values 1,2,3. The geodesic equation becomes then

$$\frac{d^2\vec{x}}{dt^2} = - \frac{c^2}{2} \vec{\nabla} h_{44} \tag{3.46}$$

and coincides with Newton's equation of motion

$$\frac{d^2\vec{x}}{dt^2} = - \vec{\nabla} \phi \tag{3.47}$$

provided that

$$h_{44} = \frac{2}{c^2}\,\phi \tag{3.48}$$

In the case of very slow motions, the leading term in the energy-momentum tensor is $T_{44} = \rho c^2$, and the trace is then $T = g^{\mu\nu} T_{\mu\nu} \simeq g^{44} T_{44} \simeq \eta^{44} T_{44} = \rho c^2$. From eq. (3.42) we have then

$$R_{44} = \frac{1}{2}\chi \rho c^2 \tag{3.49}$$

Neglecting temporal derivatives, and Γ^2 terms, eq.(3.25) gives in this limit

$$R_{44} = \partial_\kappa \Gamma_{44}{}^{\kappa} = \frac{1}{2}\nabla^2 h_{44} = \frac{1}{c^2}\nabla^2 \phi \tag{3.50}$$

(we have used (3.48)). Finally, combining (3.49,3.50) we obtain

$$\nabla^2 \phi = \frac{1}{2}\chi \rho c^4 \tag{3.51}$$

which coincides with the Poisson equation provided that we fix

$$\chi = \frac{8\pi G}{c^4} = 2.073 \times 10^{-48} \text{ c.g.s.} \tag{3.52}$$

3.- Field equations from a variational principle

The field equations (4.40) can also be deduced formally from an action using a minimum principle.

In order to formulate a theory of gravity generally covariant and locally Lorentz invariant, we must start from an action inte gral which is a scalar under both Lorentz and general coordinate transformations.

To this aim we note that the infinitesimal four-volume d^4x is a Lorentz scalar, but, under a general coordinate transformation, its transforms like $d^4x \longrightarrow d^4x' = d^4x\,|\partial x'/\partial x|$, where $|\partial x'/\partial x|$ is the Jacobian determinant of the transformation . In general, a geometrical object is called a density of weight w if its tensorial tranformation rule is multiplied by $|\partial x'/\partial x|^{w}$. For example, $\mathfrak{F}^{\mu}$ is a vector density of weight w if it transforms as

$$\xi'^{\mu} = \left|\frac{\partial x'}{\partial x}\right|^{w} \frac{\partial x'^{\mu}}{\partial x^{\nu}} \xi^{\nu} \tag{3.53}$$

The four-volume element is then a scalar density of weight $w = +1$. A true scalar action therefore is obtained integrating a scalar density of weight $w = -1$, so that the total weight is zero. Such a scalar density can be simply obtained considering the transformations of the metric tensor

$$g'_{\mu\nu} = \frac{\partial x^{\alpha}}{\partial x'^{\mu}} \frac{\partial x^{\beta}}{\partial x'^{\nu}} g_{\alpha\beta} \tag{3.54}$$

from which, by taking the determinant, and putting $g = \det g_{\mu\nu}$, we have

$$g' = \left|\frac{\partial x}{\partial x'}\right|^{2} g = \left|\frac{\partial x'}{\partial x}\right|^{-2} \cdot g \tag{3.55}$$

$$\sqrt{-g'} = \left|\frac{\partial x'}{\partial x}\right|^{-1} \sqrt{-g} \tag{3.56}$$

Therefore g has weight $w = -2$, and $\sqrt{-g}$ has $w = -1$. The required scalar density for the action can be obtained then by multiplying a scalar Lagrangian by $\sqrt{-g}$, and the action can be written as follows

$$S = \int d^4x \sqrt{-g} \left(\mathcal{L}_g + \mathcal{L}_m \right) \tag{3.57}$$

where $\mathcal{L}_g$ is the Lagrangian for the gravitational field, and $\mathcal{L}_m$ for the matter fields.

It should be noted that another density of weight -1 is the totally antisymmetric tensor $\varepsilon^{\mu\nu\alpha\beta}$ (see Appendix A), which is contained, for example, in the matter Lagrangian for a spin 3/2 particle (see Chapter XIII); in this case the Lagrangian as already $w = -1$, and the additional factor $\sqrt{-g}$ is not needed.

Note also that, in the expression for the total action, we have not written explicitly an interaction Lagrangian. This because the coupling between matter and gravitation is already contained in the matter Lagrangian , where, following a "minimal coupling" procedure (suggested by the principle of equivalence), scalar products are performed with the Riemanniann metric $g_{\mu\nu}$, and the partial derivatives are replaced by the covariant ones.

The field Lagrangian $\mathcal{L}_g$ must be a scalar containing the gravitational potential $g_{\mu\nu}$ and its derivatives. In general relativity one chooses

$$\mathcal{L}_g = -\frac{R}{2\chi} = -\frac{1}{2\chi} g^{\mu\nu} R_{\mu\nu} \qquad (3.58)$$

where R is the curvature scalar, and this choice leads, as we shall see, to the field equations (3.40) (other choices are however possible, which lead to gravitational theories other than general relativity; the only fundamental constraint is of course the agreement with experimental data).

In order to justify the presence of the minus sign in front of the scalar curvature, in eq.(3.58), it must be remembered that the choice of the sign of a Lagrangian density $\mathcal{L}$ is determined by the requirement that the corresponding action $S = \int dt \int d^3x\, \mathcal{L}$ have a minimum. Considering the temporal derivatives of $g_{\mu\nu}$ appearing in R, it can be easily shown that, with our metric conventions and with the definition (3.25) of $R_{\mu\nu}$, they are contained in a term which is always negative. Therefore, as $\chi > 0$, the quantity $R/2\chi$ would decrease without limit, by a sufficiently rapid change of $g_{\mu\nu}$ with t. This is avoided chosing $-R/2\chi$ as the gravitational Lagrangian density.

The variation of the field action with respect to $g^{\mu\nu}$ gives

$$\delta S_g = \frac{\delta S_g}{\delta g^{\mu\nu}} \delta g^{\mu\nu} = -\frac{1}{2\chi} \int d^4x\, \delta\left(\sqrt{-g}\, g^{\mu\nu} R_{\mu\nu}\right) =$$

$$= -\frac{1}{2\chi} \int d^4x \left(\sqrt{-g}\, R_{\mu\nu}\, \delta g^{\mu\nu} + \sqrt{-g}\, g^{\mu\nu}\, \delta R_{\mu\nu} - \frac{1}{2} R \sqrt{-g}\, g_{\mu\nu}\, \delta g^{\mu\nu}\right) =$$

$$= -\frac{1}{2\chi} \int d^4x \sqrt{-g} \left(R_{\mu\nu} - \frac{1}{2} R g_{\mu\nu}\right) \delta g^{\mu\nu} - \frac{1}{2\chi} \int d^4x \sqrt{-g}\, g^{\mu\nu}\, \delta R_{\mu\nu} \qquad (3.59)$$

(we have used (3.27) for $\delta\sqrt{-g}$).

In order to evaluate the last integral, we note that $\delta \Gamma_{\mu\nu}{}^{\alpha}$, being a difference (infinitesimal) of two connections, is a tensor (the fact that the difference of two connections transforms like a tensor can be directly verified from the transformation rules (2.38)). Therefore the variation $\delta R_{\mu\nu}$ can be performed equivalently in a locally inertial frame ($\Gamma = 0$) where we obtain (see 3.25)

$$g^{\mu\nu}\delta R_{\mu\nu} = g^{\mu\nu}\left(\partial_\alpha \delta \Gamma_{\mu\nu}{}^{\alpha} - \partial_\mu \delta \Gamma_{\alpha\nu}{}^{\alpha}\right) \tag{3.60}$$

Putting

$$V^\alpha = g^{\mu\nu}\delta \Gamma_{\mu\nu}{}^{\alpha} - g^{\nu\alpha}\delta \Gamma_{\nu\beta}{}^{\beta} \tag{3.61}$$

eq.(3.60) becomes

$$g^{\mu\nu}\delta R_{\mu\nu} = \partial_\alpha V^\alpha \tag{3.62}$$

because, in the coordinate system we are considering, the first derivatives of $g_{\mu\nu}$ are vanishing. In a more general frame this equation must be written, using (3.29)

$$g^{\mu\nu}\delta R_{\mu\nu} = V^\alpha{}_{;\alpha} = \partial_\alpha V^\alpha + \Gamma_{\alpha\nu}{}^{\alpha} V^\nu =$$

$$= \partial_\alpha V^\alpha + V^\alpha \partial_\alpha (\ln \sqrt{-g}) = \frac{1}{\sqrt{-g}}\partial_\alpha(\sqrt{-g}\, V^\alpha) \tag{3.63}$$

and then

$$\int d^4x \sqrt{-g}\, g^{\mu\nu}\delta R_{\mu\nu} = \int d^4x\, \partial_\alpha(\sqrt{-g}\, V^\alpha) \tag{3.64}$$

This integral, performed over a four-volume Ω , does not contribute to the action because, using the Gauss theorem[1], it may be transformed into a flux integral for the vector V^α over the boundary $F(\Omega)$ of the four-volume

$$\int_\Omega d^4x\, \partial_\alpha(\sqrt{-g}\, V^\alpha) = \oint_{F(\Omega)} dS_\alpha \sqrt{-g}\, V^\alpha \tag{3.65}$$

(dS_α is the infinitesimal element of a three dimensional hypersurface). This last integral is vanishing because, to obtain the field equations, the variation of the action is performed so that the field variation $\delta\Gamma$ is zero on the boundary (see Appendix B).

The variation of the matter action with respect to $g^{\mu\nu}$ defines the so-called "dynamical" energy-momentum tensor $T_{\mu\nu}$

$$\delta S_m = \frac{\delta S_m}{\delta g^{\mu\nu}}\delta g^{\mu\nu} = \int d^4x\, \delta(\sqrt{-g}\,\mathcal{L}_m) =$$

$$= \frac{1}{2}\int d^4x \sqrt{-g}\, T_{\mu\nu}\,\delta g^{\mu\nu} \tag{3.66}$$

where

$$T_{\mu\nu} = \frac{2}{\sqrt{-g}} \frac{\delta(\sqrt{-g}\,\mathcal{L}_m)}{\delta g^{\mu\nu}} = T_{\nu\mu} \tag{3.67}$$

Performing explicitly the variation we obtain

$$\delta S_m = \int d^4x \left[\frac{\partial(\sqrt{-g}\,\mathcal{L}_m)}{\partial g^{\mu\nu}} \delta g^{\mu\nu} + \frac{\partial(\sqrt{-g}\,\mathcal{L}_m)}{\partial(\partial_\alpha g^{\mu\nu})} \delta(\partial_\alpha g^{\mu\nu}) \right] =$$

$$= \int d^4x \left[\frac{\partial(\sqrt{-g}\,\mathcal{L}_m)}{\partial g^{\mu\nu}} - \partial_\alpha \frac{\partial(\sqrt{-g}\,\mathcal{L}_m)}{\partial(\partial_\alpha g^{\mu\nu})} \right] \delta g^{\mu\nu} +$$

$$+ \int d^4x\, \partial_\alpha \left[\frac{\partial(\sqrt{-g}\,\mathcal{L}_m)}{\partial(\partial_\alpha g^{\mu\nu})} \delta g^{\mu\nu} \right] . \tag{3.68}$$

Again the last integral does not contribute, being the integral of a total divergence (it becomes a flux integral over the boundary, where $\delta g^{\mu\nu} = 0$) and then one obtains, comparing (3.66,3.68)

$$\frac{1}{2}\sqrt{-g}\, T_{\mu\nu} = \frac{\partial(\sqrt{-g}\,\mathcal{L}_m)}{\partial g^{\mu\nu}} - \partial_\alpha \frac{\partial(\sqrt{-g}\,\mathcal{L}_m)}{\partial(\partial_\alpha g^{\mu\nu})} \tag{3.69}$$

It is well known that imposing on the action the invariance under infinitesimal global translations in space-time, in special relativity, one obtains the energy-momentum conservation law (see the Appedix B). The expression (3.69) may be correctly interpreted as the general relativistic definition of the energy-momentum tensor, because the matter action is invariant under local infinitesimal translations, $\delta x^\mu = \xi^\mu(x)$, provided that the tensor (3.69) is covariantly conserved, i.e. $T_\mu{}^\nu{}_{;\nu} = 0$.

In fact, considering the infinitesimal transformation

$$\delta x^\mu = \xi^\mu \tag{3.70}$$

the corresponding variation of the metric tensor is (remember eq.(2.33))

$$\delta g^{\mu\nu} = - \xi^{\alpha} \partial_{\alpha} g^{\mu\nu} + g^{\alpha\nu} \partial_{\alpha} \xi^{\mu} + g^{\mu\alpha} \partial_{\alpha} \xi^{\nu} \qquad (3.71)$$

which can also be rewritten as (see Chapter VIII, Appendix A)

$$\delta g^{\mu\nu} = \xi^{\mu;\nu} + \xi^{\nu;\mu} \qquad (3.72)$$

(using the explicit definition of covariant derivatives and Christoffel symbols). The infinitesimal variation of the action (3.66) is then

$$\delta S_m = \int d^4x \sqrt{-g}\, T_{\mu\nu}\, \xi^{\mu;\nu} \qquad (3.73)$$

(remember that $T_{\mu\nu} = T_{\nu\mu}$). Therefore

$$\delta S_m = \int d^4x \sqrt{-g}\, (T_{\mu}{}^{\nu} \xi^{\mu})_{;\nu} -$$

$$- \int d^4x \sqrt{-g}\, T_{\mu}{}^{\nu}{}_{;\nu}\, \xi^{\mu} \qquad (3.74)$$

The first integral can be transformed using the property (3.29) and the Gauss theorem, and its contribution is vanishing

$$\int d^4x \sqrt{-g} \frac{1}{\sqrt{-g}} \partial_{\nu} (\sqrt{-g}\, T_{\mu}{}^{\nu} \xi^{\mu}) =$$

$$= \oint dS_{\nu} \sqrt{-g}\, T_{\mu}{}^{\nu} \xi^{\mu} = 0 \qquad (3.75)$$

(modulo surface terms coming from the boundary of the four-dimensional integration interval). Therefore the action is invariant, $\delta S_m = 0$, for local infinitesimal translations, if

$$T_{\mu}{}^{\nu}{}_{;\nu} = 0 \qquad (3.76)$$

(in flat space-time this equation is reduced to $\partial_{\nu} T_{\mu}{}^{\nu} = 0$).

Coming back to the gravitational action (3.57), using (3.59) and (3.66), and imposing that $\delta S = \delta S_g + \delta S_m = 0$ one obtains the Einstein field equations

$$R_{\mu\nu} - \frac{1}{2} g_{\mu\nu} R = \chi T_{\mu\nu} \qquad (3.77)$$

which are exactly the postulated gravitational equations (3.40), if the free parameter χ of the action is choosen to be $\chi = 8\pi G/c^4$.

These equations are nonlinear, therefore the superposition principle is no longer valid for the gravitational field (unlike the case of the electromagnetic field in special relativity). This principle holds only in the weak field limit, where the eq.(3.77) can be linearized in first approximation (see Chapter VIII, Sect.1).

Other two very important features of the Einstein field equations, which at this point should be stressed, are the following:

a) the conservation equation $T^{\mu\nu}{}_{;\nu} = 0$ follows from the field equations themselves (remember the contracted Bianchi identity $G^{\mu\nu}{}_{;\nu} = 0$);

b) the energy-momentum tensor $T^{\mu\nu}$ represents the total energy density but the gravitational one, that is in $T^{\mu\nu}$ are included all the fields which contribute to the energy content of space-time, except the gravitational field itself.

Consider the point a) (the implications of the point b) will be discussed in detail in the Appendix B of Chapter VIII).

The fact that $T_\mu{}^\nu{}_{;\nu} = 0$ may be deduced directly from the field equations means that the field equations contain also the equations of motion for the matter producing the field (in fact, as we shall see explicitly in Chapter X, Sect.4, they follow from the condition $T_\mu{}^\nu{}_{;\nu} = 0$ which expresses the generalized conservation law of the matter energy-momentum in the presence of gravity).

In the electromagnetic theory the situation is different, as from the Maxwell field equations one can deduce the conservation of the electromagnetic current, but not the equations of motion of charges, i.e. the Lorentz equations: therefore one may assign the charge and current distributions, and then calculate the fields solving exactly the field equations. In the gravitational case, on the contrary, the matter distribution and its motions cannot be prescribed independently from the gravitational field produced by them. This characteristic aspect of gravitation is to be ascribed to the fact that the energy plays the double role of gravitational source and inertial mass.

It must be stressed, moreover, that the field equations

do not determine completely the distribution and motion of matter, as the ten equations (3.77) are related by the four conditions $G_\mu{}^\nu{}_{;\nu} = 0$, and then only six independent equations are left to determine the ten components of the metric tensor (this point will be discussed in detail in Section 5 of this Chapter).

In conclusion of this Section we note that the tensor $T^{\mu\nu}$ is called dynamical because it is the source of the gravitational field, as shown in eq.(3.77). Given a matter field, however, one can define also the so-called canonical energy-momentum tensor: starting with the matter Lagrangian, performing a variation not with respect to the metric, but with respect to x^μ and the matter field ψ, and considering in special relativity an infinitesimal global translation, one is led to define the canonical tensor $\theta_\mu{}^\nu$ (see the Appendix B).

This tensor is conserved, $\partial_\nu \theta_\mu{}^\nu = 0$, but is not symmetric in general; it can be symmetrized, however, by the addition of suitable divergence-free terms (Belinfante-Rosenfeld procedure[2]) (see Chapter X, Sect.3 for the possibility of using the canonical tensor as a source of gravity).

4.- Maxwell's equations in a Riemann space-time

If we introduce the coupling between electromagnetic and gravitational fields according to the minimal coupling procedure (i.e. replacing partial derivatives with covariant ones), then the field strenght tensor $F_{\mu\nu} = \partial_\mu A_\nu - \partial_\nu A_\mu$, where A_μ is the usual potential vector, is not modified: in fact, using the definition of covariant derivative (2.50),

$$F'_{\mu\nu} = A_{\nu;\mu} - A_{\mu;\nu} = F_{\mu\nu} + 2\,\Gamma_{[\mu\nu]}{}^\alpha A_\alpha = F_{\mu\nu} \qquad (3.78)$$

if the connection is symmetric. Therefore the first couple of Maxwell equations is unchanged, because the identity

$$\partial_{[\mu} F_{\alpha\beta]} = \partial_{[\mu}\partial_\alpha A_{\beta]} = 0 \qquad (3.79)$$

is still satisfied (being a consequence of the definition of $F_{\mu\nu}$). We remember, incidentally, that

$$T_{[\mu\nu\alpha]} = \frac{1}{3!}\left(T_{\mu\nu\alpha} + T_{\nu\alpha\beta} + T_{\alpha\mu\nu} - T_{\mu\alpha\nu} - T_{\nu\beta\alpha} - T_{\alpha\nu\mu}\right) \qquad (3.80)$$

Even more important, the electromagnetic theory is stil invariant under the local gauge transformations

$$A_\mu \to A'_\mu = A_\mu + \partial_\mu \Lambda \qquad (3.81)$$

(where Λ is a scalar function) also in the presence of gravity. In fact, the special relativistic action for the electromagnetic field coupled to an electric current density J^μ

$$S = \int d^4x \left(-\frac{1}{16\pi} F_{\mu\nu} F^{\mu\nu} + \frac{1}{c} J^\mu A_\mu \right) \qquad (3.82)$$

in the presence of a gravitational field becomes simply, according to the minimal coupling procedure,

$$S = \int d^4x \sqrt{-g} \left(-\frac{1}{16\pi} F_{\mu\nu} F^{\mu\nu} + \frac{1}{c} J'^\mu A_\mu \right) \qquad (3.83)$$

where J'^μ is the general relativistic definition of the current density (see 3.88) and where, of course, indices are raised and lowered, and scalar products are performed, using the metric tensor $g_{\mu\nu}$ instead of $\eta_{\mu\nu}$.

Under the local infinitesimal gauge transformations $\delta A_\mu = \partial_\mu \Lambda$, we have $\delta F_{\mu\nu} = \partial_{[\mu}\partial_{\nu]} \Lambda = 0$, and the variation of the action is then

$$\delta S = \int d^4x \sqrt{-g} \, \frac{1}{c} J'^\mu \partial_\mu \Lambda =$$

$$= \frac{1}{c} \int d^4x \, \partial_\mu (\sqrt{-g} \, J'^\mu \Lambda) - \frac{1}{c} \int d^4x \, \Lambda \, \partial_\mu (\sqrt{-g} \, J'^\mu) \qquad (3.84)$$

As already shown in the previous Section, the first integral may be transformed, using the Gauss theorem, into a flux integral over the boundary of the four-dimensional volume. The action is then gauge invariant, $\delta S = 0$ (modulo surface terms), if the general relativistic current J'^μ is covariantly conserved,

$$J'^\mu{}_{;\mu} = 0 \qquad (3.85)$$

In fact, using (3.29), this condition can be rewritten

$$\partial_\mu J'^\mu + \Gamma_{\mu\nu}{}^\mu J'^\nu = \partial_\mu J'^\mu + \partial_\nu (\ln \sqrt{-g}) J'^\nu =$$

$$= \partial_\mu J'^\mu + \frac{J'^\nu}{\sqrt{-g}} \partial_\nu \sqrt{-g} = \frac{1}{\sqrt{-g}} \partial_\mu (\sqrt{-g} \, J'^\mu) = 0 \qquad (3.86)$$

In order to show explicitly that this condition is satisfied, we remember that J'^{μ}, under a general coordinate transformations, must transform like a controvariant vector. As dx^{μ} is a vector, then also $dq\,dx^{\mu}$ is a vector, where dq is the electric charge contained inside an infinitesimal volume d^3x, because the electric charge q is a scalar. Putting $dq = \rho\, d^3x$, where ρ is the charge density, then $\rho\, d^3x\, dx^{\mu}$ is a vector, and also

$$\rho \frac{dx^{\mu}}{dt} d^3x\, dt = \rho \frac{dx^{\mu}}{dt} d^4x \qquad (3.87)$$

is a vector. The four-dimensional volume-element d^4x however is not a scalar under a general coordinate transformation and, as shown in the previous Section, to obtain a scalar we have to multiply it by $\sqrt{-g}$. The general relativistic controvariant current J'^{μ} (such that $J'^{\mu}\sqrt{-g}\, d^4x$ is a vector) is then

$$J'^{\mu} = \frac{\rho}{\sqrt{-g}} \frac{dx^{\mu}}{dt} = \frac{\rho_0}{\sqrt{-g}} \frac{dt}{d\tau} \frac{dx^{\mu}}{dt} = \frac{1}{\sqrt{-g}} J^{\mu} \qquad (3.88)$$

where ρ_0 is the proper charge density, i.e. the charge density measured in the system at rest with the charge, and J^{μ} is the special relativistic expression for the current density, which satisfies $\partial_{\mu} J^{\mu} = 0$ (charge conservation). The condition (3.86) is then satisfied, as $\partial_{\mu}(\sqrt{-g}\, J'^{\mu}) = \partial_{\mu} J^{\mu}$, and the action is gauge invariant.

The second pair of Maxwell's equations in a gravitational field can be obtained directly from the action (3.83), performing the variation with respect to the potential vector A_{μ}, and imposing $\delta S = 0$. Putting

$$\mathcal{L} = -\frac{\sqrt{-g}}{16\pi}(\partial_{\mu} A_{\nu} - \partial_{\nu} A_{\mu})(\partial^{\mu} A^{\nu} - \partial^{\nu} A^{\mu}) + \frac{\sqrt{-g}}{c} J'^{\mu} A_{\mu} \qquad (3.89)$$

we obtain

$$\delta S = \int d^4x \left\{ \frac{\partial \mathcal{L}}{\partial A_{\mu}} \delta A^{\mu} + \frac{\partial \mathcal{L}}{\partial(\partial_{\nu} A_{\mu})} \delta \partial_{\mu} A_{\mu} \right\} =$$

$$= \int d^4x \left\{ \frac{\partial \mathcal{L}}{\partial A_{\mu}} - \partial_{\nu} \frac{\partial \mathcal{L}}{\partial(\partial_{\nu} A_{\mu})} \right\} \delta A_{\mu} + \int d^4x\, \partial_{\nu} \left[\frac{\partial \mathcal{L}}{\partial(\partial_{\nu} A_{\mu})} \delta A_{\mu} \right]$$

$$(3.90)$$

Again the last integral does not contributes to the variation, as it becomes a flux integral over the boundary of the four-volume, where δA_μ is vanishing (see the Appendix B); imposing $\delta S = 0$ we obtain then the Eulero-Lagrange equations

$$\frac{\partial \mathcal{L}}{\partial A_\mu} = \partial_\nu \frac{\partial \mathcal{L}}{\partial(\partial_\nu A_\mu)} \tag{3.91}$$

From the Lagrangian density (3.89) we have

$$\frac{\partial \mathcal{L}}{\partial A_\mu} = \frac{\sqrt{-g}}{c} J'^\mu$$

$$\frac{\partial \mathcal{L}}{\partial(\partial_\nu A_\mu)} = \frac{4\sqrt{-g}}{16\pi} F^{\mu\nu} = \frac{\sqrt{-g}}{4\pi} F^{\mu\nu} \tag{3.92}$$

$$\partial_\nu \frac{\partial \mathcal{L}}{\partial(\partial_\nu A_\mu)} = \frac{1}{4\pi} \partial_\nu (\sqrt{-g}\, F^{\mu\nu})$$

and from (3.91) we obtain the following generalized Maxwell equations in a curved space-time:

$$\frac{1}{\sqrt{-g}} \partial_\nu (\sqrt{-g}\, F^{\mu\nu}) = \frac{4\pi}{c} J'^\mu \tag{3.93}$$

Using the symmetry of the connection, and eq.(3.29), we have

$$F^{\mu\nu}{}_{;\nu} = \partial_\nu F^{\mu\nu} + \Gamma_{\nu\alpha}{}^{\mu} F^{\alpha\nu} + \Gamma_{\nu\alpha}{}^{\nu} F^{\mu\alpha} =$$

$$= \partial_\nu F^{\mu\nu} + \partial_\alpha (\ln(\sqrt{-g})\, F^{\mu\alpha} = \partial_\nu F^{\mu\nu} + \frac{F^{\mu\alpha}}{\sqrt{-g}} \partial_\alpha \sqrt{-g} =$$

$$= \frac{1}{\sqrt{-g}} \partial_\nu (\sqrt{-g}\, F^{\mu\nu}) \tag{3.94}$$

(this result is valid for any antisymmetric tensor $F_{\mu\nu}$). The Maxwell equations can also be rewritten then

$$F^{\mu\nu}{}_{;\nu} = \frac{4\pi}{c} J'^\mu \tag{3.95}$$

and, as expected, they correspond to the electromagnetic equations of special relativity in which the partial derivatives have been replaced by the covariant ones, and the current J^μ by the generally covariant current vector $J'^\mu = J^\mu / \sqrt{-g}$.

It should be mentioned that, even in the absence of matter, $J'^{\mu} = 0$, the electromagnetic field itself is a source of gravity and bends space-time. From the electromagnetic free-field Lagrangian density

$$\mathcal{L} = -\frac{\sqrt{-g}}{16\pi} F_{\mu\nu} F^{\mu\nu} = -\frac{\sqrt{-g}}{16\pi} F_{\mu\nu} F_{\alpha\beta} g^{\mu\alpha} g^{\nu\beta} \qquad (3.96)$$

applying the formula (3.69) we obtain in fact the following dynamical energy-momentum tensor for the electromagnetic field

$$T_{\mu\nu} = \frac{1}{4\pi} \left(F_{\mu\alpha} F_{\nu}{}^{\alpha} - \frac{1}{4} g_{\mu\nu} F_{\alpha\beta} F^{\alpha\beta} \right) \qquad (3.97)$$

which is nonvanishing, and can be inserted into the right-hand side of Einstein's equations (3.77).

In this case $T_{\mu\nu}$ is traceless, $T = g^{\mu\nu} T_{\mu\nu} = 0$, and then the field equations, which can be equivalently written in the form (3.42), are reduced to

$$R_{\mu\nu} = \chi T_{\mu\nu} \qquad (3.98)$$

Finally we note that, for a particle subject to the action of a gravitational and electromagnetic force, we must introduce the Lorentz term into the geodesic equation, and we obtain the following equation of motion

$$\frac{Du^{\mu}}{Ds} = \frac{d^2 x^{\mu}}{ds^2} + \Gamma_{\alpha\beta}{}^{\mu} u^{\alpha} u^{\beta} = \frac{e}{mc} F^{\mu}{}_{\nu} u^{\nu} \qquad (3.99)$$

5.- The initial values problem in general relativity

The Einstein field equations (3.77) are ten, second order, partial differential equations for the ten components of $g_{\mu\nu}$. According to the usual Chauchy problem one should expect then that once the values of $g_{\mu\nu}$ and $\partial_t g_{\mu\nu}$ are assigned on the space-like initial hypersurface $t = t_0$, the temporal evolution of $g_{\mu\nu}$ is fixed, and then the metric can be calculated for each value of $\vec{x}$ and t.

From a physical point of view we know, however, that the metric tensor never may be determined univocally, as it can be always subject to an arbitrary change of coordinates.

From a mathematical point of view, the ten field equations $G_{\mu\nu} = \chi T_{\mu\nu}$ are related by the four conditions following from the contracted Bianchi identities, $G_{\mu}{}^{\nu}{}_{;\nu} = 0$, and then only six independent equations are left to determine the ten unknown components of $g_{\mu\nu}$. We can impose then four "gauge" conditions to fix the metric tensor, for example the so-called "harmonic gauge" $\partial_{\nu}(\sqrt{-g}\, g^{\mu\nu}) = 0$.

But let us consider the Chauchy problem. If we could deduce, from the field equations, the value of $\partial^2 g_{\mu\nu}/\partial t^2$, then, using the knowledge of $g_{\mu\nu}$ and its first temporal derivative at a given initial time t_0, then the metric would be determined everywhere. The field equations, however, contain only the second temporal derivatives of g_{ik} (Latin indices run from 1 to 3), as one can see directly from the explicit definition of the Riemann tensor, eq.(3.7) (the second temporal derivatives are contained in $R_{i4k4} \propto \propto \partial^2 g_{ik}/\partial t^2$).

On the other hand, consider the Bianchi identity, which may be written

$$\partial_4 G^{\mu 4} = -\partial_i G^{\mu i} - \Gamma_{\nu\alpha}{}^{\mu} G^{\alpha\nu} - \Gamma_{\nu\alpha}{}^{\nu} G^{\mu\alpha} \tag{3.100}$$

As the right-hand side of these equations contains at most the second temporal derivatives of $g_{\mu\nu}$, then $G^{\mu 4}$ must contain only first order temporal derivatives. Therefore the four field equations

$$G^{\mu 4} = \chi T^{\mu 4} \tag{3.101}$$

do not determine the temporal evolution of $g_{\mu\nu}$, but represent only a condition on the initial data. The dynamical equations for the gravitational field are the six equations

$$G^{ik} = \chi T^{ik} \tag{3.102}$$

from which the value of $\partial^2 g_{ik}/\partial t^2$ can be obtained, but $\partial^2 g_{i4}/\partial t^2$ is left undetermined.

The problem of the initial data can be solved removing this ambiguity with a gauge condition which fixes the coordinate system. For example, in the harmonic gauge one has

$$\frac{\partial^2}{\partial t^2}(\sqrt{-g}\, g^{\mu 4}) = -\frac{\partial}{\partial t}\partial_i(\sqrt{-g}\, g^{\mu i}) \tag{3.103}$$

and then all the ten values of $\partial^2 g_{\mu\nu}/\partial t^2$ are determined by these four additional conditions, plus the six dynamical equations (3.102).

This method of solving the initial value problem can be used, in practice, to generate numerical solutions, starting from the initial data, by means of a computer program.

In conclusion it is interesting to remark that if the initial values are assigned on a null hypersurface, i.e. a three-dimensional surface whose normal vector V^μ is light-like, $V^\mu V_\mu = 0$, then the field equations represent only a set of conditions on the initial data, and none of them is a dynamical equation, in the sense discussed before.

APPENDIX A

The totally antisymmetric tensor in curvilinear coordinates

In special relativity, the components of the totally antisymmetric tensor $\varepsilon^{\mu\nu\alpha\beta}$ are : +1 if $(\mu, \nu, \alpha, \beta)$ are equal to (1,2,3,4) or to a permutation of even degree of these four numbers; -1 if $(\mu, \nu, \alpha, \beta)$ are a permutation of odd degree of (1,2,3,4); finally, $\varepsilon^{\mu\nu\alpha\beta} = 0$ if at least two of the four indices are equal.

Under a general transformation of coordinates, the symbol $\varepsilon^{\mu\nu\alpha\beta}$ does not behave like a true tensor, but like a density of weight w = -1 (remember the definition of tensor density given in the Sect.3 of this Chapter).

In fact, given any 4×4 matrix $K_\alpha{}^\beta$, its determinant K satisfies the general relation

$$\varepsilon^{\mu\nu\alpha\beta} K = \varepsilon^{\lambda\rho\sigma\delta} K_\lambda{}^\mu K_\rho{}^\nu K_\sigma{}^\alpha K_\delta{}^\beta \qquad (A.1)$$

which follows from the definition of determinant. In the case of a coordinate transformation $x^\mu \to x'^\mu$, replacing the matrix $K_\alpha{}^\beta$ with $\partial x'^\beta / \partial x^\alpha$, one obtains, from (A.1), the following transformation rule

$$\varepsilon'^{\mu\nu\alpha\beta} = \left| \frac{\partial x'}{\partial x} \right|^{-1} \varepsilon^{\lambda\rho\sigma\delta} \frac{\partial x'^\mu}{\partial x^\lambda} \frac{\partial x'^\nu}{\partial x^\rho} \frac{\partial x'^\alpha}{\partial x^\sigma} \frac{\partial x'^\beta}{\partial x^\delta} \qquad (A.2)$$

where $|\partial x'/\partial x| = \det(\partial x'^\mu / \partial x^\nu)$ is the Jacobian determinant of the transformation. According to the definition of Sect.3, this equation shows that $\varepsilon^{\mu\nu\alpha\beta}$ behaves like a tensor density of rank four and weight w = -1.

A fourth rank totally antisymmetric controvariant tensor can be obtained then by multiplying $\varepsilon^{\mu\nu\alpha\beta}$ by a scalar density of weight w = +1. Remembering that $\sqrt{-g}$ has w = -1, we are led then to introduce the following generalized symbol

$$\eta^{\mu\nu\alpha\beta} = \frac{1}{\sqrt{-g}} \varepsilon^{\mu\nu\alpha\beta} \qquad (A.3)$$

which transforms correctly as a controvariant tensor under general coordinate transformations, and which must replace the special relativistic symbol $\varepsilon^{\mu\nu\alpha\beta}$ in the case of a curvilinear coordinate system.

Following the same procedure, one can show easily that $\varepsilon_{\mu\nu\alpha\beta}$ is a tensor density of weight +1, as

$$\varepsilon'_{\mu\nu\alpha\beta} = \left|\frac{\partial x'}{\partial x}\right| \varepsilon_{\lambda\rho\sigma\delta} \frac{\partial x^{\lambda}}{\partial x'^{\mu}} \frac{\partial x^{\rho}}{\partial x'^{\nu}} \frac{\partial x^{\sigma}}{\partial x'^{\alpha}} \frac{\partial x^{\delta}}{\partial x'^{\beta}} \qquad (A.4)$$

and the corresponding generalized totally antisymmetric tensor, transforming covariantly under a general coordinate transformation, is then

$$\eta_{\mu\nu\alpha\beta} = \sqrt{-g}\, \varepsilon_{\mu\nu\alpha\beta} \qquad (A.5)$$

Using the antisymmetry properties of the ε symbols, we obtain the following multiplication rules for two totally antisymmetric tensors:

$$\eta^{\mu\nu\alpha\beta}\, \eta_{\mu\nu\alpha\beta} = -4! \qquad (A.6)$$

$$\eta^{\mu\nu\alpha\beta}\, \eta_{\lambda\nu\alpha\beta} = -3!\, \delta^{\mu}_{\lambda} \qquad (A.7)$$

$$\eta^{\mu\nu\alpha\beta}\, \eta_{\lambda\rho\alpha\beta} = -2\, \delta^{\mu\nu}_{\lambda\rho} \qquad (A.8)$$

$$\eta^{\mu\nu\alpha\beta}\, \eta_{\lambda\rho\sigma\beta} = -\delta^{\mu\nu\alpha}_{\lambda\rho\sigma} \qquad (A.9)$$

$$\eta^{\mu\nu\alpha\beta}\, \eta_{\lambda\rho\sigma\gamma} = -\delta^{\mu\nu\alpha\beta}_{\lambda\rho\sigma\gamma} \qquad (A.10)$$

where

$$\delta^{\mu\nu}_{\lambda\rho} = \begin{vmatrix} \delta^{\mu}_{\lambda} & \delta^{\nu}_{\lambda} \\ \delta^{\mu}_{\rho} & \delta^{\nu}_{\rho} \end{vmatrix} = \delta^{\mu}_{\lambda}\delta^{\nu}_{\rho} - \delta^{\nu}_{\lambda}\delta^{\mu}_{\rho} \qquad (A.11)$$

$$\delta^{\mu\nu\alpha}_{\lambda\rho\sigma} = \begin{vmatrix} \delta^{\mu}_{\lambda} & \delta^{\nu}_{\lambda} & \delta^{\alpha}_{\lambda} \\ \delta^{\mu}_{\rho} & \delta^{\nu}_{\rho} & \delta^{\alpha}_{\rho} \\ \delta^{\mu}_{\sigma} & \delta^{\nu}_{\sigma} & \delta^{\alpha}_{\sigma} \end{vmatrix} = \delta^{\mu}_{\lambda}\delta^{\nu\alpha}_{\rho\sigma} - \delta^{\nu}_{\lambda}\delta^{\mu\alpha}_{\rho\sigma} + \delta^{\alpha}_{\lambda}\delta^{\mu\nu}_{\rho\sigma} \qquad (A.12)$$

$$\delta^{\mu\nu\alpha\beta}_{\lambda\rho\sigma\gamma} = \begin{vmatrix} \delta^{\mu}_{\lambda} & \delta^{\nu}_{\lambda} & \delta^{\alpha}_{\lambda} & \delta^{\beta}_{\lambda} \\ \delta^{\mu}_{\rho} & \delta^{\nu}_{\rho} & \delta^{\alpha}_{\rho} & \delta^{\beta}_{\rho} \\ \delta^{\mu}_{\sigma} & \delta^{\nu}_{\sigma} & \delta^{\alpha}_{\sigma} & \delta^{\beta}_{\sigma} \\ \delta^{\mu}_{\gamma} & \delta^{\nu}_{\gamma} & \delta^{\alpha}_{\gamma} & \delta^{\beta}_{\gamma} \end{vmatrix} =$$

$$= \delta^{\mu}_{\lambda}\, \delta^{\nu\alpha\beta}_{\rho\sigma\gamma} - \delta^{\nu}_{\lambda}\, \delta^{\mu\alpha\beta}_{\rho\sigma\gamma} + \delta^{\alpha}_{\lambda}\, \delta^{\mu\nu\beta}_{\rho\sigma\gamma} -$$

$$- \delta^{\beta}_{\lambda}\, \delta^{\mu\nu\alpha}_{\rho\sigma\gamma} \qquad (A.13)$$

In order to deduce (3.27) we note that $\eta_{\mu\nu\alpha\beta}$, being a fourth rank covariant tensor, can be related to the corresponding controvariant tensor of eq.(A.3) by lowering indices with the metric tensor,

$$\eta_{\mu\nu\alpha\beta} = g_{\mu\lambda}\, g_{\nu\rho}\, g_{\alpha\sigma}\, g_{\beta\delta}\, \eta^{\lambda\rho\sigma\delta} \qquad (A.14)$$

Differentiating both sides of this equation we have then

$$\varepsilon_{\mu\nu\alpha\beta}\, d(\sqrt{-g}) = dg_{\mu\lambda}\, \eta^{\lambda}{}_{\nu\alpha\beta} + dg_{\nu\rho}\, \eta_{\mu}{}^{\rho}{}_{\alpha\beta} +$$

$$+ dg_{\alpha\sigma}\, \eta_{\mu\nu}{}^{\sigma}{}_{\beta} + dg_{\beta\delta}\, \eta_{\mu\nu\alpha}{}^{\delta} - \varepsilon_{\mu\nu\alpha\beta}\, d(\sqrt{-g}) \qquad (A.15)$$

(we have used the fact that

$$g_{\mu\lambda}\, g_{\nu\rho}\, g_{\alpha\sigma}\, g_{\beta\delta}\, \varepsilon^{\lambda\rho\sigma\delta} = g\, \varepsilon_{\mu\nu\alpha\beta} \qquad (A.16)$$

according to the property (A.1)).

Multiplying (A.15) by $\eta^{\mu\nu\alpha\beta}$, and using eqs.(A.6,7) we obtain

$$- \frac{4!}{\sqrt{-g}}\, d(\sqrt{-g}) = -3!\, 4\, g^{\mu\lambda}\, dg_{\mu\lambda} + \frac{4!}{\sqrt{-g}}\, d(\sqrt{-g}) \qquad (A.17)$$

Therefore

$$d(\sqrt{-g}) = \frac{1}{2}\sqrt{-g}\, g^{\mu\nu}\, d g_{\mu\nu} \qquad (A.18)$$

Finally, remembering that

$$g_{\mu\alpha}\, g^{\mu\beta} = \delta_\alpha{}^\beta \qquad (A.19)$$

one has

$$g_{\mu\nu}\, g^{\mu\nu} = \delta_\nu{}^\nu = 4 \qquad (A.20)$$

Differentiating we have

$$g_{\mu\nu}\, d g^{\mu\nu} + g^{\mu\nu}\, d g_{\mu\nu} = 0 \qquad (A.21)$$

Eq.(A.18) may also be rewritten then

$$d(\sqrt{-g}) = \frac{1}{2}\sqrt{-g}\, g^{\mu\nu}\, d g_{\mu\nu} = -\frac{1}{2}\sqrt{-g}\, g_{\mu\nu}\, d g^{\mu\nu} \qquad (A.22)$$

and this coincides with the formula (3.27), used in the Sections 1 and 3 of this Chapter.

APPENDIX B

Four-dimensional variational formalism: the canonical energy-momentum tensor and the spin density tensor.

Consider a matter field ψ , described by special relativistic Lagrangian density $\mathcal{L}(\psi, \partial_\mu \psi, x^\mu)$. The field Lagrangian is then

$$L(\psi, \partial\psi, x) = \int_{-\infty}^{+\infty} d^3x\, \mathcal{L}(\psi, \partial\psi, x) \qquad (B.1)$$

and the action

$$S = \int_{t_1}^{t_2} dt\, L = \int_{t_1}^{t_2} dt \int_{-\infty}^{+\infty} d^3x\, \mathcal{L} = \int_\Omega d^4x\, \mathcal{L}(\psi, \partial\psi, x) \qquad (B.2)$$

The field equations can be obtained, according to the minimal action principle, performing the variation and imposing $\delta S = 0$, keeping however fixed the boundary of the integration interval.

The variation of the action may be written, in general,

$$\delta S = \int_\Omega d^4x \left[\frac{\partial \mathcal{L}}{\partial \psi}\delta\psi + \frac{\partial \mathcal{L}}{\partial(\partial_\nu \psi)}\delta\partial_\nu\psi + \frac{\partial \mathcal{L}}{\partial x^\nu}\delta x^\nu\right] =$$

$$= \int_\Omega d^4x \left[\frac{\partial \mathcal{L}}{\partial \psi} - \partial_\nu \frac{\partial \mathcal{L}}{\partial(\partial_\nu \psi)}\right]\delta\psi + \int_\Omega d^4x\, \partial_\nu J^\nu \qquad (B.3)$$

where

$$J^\nu = \frac{\partial \mathcal{L}}{\partial(\partial_\nu \psi)}\delta\psi + \mathcal{L}\,\delta x^\nu \qquad (B.4)$$

The last term in eq.(B.3) is the integral of a four-divergence over the four-dimensional volume Ω, bounded by the two hypersurfaces $t = t_1$ and $t = t_2$, and spatially extended to infinity . Applying the Gauss theorem, this becomes the flux integral of J^μ over the close boundary $F(\Omega)$ of th volume Ω :

$$\int_\Omega d^4x\, \partial_\nu J^\nu = \oint_{F(\Omega)} J^\nu dS_\nu \qquad (B.5)$$

(dS_ν is the infinitesimal four-dimensional vector for a three-dimensional volume-element). But at spatial infinity the fields are vanishing, and the variations $\delta\psi$ and δx are constrained to be zero at the temporal boundaries. Therefore the integral (B.5) vanishes, giving no contribution to the variation of the action.

By imposing on the action to have an extremum, $\delta S=0$, for an arbitrary variation $\delta\psi$, we obtain then, from eq. (B.13), the so-called Eulero-Lagrange field equations

$$\frac{\partial \mathcal{L}}{\partial \psi} = \partial_\nu \frac{\partial \mathcal{L}}{\partial(\partial_\nu \psi)} \qquad (B.6)$$

In order to express the variation of the action as a function of its generalized coordinates (the field ψ in this case) we impose on the variation to be vanishing only at one temporal boundary. Using the Eulero Lagrange equations as a constraint, we obtain then, from eq.(B.3),

$$\delta \mathcal{L} = \partial_\nu J^\nu \qquad (B.7)$$

and we can relate then the invariance properties of the action to the conservation of the four-dimensional current density J^μ.

Consider, for example, an infinitesimal global translation

$$\delta x^\mu = \varepsilon^\mu \quad , \qquad \delta\psi = -\varepsilon^\mu \partial_\mu \psi \tag{B.8}$$

where ε^μ are four infinitesimal constant parameters. Eq. (B.7) becomes in this case, using the definition (B.4),

$$\delta\mathcal{L} = \varepsilon^\mu \partial_\mu \left[\mathcal{L}\, \delta_\mu{}^\nu - \frac{\partial \mathcal{L}}{\partial(\partial_\nu \psi)} \partial_\mu \psi \right] \tag{B.9}$$

The requirement that the action be invariant under global translations leads then to the following conservation law

$$\partial_\nu \, \theta_\mu{}^\nu = 0 \tag{B.10}$$

where $\theta_\mu{}^\nu$ is the so-called canonical energy-momentum tensor

$$\theta_\mu{}^\nu = \mathcal{L}\, \delta_\mu{}^\nu - \frac{\partial \mathcal{L}}{\partial(\partial_\nu \psi)} \partial_\mu \psi \tag{B.11}$$

The corresponding energy-momentum four-vector P_μ is obtained by integrating the energy-momentum tensor over a space-like hypersurface Σ (in this Section we use units such that $\hbar = c = 1$)

$$P_\mu = \int \theta_\mu{}^\nu \, dS_\nu \tag{B.12}$$

This vector is conserved, i.e. its definition is independent from the choice of the hypersurface, provided that eq.(B.10) holds. Consider in fact two different space-like hypersurfaces Σ_1 and Σ_2. Then to the difference

$$\Delta P_\mu = \int_{\Sigma_2} \theta_\mu{}^\nu dS_\nu - \int_{\Sigma_1} \theta_\mu{}^\nu dS_\nu \tag{B.13}$$

we can add the integral of $\theta_{\mu\nu}$ over the time-like hypersurface Σ_3, connecting Σ_1 and Σ_2 at spatial infinity where the fields are vanishing. Therefore we can write, denoting with $F(\Omega)$ the closed hypersurface formed with $\Sigma_1, \Sigma_2, \Sigma_3$,

$$\Delta P_\mu = \oint_{F(\Omega)} \theta_\mu{}^\nu dS_\nu \tag{B.14}$$

and, using the Gauss theorem, eq.(B10) implies then the conservation of the four-momentum vector

$$\Delta P_\mu = \int_\Omega d^4x\, \partial_\nu \theta_\mu{}^\nu = 0 \tag{B.15}$$

If the action is invariant under global Lorentz rotations, one is led to the angular momentum conservation law, and to the definition of the canonical spin density tensor.

Consider in fact a global Lorentz transformation represented by the matrix $\Lambda^\mu{}_\nu$

$$x'^\mu = \Lambda^\mu{}_\nu x^\nu \tag{B.16}$$

such that

$$\eta_{\mu\nu} \Lambda^\mu{}_\alpha \Lambda^\nu{}_\beta = \eta_{\alpha\beta} \tag{B.17}$$

For a transformation infinitely closed to the identity we have

$$\Lambda^\mu{}_\nu = \delta^\mu{}_\nu + \omega^\mu{}_\nu \tag{B.18}$$

and then

$$\delta x^\mu = x'^\mu - x^\mu = \omega^\mu{}_\nu x^\nu \tag{B.19}$$

where $\omega_{\mu\nu} = -\omega_{\nu\mu}$(the antisymmetry of ω follows from B.17) are the six constant infinitesimalparameters of the transformation. The corresponding infinitesimal variation of the field is then

$$\begin{aligned} \delta\psi &= \frac{1}{2}\omega^{\mu\nu} S_{\mu\nu}\psi - \delta x^\mu \partial_\mu \psi = \\ &= \frac{1}{2}\omega^{\mu\nu}\left(S_{\mu\nu} - x_\nu\partial_\mu + x_\mu\partial_\nu\right)\psi \end{aligned} \tag{B.20}$$

where $S_{\mu\nu} = -S_{\nu\mu}$ are the six Lorentz generators for the field ψ. If the action is globally Lorentz invariant, then $\delta\mathcal{L} = 0$ and from eqs.(B.4,7,19,20) we get

$$\begin{aligned} \partial_\nu J^\nu = \partial_\nu \Big\{ &\frac{\partial\mathcal{L}}{\partial(\partial_\nu\psi)}\left[\frac{1}{2}\omega_{\alpha\beta}\left(S^{\alpha\beta} - x^\beta\partial^\alpha + x^\alpha\partial^\beta\right)\psi\right] + \\ &+ \mathcal{L}\,\frac{1}{2}\omega_{\alpha\beta}\left(\eta^{\alpha\nu}x^\beta - \eta^{\beta\nu}x^\alpha\right)\Big\} = \end{aligned}$$

$$= \frac{1}{2}\omega_{\alpha\beta}\partial_\nu\left\{\frac{\partial\mathcal{L}}{\partial(\partial_\nu\psi)} S^{\alpha\beta}\psi + \left[\mathcal{L}\eta^{\alpha\nu} - \frac{\partial\mathcal{L}}{\partial(\partial_\nu\psi)}\partial^\alpha\psi\right]x^\beta - \right.$$

$$\left. - \left[\mathcal{L}\eta^{\beta\nu} - \frac{\partial\mathcal{L}}{\partial(\partial_\nu\psi)}\partial^\beta\psi\right]x^\alpha\right\} =$$

$$= \omega_{\alpha\beta}\partial_\nu J^{\alpha\beta\nu} \qquad (B.21)$$

where, using (B.11),

$$J^{\alpha\beta\nu} = -S^{\alpha\beta\nu} + \frac{1}{2}\left(x^\beta\theta^{\alpha\nu} - x^\alpha\theta^{\beta\nu}\right) = -J^{\beta\alpha\nu} \qquad (B.22)$$

and

$$S^{\alpha\beta\nu} = -\frac{1}{2}\frac{\partial\mathcal{L}}{\partial(\partial_\nu\psi)} S^{\alpha\beta}\psi = -S^{\beta\alpha\nu} \qquad (B.23)$$

As $\theta_\mu{}^\nu$ is a four-momentum density, see (B.12), the last term in eq.(B.22) represents the density of orbital angular momentum. Therefore $J^{\alpha\beta\nu}$ corresponds to the total angular momentum density, provided that $S^{\alpha\beta\nu}$ is identified with the intrinsic spin density tensor of the field ψ .

In the case of a Dirac spinor, for example, the Lagrangian density is

$$\mathcal{L} = \frac{i}{2}\bar{\psi}\gamma^\nu\partial_\nu\psi - \frac{1}{2}m\bar{\psi}\psi + h.c. \qquad (B.24)$$

(where h.c. means hermitian conjugate) and the Lorentz generators are

$$S^{\alpha\beta} = \frac{1}{4}\left(\gamma^\alpha\gamma^\beta - \gamma^\beta\gamma^\alpha\right) \qquad (B.25)$$

where γ^μ are the Dirac matrices (see Chapter X, Sect.5). The canonical spin density tensor is then, from (B.23),

$$S^{\alpha\beta\nu} = -\frac{i}{8}\bar{\psi}\gamma^\nu\gamma^{[\alpha}\gamma^{\beta]}\psi + h.c. =$$

$$= -\frac{i}{8}\bar{\psi}\gamma^\nu\gamma^{[\alpha}\gamma^{\beta]}\psi - \frac{i}{8}\bar{\psi}\gamma^{[\alpha}\gamma^{\beta]}\gamma^\nu\psi \qquad (B.26)$$

and using the algebra of the Dirac matrices (see eqs.(10.155-157) for an explicit computation) one obtains

$$S^{\alpha\beta\nu} = -\frac{i}{4}\bar{\psi}\gamma^{[\nu}\gamma^{\alpha}\gamma^{\beta]}\psi = S^{[\alpha\beta\nu]} \qquad (B.27)$$

(exactly the same expression will be deduced, in Chapter X, using the field equations of the Einstein-Cartan theory, see eq.(10.176)).

The conservation law, following from (B.21),

$$\partial_\nu J^{\alpha\beta\nu} = 0 \qquad (B.28)$$

implies the conservation of the total angular momentum $M^{\alpha\beta} = - M^{\beta\alpha}$, defined as the integral of $J^{\alpha\beta\nu}$ over a space-like hypersurface Σ

$$M^{\alpha\beta} = \int_\Sigma J^{\alpha\beta\nu}\, dS_\nu \qquad (B.29)$$

In fact, using the Gauss theorem, we can show that the definition of $M^{\alpha\beta}$ is independent from the choice of Σ , provided that (B.28) is satisfied, just like in the case of the energy-momentum conservation.

In conclusion, we wish to stress that the angular momentum conservation law, (B.28), can be rewritten, using (B.10), as

$$\partial_\nu J^{\alpha\beta\nu} = -\partial_\nu S^{\alpha\beta\nu} + \frac{1}{2}\left(\delta_\nu^\beta\,\theta^{\alpha\nu} - \delta_\nu^\alpha\,\theta^{\beta\nu}\right) = 0 \qquad (B.3$$

Therefore

$$\theta^{[\alpha\beta]} = \partial_\nu S^{\alpha\beta\nu} \qquad (B.31)$$

that is the antisymmetric part of the canonical energy-momentum tensor is related to the divergence of the spin current density. Therefore the canonical stress-tensor $\theta_{\mu\nu}$ for a scalar field (representing spinless particles, $S_{\alpha\beta} = 0$) is symmetric, but in general $\theta_{\mu\nu} \neq \theta_{\nu\mu}$ in the case of a spinning matter field ($S_{\alpha\beta} \neq 0$).

Using eq.(B.31), the canonical tensor $\theta_{\mu\nu}$ can however be symmetrized in the following way (Belinfante-Rosenfeld procedure)

$$\theta_{\alpha\beta} \longrightarrow T_{\alpha\beta} = \theta_{\alpha\beta} - \partial^\nu\left(S_{\alpha\beta\nu} - S_{\beta\nu\alpha} + S_{\nu\alpha\beta}\right) \qquad (B.32)$$

(the reader can easily verify that $T_{\alpha\beta}$ is symmetric and conserved, $\partial^{\beta} T_{\alpha\beta} = 0$).

As we shall see in Chapter X, Sect.3, in the framework of the Einstein-Cartan theory of gravity, where the spin density like the energy density is a dynamical source of gravitation, the covariant generalization of the special relativistic conservation law (B.31) follows directly from the gravitational field equations (see 10.57) , and the symmetrized tensor $T_{\alpha\beta}$ of eq. (B.32) can be identified with the dynamical energy-momentum tensor (see 10.53), obtained by varying the matter Lagrangian with respect to $g^{\alpha\beta}$.

REFERENCES

1) As regards the four-dimensional expression of the theorems of Gauss and Stokes, see for example L.Landau and E.Lifchitz: "Theorie du Champ" (ed.MIR, Moscow 1966) § 6, p.25 and § 83, p.292

2) A complete discussion of the notion of canonical energy-momentum tensor for spinning matter fields in the context of the gravitational theory may be found in a paper written by F.W.Hehl: Rep.Math.Phys.9,55 (1976)

CHAPTER IV

THE THREE CLASSICAL TESTS OF EINSTEIN'S THEORY

1.- The Schwarzschild solution

The first solution of the Einstein equations was obtained by Schwarzschild for a static and spherically symmetric field, which is a good approximation for the field produced by our sun.

The most general spherically symmetric line-element is

$$ds^2 = -U(r)\,dr^2 - V(r)\,(r^2 d\theta^2 + r^2 \sin^2\theta\, d\varphi^2) + W(r)\,dt^2 \tag{4.1}$$

where U,V and W are functions only of r, as we are considering a static field. Putting $r'^2 = r^2 V(r)$ we can write

$$ds^2 = -U_1(r')\,dr'^2 - r'^2(d\theta^2 + \sin^2\theta\, d\varphi^2) + W_1(r')\,dt^2 \tag{4.2}$$

where U_1 and W_1 are arbitrary functions of r'. Neglecting primes, this can be rewritten in the useful form

$$ds^2 = -e^{\lambda} dr^2 - r^2 d\theta^2 - r^2 \sin^2\theta\, d\varphi^2 + e^{\nu} dt^2 \tag{4.3}$$

with $\lambda = \lambda(r)$ and $\nu = \nu(r)$. Defining spherical polar coordinates $x^1 = r$, $x^2 = \theta$, $x^3 = \varphi$, $x^4 = t$ (we put c = 1), the fundamental metric tensor is then

$$g_{11} = -e^{\lambda}, \quad g_{22} = -r^2, \quad g_{33} = -r^2\sin^2\theta, \quad g_{44} = e^{\nu}$$

$$g_{\mu\nu} = 0 \quad \text{for } \mu \neq \nu \tag{4.4}$$

and its determinant is given by

$$g = -r^4 \sin^2\theta\, e^{\nu+\lambda} \tag{4.5}$$

The elements of the matrix corresponding to the inverse metric tensor are $g^{\mu\nu} = (\text{cofactor } g_{\mu\nu})/g$, so that

$$g^{11} = \frac{\text{cofactor } g_{11}}{g} = \frac{r^4 \sin^2\theta\, e^{\nu}}{g} = -\frac{1}{e^{\lambda}} = \frac{1}{g_{11}} \tag{4.6}$$

and so on, i.e.

$$g^{11} = -e^{-\lambda}, \quad g^{22} = -\frac{1}{r^2}, \quad g^{33} = -\frac{1}{r^2 \sin^2\theta}, \quad g^{44} = e^{-\nu} \qquad (4.7)$$

In order to write explicitly the Einstein equations (3.77), which outside the gravitating source are reduced to

$$R_{\mu\nu} = 0 \qquad (4.8)$$

(because $R = -\chi T$, and $T_{\mu\nu} = 0$ in vacuum), we must calculate explicitly the Christoffel symbols, and then apply the definition

$$R_{\mu\nu} = \partial_\sigma \Gamma_{\mu\nu}{}^{\sigma} - \partial_\nu \Gamma_{\mu\sigma}{}^{\sigma} + \Gamma_{\mu\nu}{}^{\alpha} \Gamma_{\alpha\sigma}{}^{\sigma} - \Gamma_{\mu\sigma}{}^{\alpha} \Gamma_{\alpha\nu}{}^{\sigma} \qquad (4.9)$$

Remembering that (see 2.60)

$$\Gamma_{\mu\nu}{}^{\alpha} = \frac{1}{2} g^{\alpha\beta} (\partial_\mu g_{\beta\nu} + \partial_\nu g_{\beta\mu} - \partial_\beta g_{\mu\nu}) \qquad (4.10)$$

we have three possible cases. If all the thre indices of Γ are equal then (no summation over μ in this case)

$$\Gamma_{\mu\mu}{}^{\mu} = \frac{1}{2} g^{\mu\mu} \partial_\mu g_{\mu\mu} = \frac{1}{2} \partial_\mu (\ln g_{\mu\mu}) \qquad (4.11)$$

If two indices are equal, then

$$\Gamma_{\mu\mu}{}^{\nu} = \frac{1}{2} g^{\nu\alpha} (\partial_\mu g_{\alpha\mu} + \partial_\mu g_{\alpha\mu} - \partial_\alpha g_{\mu\mu}) \qquad (4.12)$$

But since in our case $g_{\mu\nu}$ is diagonal we have only the terms with $\alpha = \nu \neq \mu$, so that

$$\Gamma_{\mu\mu}{}^{\nu} = \frac{1}{2} g^{\nu\nu} (-\partial_\nu g_{\mu\mu}) = -\frac{1}{2} g^{\nu\nu} \partial_\nu g_{\mu\mu} \qquad (4.13)$$

In the case of $\Gamma_{\mu\nu}{}^{\nu}$ we have already shown in the previous Chapter that

$$\Gamma_{\mu\nu}{}^{\nu} = \partial_\mu (\ln \sqrt{-g}) \qquad (4.14)$$

Finally, if all the three indices are different, Γ in this case is vanishing.

Now the computation of Γ is easy and we obtain that the only nonvanishing components are the following (a prime denotes the derivative with respect to r)

$$\Gamma_{11}{}^{1} = \frac{\lambda'}{2} \ , \quad \Gamma_{22}{}^{1} = -r\, e^{-\lambda}$$

$$\Gamma_{12}{}^{2} = r^{-1} = \Gamma_{21}{}^{2} \ , \quad \Gamma_{13}{}^{3} = r^{-1} = \Gamma_{31}{}^{3}$$

$$\Gamma_{14}{}^{4} = \frac{1}{2}\nu' = \Gamma_{41}{}^{4} \ , \quad \Gamma_{23}{}^{3} = \cot\vartheta = \Gamma_{32}{}^{3}$$

$$\Gamma_{33}{}^{1} = -r \sin^2\vartheta\, e^{-\lambda} \ , \quad \Gamma_{33}{}^{2} = -\sin\vartheta\cos\vartheta$$

$$\Gamma_{44}{}^{1} = \frac{1}{2}\nu' e^{\nu-\lambda} \tag{4.15}$$

(note that we have used the symmetry of the Christoffel connection, $\Gamma_{\mu\nu}{}^{\alpha} = \Gamma_{\nu\mu}{}^{\alpha}$).

These values are to be inserted in eq.(4.9), which can also be rewritten as

$$R_{\mu\nu} = \partial_\alpha \Gamma_{\mu\nu}{}^{\alpha} - \partial_\mu \partial_\nu (\ln\sqrt{-g}) - \Gamma_{\mu\sigma}{}^{\alpha}\Gamma_{\alpha\nu}{}^{\sigma} + \Gamma_{\mu\nu}{}^{\alpha}\partial_\alpha(\ln\sqrt{-g}) \tag{4.16}$$

so that we find explicitly

$$-R_{11} = -\frac{\partial}{\partial r}\Gamma_{11}{}^{1} + (\Gamma_{11}{}^{1})^2 + (\Gamma_{12}{}^{2})^2 + (\Gamma_{13}{}^{3})^2 + (\Gamma_{14}{}^{4})^2 + \frac{\partial^2}{\partial r^2}(\ln\sqrt{-g}) - \Gamma_{11}{}^{1}\frac{\partial}{\partial r}(\ln\sqrt{-g}) \tag{4.17a}$$

$$-R_{22} = -\frac{\partial}{\partial r}\Gamma_{22}{}^{1} + 2\Gamma_{22}{}^{1}\Gamma_{21}{}^{2} + (\Gamma_{23}{}^{3})^2 + \frac{\partial^2}{\partial\theta^2}(\ln\sqrt{-g}) - \Gamma_{22}{}^{1}\frac{\partial}{\partial r}(\ln\sqrt{-g}) \tag{4.17b}$$

$$-R_{33} = -\frac{\partial}{\partial r}\Gamma_{33}{}^{1} - \frac{\partial}{\partial\theta}\Gamma_{33}{}^{2} + 2\Gamma_{33}{}^{1}\Gamma_{31}{}^{3} + 2\Gamma_{33}{}^{2}\Gamma_{32}{}^{3} - \Gamma_{33}{}^{1}\frac{\partial}{\partial r}(\ln\sqrt{-g}) - \Gamma_{33}{}^{2}\frac{\partial}{\partial\theta}(\ln\sqrt{-g}) \tag{4.17c}$$

$$-R_{44} = -\frac{\partial}{\partial r}\Gamma_{44}{}^{1} + 2\Gamma_{44}{}^{1}\Gamma_{41}{}^{4} - \Gamma_{44}{}^{1}\frac{\partial}{\partial r}(\ln\sqrt{-g}) \tag{4.17d}$$

(all the other non-diagonal terms are identically vanishing). Using eq.(4.15) now we obtain

$$-R_{44} = e^{\nu-\lambda}\left(-\frac{\nu''}{2} - \frac{\nu'}{r} - \frac{\nu'^2}{4} + \frac{\lambda'\nu'}{4}\right) \qquad (4.18)$$

$$-R_{11} = -\frac{\lambda'}{r} - \frac{1}{4}\lambda'\nu' + \frac{1}{2}\nu'' + \frac{1}{4}\nu'^2 \qquad (4.19)$$

$$-R_{22} = e^{-\lambda}\left[1 + \frac{r}{2}(\nu' - \lambda')\right] - 1 \qquad (4.20)$$

$$-R_{33} = \sin^2\vartheta\, e^{-\lambda}\left[1 + \frac{1}{2}r(\nu' - \lambda')\right] - \sin^2\vartheta \qquad (4.21)$$

As we are interested in the gravitational field outside the body, we have to use the field equations for empty space, $R_{\mu\nu} = 0$, and then we are left with the following three equations only

$$\frac{\nu''}{2} + \frac{\nu'}{r} + \frac{\nu'^2}{4} - \frac{\lambda'\nu'}{4} = 0$$
$$\frac{\nu''}{2} + \frac{\nu'^2}{4} - \frac{\lambda'}{r} - \frac{\lambda'\nu'}{4} = 0 \qquad (4.22)$$
$$e^{-\lambda}\left[1 + \frac{1}{2}r(\nu' - \lambda')\right] - 1 = 0$$

because $R_{33} = 0$ is equivalent to $R_{22} = 0$. By subtracting the first two equations we have

$$\nu' + \lambda' = 0 \qquad (4.23)$$

so that

$$\nu + \lambda = \text{const} \qquad (4.24)$$

By imposing the boundary condition that, at $r = \infty$, λ and ν must be vanishing (because the gravitational field, which represents a modification of the flat space-time metric, goes to zero as we go to an infinite distance from the source, and then in this limit ds^2 must be reduced to the Mikowski form) we obtain

$$\lambda = -\nu \qquad (4.25)$$

The third equation (4.22) becomes then

$$e^{\nu}(1 + r\nu') = 1$$

or

$$(e^{\nu} r)' = 1 \tag{4.26}$$

and its integration gives

$$e^{\nu} r = r - 2m \tag{4.27}$$

where -2m is an integration constant that, as we will see, is related to the mass of the central source.

The result is then

$$e^{\nu} = e^{-\lambda} = 1 - 2\frac{m}{r} \tag{4.28}$$

and the proper-time interval (4.3) becomes

$$ds^2 = \left(1 - 2\frac{m}{r}\right) dt^2 - \frac{dr^2}{1 - 2\frac{m}{r}} - r^2 \left(d\theta^2 + \sin^2\theta \, d\varphi^2\right) \tag{4.29}$$

which is the famous solution of the vacuum field equations known as the "Schwarzschild line-element". This metric describes the exterior gravitational field outside a central, spherically symmetric body, and does not depend on the distribution of matter in the interior of the body, like in the Newton theory of gravitation.

In order to interpret the constant of integration m, we can note that, in the weak field limit, it must be $e^{\nu} = g_{44} = 1 + h_{44} = 1 + 2\phi / c^2$ (see eq.3.48), where ϕ is the Newton potential. For a central body of total mass M we have $\phi = - GM / r$, and then, in the limit $r \gg m$, one must obtain

$$g_{44} = e^{\nu} = 1 - \frac{2GM}{rc^2} \tag{4.30}$$

that is, from (4.28)

$$m = \frac{GM}{c^2} \tag{4.31}$$

Therefore m is the gravitational mass of the source in relativistic units; it has the dimension of a length, and for this reason it is usually called the "gravitational radius" of the central body (for instance the value of m for the sun is about 1.475 Km, for the earth about 0.5 cm).

2.- Planetary motions

In this Section we will discuss the motion of a test body in the Schwarzschild field, that is the motion of a particle moving freely in a curved space-time whose metric is given by eq. (4.29).

As we have seen in Chapter II, the motion of a free structureless particle is described by the geodesic equation

$$\frac{d^2x^\alpha}{ds^2} + \Gamma_{\mu\nu}{}^{\alpha} \frac{dx^\mu}{ds} \frac{dx^\nu}{ds} = 0 \qquad (4.32)$$

Taking first $x^2 = \vartheta$ we have

$$\frac{d^2\vartheta}{ds^2} + 2\Gamma_{12}{}^{2} \frac{dx^1}{ds} \frac{dx^2}{ds} + \Gamma_{33}{}^{2} \frac{dx^3}{ds} \frac{dx^3}{ds} = 0 \qquad (4.33)$$

(we have used (4.15)), and then

$$\frac{d^2\vartheta}{ds^2} + \frac{2}{r} \frac{dr}{ds} \frac{d\vartheta}{ds} - \cos\vartheta \sin\vartheta \left(\frac{d\varphi}{ds}\right)^2 = 0 \qquad (4.34)$$

Choosing for simplicity, and without loss of generality, the initial conditions

$$\vartheta = \pi/2 \quad , \qquad \frac{d\vartheta}{ds} = 0 \qquad (4.35)$$

then eq.(4.34) provides also $d^2\vartheta/ ds^2 = 0$. This means that the motion is confined to the $\vartheta = \pi/2$ plane, and we may simplify all the remaining equations putting everywhere $\vartheta = \pi/2 =$ constant.

We find: for $x^1 = r$

$$\frac{d^2r}{ds^2} + \frac{1}{2}\lambda' \left(\frac{dr}{ds}\right)^2 - r e^{-\lambda}\left(\frac{d\varphi}{ds}\right)^2 + \frac{\nu'}{2} e^{\nu-\lambda} \left(\frac{dt}{ds}\right)^2 = 0 \qquad (4.36)$$

for $x^3 = \varphi$

$$\frac{d^2\varphi}{ds^2} + \frac{2}{r} \frac{dr}{ds} \frac{d\varphi}{ds} = 0$$

or

$$\frac{d}{ds}\left(r^2 \frac{d\varphi}{ds}\right) = 0 \qquad (4.37)$$

for $x^4 = t$

$$\frac{d^2t}{ds^2} + \nu' \frac{dr}{ds}\frac{dt}{ds} = 0$$

or

$$\frac{d}{ds}\left(e^{\nu}\frac{dt}{ds}\right) = 0 \qquad (4.38)$$

The integration of these last two geodesic equations is immediate, and gives

$$r^2 \frac{d\varphi}{ds} = h \qquad (4.39)$$

$$\frac{dt}{ds} = k e^{-\nu} = \frac{k}{\gamma} \qquad (4.40)$$

where h and k are two integration constants, and we have put $e^{\nu} = \gamma = 1 - 2m/r$.

In order to obtain an equation for the radial coordinate, instead of integrating the geodesic (4.36) we can equivalently start with the line-elemnt (4.29), and we write the squared modulus $g_{\mu\nu} v^{\mu} v^{\nu} = 1$ of the four-velocity vector $v^{\mu} = dx^{\mu}/ds$:

$$1 = \gamma \left(\frac{dt}{ds}\right)^2 - \gamma^{-1}\left(\frac{dr}{ds}\right)^2 - r^2\left(\frac{d\varphi}{ds}\right)^2 \qquad (4.41)$$

(we have put $\vartheta = \pi/2$). Using eqs.(4.39,40) this becomes

$$1 = \frac{k^2}{\gamma} - \frac{h^2}{r^2} - \frac{1}{\gamma}\left(\frac{h}{r^2}\frac{dr}{d\varphi}\right)^2 \qquad (4.42)$$

Multiplying by γ , and putting $u = r^{-1}$ we find

$$\left(\frac{du}{d\varphi}\right)^2 + u^2 = \frac{k^2-1}{h^2} + \frac{2m}{h^2}u + 2mu^3 \qquad (4.43)$$

Finally, differentiating with respect to φ , we obtain two solutions: one is $du/d\varphi = 0$, corresponding to $r = $ const, i.e. to a circular orbit. The other case $du/d\varphi \neq 0$, corresponds to an orbit described by the following differential equation

$$\frac{d^2u}{d\varphi^2} + u = \frac{m}{h^2} + 3mu^2 \qquad (4.44)$$

In the Newton theory, to obtain the equation of the orbit one starts with the two-bodies Lagrangian, which, for a test particle of unit mass, is given by

$$L = \frac{1}{2}\left(\dot{r}^2 + r^2\dot{\varphi}^2\right) + \frac{GM}{r} \qquad (4.45).$$

where a dot denotes differentiation with respect to t. The Eulero-Lagrange equation for φ gives then

$$\frac{d}{dt}\frac{\partial L}{\partial \dot{\varphi}} = \frac{d}{dt}\left(r^2\dot{\varphi}\right) = 0 \Rightarrow r^2\dot{\varphi} = hc \qquad (4.47)$$

where h is a constant, and c is the light velocity. Another constant of the motion is the total energy E, which is conserved as the Lagrangian (4.45) is not explicitly time-dependent,

$$E = \frac{1}{2}\left(\dot{r}^2 + r^2\dot{\varphi}^2\right) - \frac{GM}{r} \qquad (4.48)$$

Using (4.47), putting $\dot{r} = (dr/d\varphi)\,\dot{\varphi}$, and u = r, differentiating with respect to φ and assuming $du/d\varphi \neq 0$, one obtains, from (4.48), the following equation

$$\frac{d^2u}{d\varphi^2} + u = \frac{GM}{c^2h^2} \qquad (4.49)$$

This equation differs from the general relativistic equation of the orbit (4.44) by the nonlinear term $3mu^2$. Comparing the two equation we obtain again $m = GM/c^2$.

The general relativistic correction is very small, and for ordinary speeds it induces a neglegible deviation from the Newton orbit (4.49). In fact, the ratio between $3mu^2$ and m/h^2 is $3h^2u^2$, that is $3u^2r^4\dot{\varphi}^2/c^2 = 3(r\dot{\varphi}/c)^2$; it represents then three times the square of the transverse velocity of the planet measured in units of c, and in the case of the earth, for example, being the orbital velocity ~ 30 Km/sec, the ratio is $\sim 10^{-8}$.

Because the additional term, however, the planetary orbits are no longer closed, and a precession of the perihelion is produced.

In order to calculate explicitly the perihelion advance from (4.44), we can adopt an iterative procedure, expanding the solution in power series of m/h^2

$$u \simeq u^{(0)} + u^{(1)} + \cdots\cdots \qquad (4.50)$$

and taking as the zero-order unperturbed solution the solution of the Newton equation (4.49), obtained by putting

$3mu^2 = 0$.

Therefore

$$u^{(0)} = \frac{m}{h^2}\left[1 + e\cos(\varphi - \varphi_0)\right] \tag{4.51}$$

where e and φ_0 are integration constants. For $0 < e < 1$ this solution represents an ellipse of eccentricity e, perihelion longitude φ_0, and major semiaxis a given by

$$a = \frac{L}{1 - e^2} \tag{4.52}$$

where $L = m/h^2$ is the so-called semilatus rectum of the orbit.

In first approximation eq.(4.44) becomes then

$$\frac{d^2 u}{d\varphi^2} + u = \frac{m}{h^2} + 3m\,u^{(0)2} \tag{4.53}$$

that is, using (4.50,51),

$$\frac{d^2u^{(1)}}{d\varphi^2} + u^{(1)} = 3\frac{m^3}{h^4} + 6\frac{m^3}{h^4}e\cos(\varphi - \varphi_0) +$$

$$+ 3\frac{m^3 e^2}{h^4}\cos^2(\varphi - \varphi_0) \tag{4.54}$$

Considering orbits with small eccentricity (like the one of Mercury), the terms of order e^2 and higher may be neglected, in first approximation. Moreover, the contribution of the constant term $3m^3/h^4$ is neglegible, and the only one that produces an observable effect is the term in $\cos(\varphi - \varphi_0)$, whose contribution increases continuosly at each revolution. Therefore we have

$$\frac{d^2u^{(1)}}{d\varphi^2} + u^{(1)} = 6\frac{m^3}{h^4}e\cos(\varphi - \varphi_0) \tag{4.55}$$

and the solution is

$$u^{(1)} = \frac{3m^3}{h^4}e\varphi\sin(\varphi - \varphi_0) \tag{4.56}$$

Including this small correction, the solution of the orbital equation becomes

$$u \simeq u^{(0)} + u^{(1)} = \frac{m}{h^2}\left[1 + e\cos(\varphi - \varphi_0) + \frac{3m^2}{h^2}e\varphi\sin(\varphi - \varphi_0)\right] \tag{4.57}$$

and putting

$$\Delta\varphi_0 = \frac{3m^2}{h^2}\varphi \tag{4.58}$$

it may be written as

$$u = \frac{m}{h^2}\left[1 + e\cos(\varphi - \varphi_0 - \Delta\varphi_0)\right] \tag{4.59}$$

(note in fact that

$$\cos(\varphi - \varphi_0 - \Delta\varphi_0) = \cos(\varphi - \varphi_0)\cos\Delta\varphi_0 + \sin(\varphi - \varphi_0)\sin\Delta\varphi_0$$
$$\simeq \cos(\varphi - \varphi_0) + \Delta\varphi_0 \sin(\varphi - \varphi_0) \tag{4.60}$$

for $\Delta\varphi_0 \ll 1$).

From eq.(4.59) we can see that while the planet moves through an angle φ, the perihelion advances for a fraction of the revolution angle equal to

$$\frac{\Delta\varphi_0}{\varphi} = \frac{3m^2}{h^2} = \frac{3m}{a(1-e^2)} = \frac{3m}{L} \tag{4.61}$$

(we have used 4.52). After a full revolution ($\varphi = 2\pi$), we have

$$\Delta\varphi_0 = 6\pi\frac{m^2}{h^2} = \frac{6\pi m}{L} = \frac{6\pi GM}{c^2 a(1-e^2)} \tag{4.62}$$

Introducing in this equation the value of m for the sun, m = 1.475 Km, and the value of L for Mercury, L = 55.3×10^6 Km, we find an extremely small angle, $\Delta\varphi_0 \simeq 0.1''$. But since this is a secular effect that increases with the number of revolutions, if we take into account that the period of revolution of Mercury around the sun is about 0.24 years, we obtain for Mercury a perihelion advance in one century of $\simeq$ 43'' . (We have considered Mercury because, for the other planets, L is greater, and the perihelion shift is smaller: for instance, we have for Venus $\simeq$ 8.6'' , for the earth $\simeq$ 3.8'' , for Mars $\simeq$ 1.35'' (per century)).

The value of 43'' coincides, up to 1% , with the residue in the motion of the perihelion of Mercury left unexplained in the Newton theory of gravitation, after that all the perturbations caused by the other planets are taken into account.

This theoretical prediction is then in good agreement with the observational data.

It should be noted, however, that if the shape of the sun is not exactly spherically symmetric, as pointed out first by Dicke, there is a further shift of the perihelion, produced by this cause, to be added to the other Newtonian effects. Therefore the agreement between the general relativistic prediction and the observational data could be not so good (some recent results seem to indicate some very small solar oblateness[1]).

In conclusion of this Section, it is interesting to compare the amount of perihelion advance predicted by the theory of general relativity, due to a dynamical gravitational effect, with the perihelion shift due to special relativistic kinematics.

Consider the special relativistic generalization of the Newton Lagrangian (4.45):

$$L = -m_0 c^2 \sqrt{1-\beta^2} + m_0 G \frac{m}{r} \tag{4.63}$$

where m_0 is the rest mass of the test particle and, using polar coordinates, the velocity is

$$\beta^2 = \frac{v^2}{c^2} = \frac{1}{c^2}\left(\dot{r}^2 + r^2\dot{\varphi}^2\right) \tag{4.64}$$

(a dot denote differentiation with respect to the time t).

From the Lagrange equation for φ

$$\frac{d}{dt}\left(\frac{m_0 c^2}{\sqrt{1-\beta^2}} r^2 \dot{\varphi}\right) = 0 \tag{4.65}$$

we have

$$r^2 \dot{\varphi} = h\sqrt{1-\beta^2} \tag{4.66}$$

where h is a constant. The other constant is the total energy which coincides with the Hamiltonian

$$H = \frac{m_0 c^2}{\sqrt{1-\beta^2}} - m_0 \frac{GM}{r} \tag{4.67}$$

We can put then

$$\frac{1}{\sqrt{1-\beta^2}} - \frac{GM}{rc^2} = \alpha = \text{const} \tag{4.68}$$

From this equation, using (4.64) and (4.66), putting $\dot{r} = \dot{\varphi}\,(dr/d\varphi)$ and $u = 1/r$, and differentiating with respect to φ, we obtain the following orbital equation

$$\frac{d^2 u}{d\varphi^2} + u = \frac{\alpha m}{h^2} + \frac{m^2}{h^2} u \qquad (4.69)$$

where $m = GM/c^2$. The constant α can be determined by imposing the initial condition that at $r = \infty$, where the potential energy is vanishing, the velocity is also vanishing, $\beta = 0$. Therefore, from (4.68), $\alpha = 1$, and the differential equation for the orbit (4.69), neglecting the special relativistic correction $m^2 u/h^2$, coincides exactly with the Newton equation (4.49).

In order to find a solution, it is convenient to rewrite (4.69) as

$$\frac{d^2 u}{d\varphi^2} + k u^2 = \frac{k^2}{p} \qquad (4.70)$$

where $k^2 = 1 - m^2/h^2$, and $p = k^2 h^2/\alpha m$. This equation may be solved exactly, and the solution is

$$u = \frac{1}{p}\left(1 + e \cos k\varphi\right) \qquad (4.71)$$

(we have chosen, for simplicity, the initial condition $\varphi_0 = 0$).

The value of the radial coordinate of a planet becomes again

$$r = \frac{p}{1+e} \qquad (4.72)$$

after an angle $k\varphi = 2\pi$, and then we have a perihelion shift per revolution

$$\Delta\varphi_0 = \frac{2\pi}{k} - 2\pi = 2\pi\left(\frac{1}{k} - 1\right) \simeq \frac{\pi m^2}{h^2} \qquad (4.73)$$

which is exactly 1/6 of the general relativistic shift (4.62). The same result can be obtained from eq.(4.69) with an iterative procedure. In the case of the planets, however, this correction is neglegible because their velocity is nonrelativistic.

3.- Deflection of light rays

In order to investigate the light propagation in a gravitational field, we can regard a light ray as a beam composed of a large number of photons, and then to study the path of test particles moving with a speed c and with vanishing rest mass.

Such a particle follows, like in special relativity, a null geodesic, $ds^2 = 0$. In the case of the Schwarzschil metric, we obtain, instead of eq. (4.41), that the squared modulus of the four-velocity satisfies now the condition

$$\gamma^{-1}\left(\frac{dr}{d\sigma}\right)^2 + r^2\left(\frac{d\varphi}{d\sigma}\right)^2 - \gamma\left(\frac{dt}{d\sigma}\right)^2 = 0 \qquad (4.74)$$

(where σ is a suitable variable which parametrizes the nul geodesic) and then, following the same procedure as in the previous Section, we are led to the following equation

$$\frac{d^2u}{d\varphi^2} + u = 3mu^2 \qquad (4.75)$$

As before the term $3mu^2$ is small with respect to the others, so that we can get an approximate solution by an iterative procedure, starting with a zeroth-order equation in which $3mu^2 = 0$, that is

$$\frac{d^2u^{(0)}}{d\varphi^2} + u^{(0)} = 0 \qquad (4.76)$$

which gives

$$u^{(0)} = \frac{1}{R}\cos(\varphi - \varphi_0) \qquad (4.77)$$

where φ_0 and R are suitable integration constants. Introducing a system of cartesian coordinates $x = r\cos\varphi$, $y = r\sin\varphi$, with origin at the center of the sun, and putting $\varphi_0 = 0$, then we can see that eq.(4.77) represents, in polar coordinates, the straight line $x = R$ parallel to the y axis (see Fig. 4.1) and R is the minimum distance between the travelling photons and the origin.

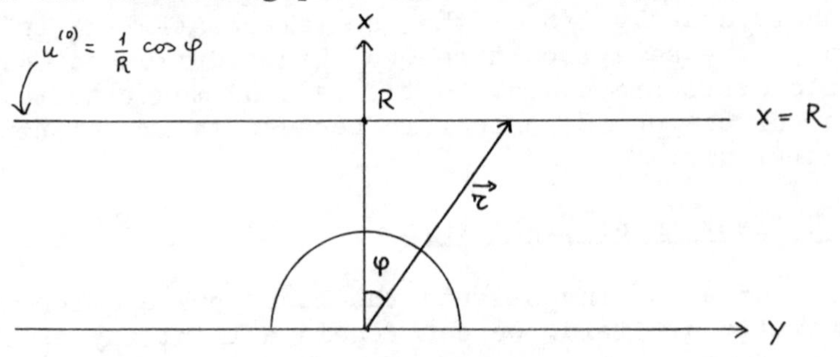

FIG. 4.1

Therefore, in this approximation, the light ray is not deflected by the gravitational field of the sun.

In a successive approximation we have, putting $u \simeq u^{(0)} + u^{(1)}$,

$$\frac{d^2 u^{(1)}}{d\varphi^2} + u^{(1)} = 3 m u^{(0)\,2} = 3\frac{m}{R^2}\cos^2\varphi \qquad (4.78)$$

A particular integral of this equation is

$$u^{(1)} = \frac{m}{R^2}\left(\cos^2\varphi + 2\sin^2\varphi\right) \qquad (4.79)$$

so that, including to first order the general relativistic effects, we have

$$u \simeq u^{(0)} + u^{(1)} = \frac{1}{R}\cos\varphi + \frac{m}{R^2}\left(\cos^2\varphi + 2\sin^2\varphi\right) \qquad (4.80)$$

Multiplying by rR we find

$$R = r\cos\varphi + \frac{m}{R}\left(r\cos^2\varphi + 2r\sin^2\varphi\right) \qquad (4.81)$$

and then, in cartesian coordinates

$$x = R - \frac{m}{R}\left(\frac{x^2 + 2y^2}{\sqrt{x^2+y^2}}\right) \qquad (4.82)$$

The second term of this equation describes a small deviation of the light ray from from the straight line $x = R$. In the limit $y \gg x$ we obtain the asymptotic equations

$$x \simeq R - \frac{m}{R}\left(\pm 2y\right) \qquad (4.83)$$

so that the total deviation angle from a straight line, given by the difference between the angular coefficients of the two asymptotes (4.83) , is (see Fig.4.2)

$$\delta = \frac{4m}{R} = 4\frac{GM}{Rc^2} \qquad (4.84)$$

A light ray coming from a distant star undergoes a maximum deflection when it passes just outside the sun surface, i.e. when the distance R is just the radius of the sun, $R \simeq 7 \times 10^{10}$ cm . Using then $m \sim 1.5$ km , we find a maximum deflection angle

$$\delta \simeq 1.75'' \qquad (4.85)$$

Fig.4.2 - The path of a light ray bent by the gravitational field

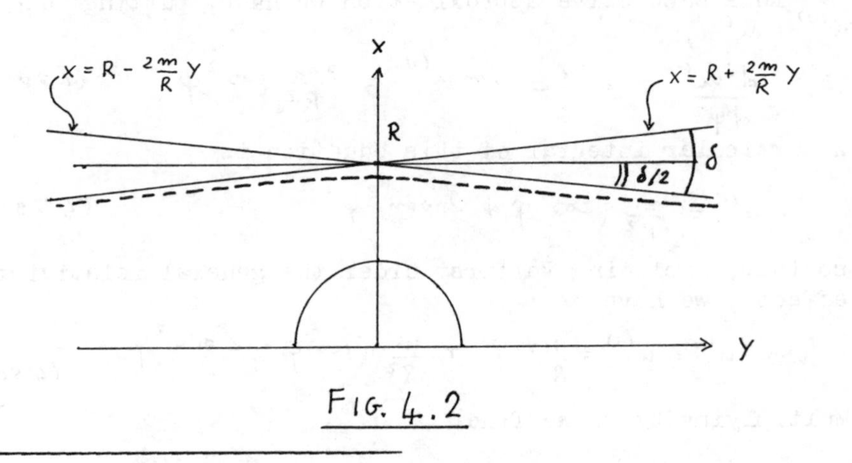

FIG. 4.2

It is important to note that this angle is just twice the value predicted by the Newtonian theory of gravitation.

In fact, from the Newton equation (4.47), we can derive the value of the constant h , for a particle travelling with a velocity c (where, of course, in this case the velocity c does not play any special role). If we consider the particle at the position of minimal distance R from the origin, we have $c = R\dot{\varphi}$, and then, from (4.47),

$$R^2 \dot{\varphi} = Rc = hc \tag{4.86}$$

that is h = R . The Newton equation (4.49) becomes then, when the speed of the test particle is c ,

$$\frac{d^2}{d\varphi^2}\left(\frac{1}{r}\right) + \frac{1}{r} = \frac{m}{R^2} \tag{4.87}$$

The solution of this equation, which can be written

$$\frac{1}{r} = \frac{m}{R^2} + \frac{1}{R}\cos\varphi \tag{4.88}$$

describes an hyperbola and becomes, in cartesian coordinates $x = r\cos\varphi$, $y = r\sin\varphi$,

$$x = R - \frac{m}{R}\sqrt{x^2+y^2} \tag{4.89}$$

For $y >> x$ we have

$$x \simeq R - \frac{m}{R} (\pm y) \qquad (4,90)$$

and calculating the angle between the asymptotes, the Newtonian deflection of light is then

$$\delta_{New} = 2\frac{m}{R} = \frac{1}{2} \delta_{Ein} \qquad (4,91)$$

It should be noted also that the deflection angle (4.84) was deduced using the corpuscolar aspect of light, but it can be shown that an identical result is obtained also considering the wave aspect, starting from the propagation of a test electromagnetic wave in the gravitational field of the sun, and using the approximation of the geometric optics.

To verify experimentally this effect, one should observe a change in the position of the stars surrounding the sun's disc during a total eclipse . The observations made up to now lead to an average value $\delta \sim 2''$ which is about 20% larger than the theoretical prediction (4.85); in view of the extreme difficulty of these observations, however, we may regard the agreement as qualitatively good in favour of Einstein's prediction.

As regards the fact that the value predicted by Einstein is twice the Newtonian one, the following "paradox" has been remarked.

Since the deflection of the light beam is doubled, the acceleration of a photon at each point is twice the value of the Newtonian acceleration, while for a massive test-particle the acceleration is the same as the Newton acceleration. To a man in a lift, for instance, descending with an acceleration m/r^2 , the tracks of ordinary particles will appear to be straight lines; but it would require an acceleration $2m/r^2$ to straighten out the light tracks. This seem to contradict the principle of equivalence.

But there is no contradiction if one has clear in mind the local validity of this principle. In fact we may put locally $\Gamma = 0$, but its derivatives are nonvanishing in a gravitational field, so that if our experiment is not sufficiently limitated in space and time, the equivalence between inertia and gravitation is no longer valid.

This can be seen easily also considering the fall of two bodies in a lift, when it is accelerated and when it is placed in a gravitational field (see Fig.4.3). The paths of the bodies are different if we imagine that the lift is not locally limited, but sufficiently large.

This argument may be applied also the experiment of light deflection. In fact, to calculate the deflection of a light ray produced by the solar gravitational field, we have introduced two different tangent planes (one before and the other behind the sun) and we have calculated the angle between them: this is obviously a nonlocal observation. To apply the principle of equivalence, we must introduce only one tangent plane at a given point, and the path of light must be referred only to this plane.

From Fig.4.4, it is easy to see that β , the angle measured by the experiment, is twice the angle α obtained introducing the tangent plane at the point of minimum distance from the central body. An since α is the angle connected with a local measure, like that performed by a man in a lift, there is no contradiction with the equivalence principle which, as we have stressed, has only local validity.

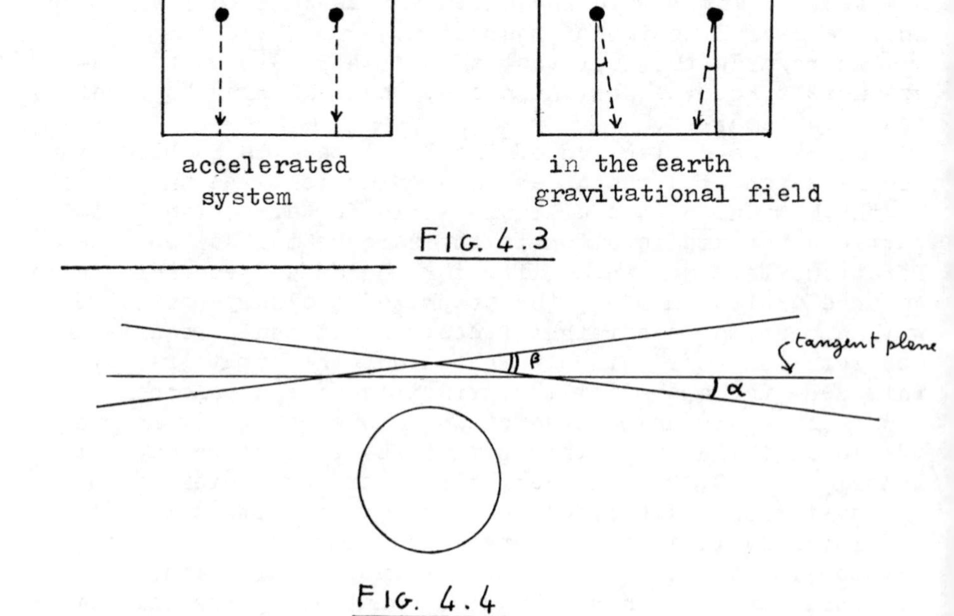

FIG. 4.3

FIG. 4.4

Finally, we note that the difference between δ_{New} and δ_{Ein} can be understood also considering the line element (4.29). In fact there the coefficient $1 - 2m/r$ appears twice, modifying the flatness of space-time in two ways . In the Newtonian limit, $g_{44} = 1 + 2\phi/c^2$ (see 3.48), but $g_{11} = 1$; therefore, in general relativity, we must add, to the Newtonian deflection, the deflection due to the non-euclidean character of the three-dimensional space, related to the presence of $1 - 2m/r$ in the coefficient of dr^2 .

We are led again to obtain a light deflection which, in the Einstein case, is twice that of the Newtonian theory, and this does not contradict the principle of equivalence, which mantains its local validity.

4.- The shift of spectral lines

Another effect predicted by the Einstein theory is the displacement of the atomic spectral lines in the presence of a gravitational field.

Consider the static gravitational line-element (4.29). The proper time $d\tau = ds/c$ is defined as the time interval between two events whose spatial separation is vanishing, $dr = d\vartheta = d\varphi = 0$. It is related then to the coordinate time dt by

$$d\tau = \sqrt{g_{44}}\, dt = \sqrt{1 - 2\frac{m}{r}}\, dt \qquad (4.32)$$

When a periodic phenomenon takes place in a curved space-time, like the emission of electromagnetic radiation by an atom, one must distinguish between "proper" and "coordinate" frequency. Consider for example the propagation of electromagnetic waves in the limit of geometric optics. In this limit the electromagnetic field can be written $f = a\exp(i\psi)$, where $a = a(\vec{x},t)$ is the amplitude of the wave, and the phase $\psi = \psi(\vec{x},t)$ is also called the eikonal function[2] . In this case the frequency of the wave can be expressed as the derivative of ψ with respect to the time, and one has a coordinate frequency, $\omega_0 = \partial\psi/\partial t$, and a proper frequency, $\omega = \partial\psi/\partial\tau$.

If the wave is emitted as a point P_1 by an atom with proper frequency ω_1, then at another point P_2 , with a different gravitational field, so that $(g_{44})_1 \neq (g_{44})_2$, one will observe a different proper frequency ω_2 , such that

$$\frac{\omega_2}{\omega_1} = \frac{(\partial\psi/\partial\tau)_2}{(\partial\psi/\partial\tau)_1} = \frac{\frac{\partial\psi}{\partial t}\left(\frac{\partial t}{\partial\tau}\right)_2}{\frac{\partial\psi}{\partial t}\left(\frac{\partial t}{\partial\tau}\right)_1} = \frac{(\sqrt{g_{44}})_1}{(\sqrt{g_{44}})_2} \qquad (4.93)$$

(as ω_0 = const). Putting $g_{44} = 1 + 2\,\phi/c^2$ we have, for $\phi_1,\ \phi_2 << c^2$,

$$\frac{\omega_2}{\omega_1} = \left(\frac{1+2\phi_1/c^2}{1+2\phi_2/c^2}\right)^{1/2} \simeq (1+\phi_1/c^2)(1-\phi_2/c^2) =$$

$$= 1 + \frac{1}{c^2}(\phi_1 - \phi_2) \qquad (4.94)$$

If the light is emitted, for example, by an atom on the surface of the sun, and observed on the earth surface, then ϕ_1 = GM/R (see eq.4.30), while ϕ_2, the gravitational potential on the earth surface, may be neglected in eq.(4.94) as $\phi_2 << \phi_1$. The observed frequency ω_2 is then

$$\omega_2 \simeq \omega_1\left(1 + \frac{\phi_1}{c^2}\right) = \omega_1\left(1 - \frac{GM}{Rc^2}\right) \qquad (4.95)$$

and we have

$$\frac{\Delta\omega}{\omega} = \frac{\omega_2 - \omega_1}{\omega_1} = -\frac{GM}{Rc^2} \qquad (4.96)$$

The frequency ω_2 measured on the earth is smaller than the proper frequency ω_1 , and introducing in this expression the values of the mass M and the radius R of the sun, we obtain that a spectral line produced on the solar surface must be displaced towards the red, compared with the corresponding line produced on the earth, by about 2×10^{-6} of its wavelength (too small to be measured).

The experimental verification of this effect is possible in the case of the so-calles "dwarf stars" whose average density exceedes that of the water by a factor of order 10^6 . The first measure was made on the faint companion of Sirius (whose mass and radius can be determined) and was in qualitative agreement with the theory.

In addition also a laboratory test of the gravitational red-shift has been performed, using the gravitational field of the earth. In this case a very weak shift, $\Delta\lambda/\lambda \sim 10^{-15}$, can be observed using the Mossbauer effect, in which the emission of γ- rays with an extremely narrow profile is allowed (they are monochromatic up to 1 part in 10^{12}). The γ- rays are then re-absorbed with a corresponding sharp resonance.

The experiment was performed in 1960 by Pound and Rebka [3] , who placed the emitter and the absorber at the opposite ends of a vertical tower 22 meters high. Gamma rays emitted at the bottom suffer a gravitational red-shift traveling to the absorber at the top, and, as a result, are less favourably observed. By moving the emitter upward at a small velocity, a compensating Doppler shift was produced which restored a resonant absorption; a measurement of the emitter velocity then allowed an easy calculation of the ratio $\Delta \nu / \nu$, and the experimental result was in very good agreement with the theoretical prediction.

REFERENCES

1) For a recent estimation of the quadrupole moment in the gravitational potential of the sun, see H.A.Hill, R.J.Bos and P.R.Gode : Phys.Rev.Lett.49, 1794 (1982)

2) As regards the propagation of light in the approximation of geometric optics see for example L.Landau and E.Lifchitz: "Theorie du Champ" (ed. MIR, Moscow 1966) § 53, p.165

3) R.V.Pound and G.A.Rebka : Phys.Rev.Lett.4,337 (1960)

CHAPTER V

ELEMENTS OF COSMOLOGY

1.- The cosmological problem

Cosmology is the science which investigates the dynamical structure and the evolution of the universe as a whole: the universe is regarded as a single unit , the behaviour of nearby matter is influenced also by the most distant regions, so that, for instance, also from the study of the structure of a galaxy one can extract informations upon the whole universe.

The basic phenomena which should provide the starting points for any theoretical cosmological model one may construct nowadays are, in our opinion, the following:

a) the homogeneity and the isotropy of the universe (on a large scale);
b) the expansion of the universe (or the Hubble law);
c) the cosmic microwave radiation background.

This last point provides informations, as we shall see, on an early hot state of the universe, and supports strongly the big-bang theory and its prediction of a much denser and hotter universe in the past. The proposition a) is also called the cosmological principle.

The modern cosmological studies began in 1917 with Einstein's famous paper in which the simplest model of universe allowed by the general relativity theory was constructed: a model homogeneous and isotropic in space, and static in time (at that time the expansion of the universe was not yet discovered).

But before introducing the relativistic models, we wish to discuss briefly the cosmological problem starting from the beginning, and to explain why the approach of Newton was unsuccessful .

If we consider, like Newton, a static and homogeneous universe, two possibilities then arise: the universe may be finite, and in this case, as Newton himself writes[1] , "If the matter of our sun and planets and all the matter of the universe were evenly scattered throughout all the heavens, and every particle had an innate gravity toward all the rest, and the whole space throughout which this matter was scattered was but finite, the matter on the outside of this space would, by its gravity, tend towards all the matter on the inside and, by consequence, fall down into the middle of the whole space and there compose one great spherical mass."

So in the hypothesis of a finite universe we should have a collapse to a central point.

Considering the other possibility, namely an universe infinitely extended, we meet the difficulty known as "Olber's paradox". It may be expressed in the following way (as written by Edmund Halley[2]) : "If the number of the fixed stars were more than finite, the whole superficies of their apparent sphere, i.e. the sky, would be luminous".

It is very easy to verify this proposition, assuming that the universe, besides being infinite, a) is static, b) it is composed of stars of similar luminisity, c) the stars are uniformly distributed. Consider in fact any large spherical shell centered on the earth: on this shell, the amount of light produced by the stars can be calculated. Then consider a shell of twice the radius: on this new shell, the number of stars will be four times larger, as the number of stars on a given shell of radius r is proportional to r^2 (it is given by $4\pi r^2 \Delta r$, if Δr is the thickness of the shell); also the light emitted by a shell is proportional then to r^2 .

On the other hand, the light that reaches the center, that is the earth, is proportional to $1/r^2$. So every shell gives the same contribution of light, and when these contributions are summed up, the sky continues to increase in brightness without limit.

Of course we can take into account that some of the light emitted is intercepted by intervening stars; even so, however, the brightness of the sky must be equal to the average brightness at the surface of a star, in apparent contradiction with our experience.

Olbers tried to resolve this paradox by suggesting that space was filled with a tenuous absorbing medium.

Although now we know that such a fog can exist, this explanation is by itself insufficient, because in the course of time the intervening gas would be heated by the absorbed radiation to become just as hot as the surface of a star: then it would radiate all the light absorbed.

To resolve this paradox, based on the hypothesis of a static, infinite and homogeneous universe, we must then neglect one of these assumptions. We could also note that there is another implicit assumption, i.e. the hypothesis that the space is euclidean; but it is easy to see that also with a non-euclidean geometry the paradox does not disappear: in fact, in this case the surface of a sphere is a function $f(r)$ different from $4\pi r^2$, but also the light received at the center is reduced by the same factor $f(r)$, so that the previous conclusions are unchanged.

This situation was resolved when Humason and Hubble discovered, in 1928, the existence of a law connecting the red-shift of a galaxy with its distance from the earth:

$$\text{red-shift} = \text{constant} \times \text{distance} \tag{5.1}$$

This relation is generally considered as an observational datum; however, one must keep in mind that observations are not directly referred to the distance, which is not measurable, but to the apparent luminosity (which is connected with the distance by some theoretical law).

According to the Hubble law (5.1), distant galaxies show a red-shift of their spectral lines which increases nearly proportionally to their distance from the observer. If we interpret the red-shift as a Doppler effect (up to now no other consistent interpretation seems to have been proposed), then the conclusion is that the universe is expanding. This fact resolves immediately the Olbers paradox; in fact the Doppler shifts of the galaxies reduce the amount of light received on the earth for two reasons: first of all the number of waves received in a second decreases, so that we have a depletion in the number of photons received in the unit of time. In addition also the energy $E = h\nu$ of each photon is smaller, so that for both these reasons the contribution to the sky brightness of the distant stars is reduced.

If the expansion is uniform, a source recedes from the observer at a rate proportional to its distance, as we shall see in a moment, and the greater is the distance, the more its contribution is reduced. Owing to the expansion of the universe, the light coming from the very distant stars is suppressed: we may conclude then that the sky is dark at night because the universe is expanding.

If the red-shift is interpreted as a Doppler effect

remembering that the Doppler shift is related to the velocity, in the nonrelativistic limit, by $\Delta\lambda/\lambda \simeq v/c$, and rewriting the Hubble law as

$$\frac{\Delta\lambda}{\lambda} = H\frac{r}{c} \tag{5.2}$$

where $H > 0$ is the so-called"Hubble constant" , we have

$$\vec{v} = H\vec{r} \tag{5.3}$$

The word "constant" means that H does not depend on the magnitude and the direction of the vector $\vec{r}$. H may depend however on time.

The velocity distribution (5.3) is isotropic: for an observer at the origin of the coordinate system, there is no preferred direction. If we consider an arbitrary point A at a distance $\vec{r}_A$ from the origin, where the matter has a velocity $\vec{v}_A$, and we perform a translation

$$\vec{r}\,' = \vec{r} - \vec{r}_A \tag{5.4}$$

we obtain

$$\vec{v}\,' = \vec{v} - \vec{v}_A = H\vec{r} - H\vec{r}_A = H\vec{r}\,' \tag{5.5}$$

that is we find the same dependence of $\vec{v}'$ on $\vec{r}'$ as $\vec{v}$ on $\vec{r}$. This means that the velocity distribution is the same for all the observers: there are no privileged positions, all points being, to this respect, equivalent; an observer moving toghether with matter sees the surrounding particles (the galaxies) going away from him.

The Hubble constant H is usually given in terms of Km/(sec x Mpc) (Mpc = 10^6 parsec, 1 parsec $\simeq 3.085\times10^{18}$ cm $\simeq$ 3.26 light-years), because the parsec is the most used astronomical unit of distance. The first measurement of H is due to Hubble, but in his work the absolute magnitude of the Cepheid variable stars was underestimated: therefore, with a distance $\vec{r}$ in eq.(5.3) too small, the Hubble value for H (350 Km/(sec x Mpc) was too large.

Subsequent measures led to larger values for the distances of the galaxies, and the value of H was lowered to 100 Km/(sec x Mpc). Nowadays the accepted value, that will be used in the following Sections, ranges from 75 to 50 Km/(sec x Mpc). This last number can also be expressed 2×10^{10} (years)$^{-1}$.

2.- Newtonian cosmology

In this Section we discuss briefly the main features of a cosmological model which takes into account the Hubble expansion law, but is based on the nonrelativistic Newtonian theory of gravitation.

In this framework, the expansion law may be expressed in the following way: the distance between any pair of material points A and B changes in time as

$$\frac{d r_{AB}}{dt} = H r_{AB} \tag{5.6}$$

so that integrating we get

$$r_{AB}(t) = r_{AB}(t_0)\, e^{\int_{t_0}^{t} H(t)\, dt} \tag{5.7}$$

To consider the corresponding variation of the matter density we take a homogeneous sphere of mass M and radius $R = R(t)$. Its density is $\rho = 3M/4\pi R^3$, from which

$$\frac{d\rho}{dt} = - \frac{3M}{\frac{4}{3}\pi R^4} \frac{dR}{dt} \tag{5.8}$$

Inserting in this equation $dR/dt = v = H r$ we have

$$\frac{d\rho}{dt} = -3H\rho \tag{5.9}$$

This equation can be obtained also starting from the matter continuity equation $\partial\rho/\partial t = - \operatorname{div}(\rho\vec{v})$, and assuming that ρ does not depend on the spatial coordinates:

$$\frac{\partial\rho}{\partial t} = -\rho \operatorname{div}\vec{v} = -\rho \operatorname{div}(H\vec{r}) = -3H\rho \tag{5.10}$$

If initially ρ is not dependent on $\vec{x}$, then also $\partial\rho/\partial t$ is constant in space, and this property holds at any subsequent time, so that $\partial\rho/\partial t = d\rho/dt$.

Eq.(5.6) describes the kinematical evolution of the universe; in order to deduce its dynamical behaviour, we consider a test particle on the surface of the sphere of radius R and mass M, and we calculate its acceleration due to the Newton gravitational force:

$$\frac{d^2R}{dt^2} = - \frac{GM}{R^2} \tag{5.11}$$

Being $v = H r$ and $M = 4\pi \rho R^3/3$ we can write

$$\frac{d^2R}{dt^2} = \frac{dv}{dt} = \frac{d}{dt}(Hr) = R\frac{dH}{dt} + H\frac{dR}{dt} = R\frac{dH}{dt} + RH^2 \quad (5.12)$$

$$\frac{d(HR)}{dt} = -\frac{4}{3}\pi G \rho R \quad (5.13)$$

and we have

$$\frac{dH}{dt} = -H^2 - \frac{4\pi}{3} G \rho \quad (5.14)$$

Eqs.(5.14,5.9) determine completely the temporal evolution of the local properties of the universe. Note that these equations do not contain neither the mass nor the radius of the arbitrarily chosen sphere.

The integration of eq.(5.11)

$$\frac{1}{2}\left(\frac{dR}{dt}\right)^2 - \frac{GM}{R} = const \quad (5.15)$$

is the expression of the energy conservation law: the first term is in fact the kinetic energy for a unit mass, the second term is its potential energy, and the constant is then the total energy for a unit mass on the surface of the sphere. To determine this constant, let us denote with H_o, ρ_o and R_o the present day values of H, ρ and R. We have then $M = 4\pi \rho_o R_o^3/3$, and

$$\left(\frac{dR}{dt}\right)_{t=t_o} = H_o R_o \quad (5.16)$$

and the constant (5.15) is

$$const = \frac{1}{2}\left(\frac{dR}{dt}\right)^2_{t=t_o} - \frac{4}{3}\pi G \rho_o \frac{R_o^3}{R_o} = \frac{1}{2} H_o^2 R_o^2 - \frac{4}{3}\pi \rho_o G R_o^2 \quad (5.17)$$

so that the conservation law can be rewritten

$$\left(\frac{dR}{dt}\right)^2 = \frac{8\pi}{3}\frac{G\rho_o R_o^3}{R} - \frac{8\pi}{3} G R_o^2 \left(\rho_o - \frac{3}{8\pi}\frac{H_o^2}{G}\right) \quad (5.18)$$

Without integrating explicitly this equation, we can easily discuss the qualitative behavior of the solution. In fact the present value of dR/dt is positive (the universe is expanding, $H_o R_o > 0$, see 5.16) and then in the past R was smaller, so that $8\pi G \rho_o R_o^3/3R$ was greater than now. Consequently in the past dR/dt was greater, and at some time $t_1 < t_0$ it was

$$\frac{dR}{dt} = +\infty \quad \text{and} \quad R = 0 \tag{5.19}$$

The future development of R(t) depends on the sign of the expression ($\rho_0 - 3H_0^2/8\pi G$). Putting

$$\rho_c = \frac{3H_0^2}{8\pi G} \tag{5.20}$$

(this is the so-called "critical density"), we may consider two cases, $\rho_0 > \rho_c$ and $\rho_0 < \rho_c$. If $\rho_0 > \rho_c$, the expression in round bracket on the right-hand side of eq.(5.18) is positive, and this means that as R increases a value will be reached for which the right-side of (5.18) become vanishing (i.e. dR/dt = 0). After that time R will start decreasing (note that the right-hand side of (5.18) cannot become negative, being the left-hand side a squared term) so that the expansion will stop and will change in contraction (see Fig.5.1) .

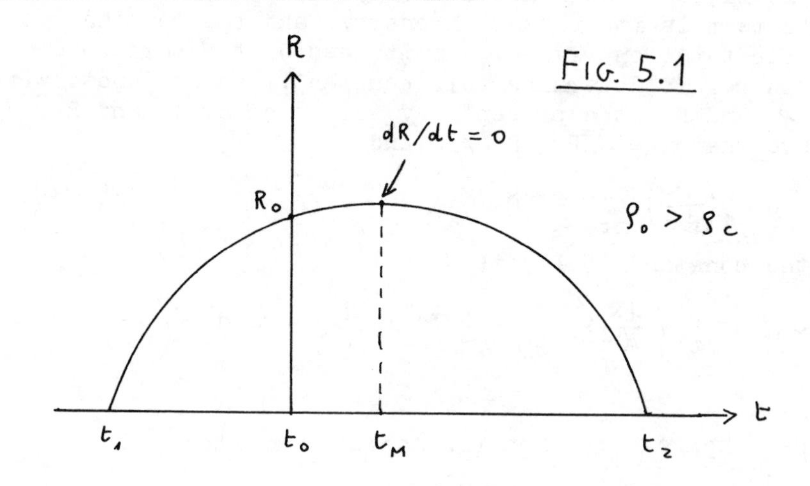

Fig.5.1

Variation in time of the distance between two points when $\rho_0 > \rho_c$: t_0 is the present time, t_1 and t_2 are the instants of infinite density and t_M is the instant of maximum expansion.

If $\rho_0 < \rho_c$, then dR/dt remains always positive (the right-hand side of (5.18) never goes to zero) and the expansion continues without end (see Fig.5.2); in the limit $t \rightarrow \infty$ and $R \rightarrow \infty$ the expansion velocity approaches the constant value

$$\left(\frac{dR}{dt}\right)_{t\rightarrow\infty} = \left[\frac{8\pi}{3} G R_0^2 (\rho_c - \rho_0)\right]^{1/2} \qquad (5,21)$$

In the first case ($\rho_0 > \rho_c$) the Hubble parameter H decrease from H_0 to $H(t_M) = 0$, and $H(t_2) = -\infty$. In the second case ($\rho_0 < \rho_c$) H decreases but remains always positive, and for $t \rightarrow \infty$ we have dR/dt = const, $R \propto t$ and $H = 1/t$.

Therefore, what prediction can be obtained for the future evolution of our universe?

The answer to this question depends on the present values of H_0 and ρ_0 . Up to now they are known only approximately. With $H_0 \sim 50$-75 Km/(secxMpc) we have, for the critical density, $5 \times 10^{-30} \leq \rho_c \leq 10^{-29}$ g/cm^3. As regards ρ_0, we can only say that it cannot be less than 3×10^{-31} g/cm^3 . This lower limit is determined by the visible matter present in the galaxies and does not take into account the intergalactic matter, or the presence of a neutrino background or other kinds of dark matter.

Therefore the value of ρ_0 could be much higher than this, near to the critical density or also greater. There is no sure answer then at present to the question whether our universe will expand indefinitely (open) or will recontract again (closed universe).

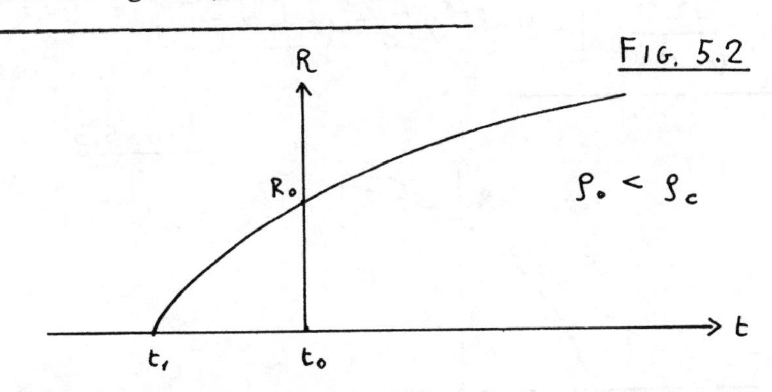

Fig.5.2. Variation in time of the distance between two points when $\rho_0 < \rho_c$.

However, going back in time, we can calculate when the expansion began, that is the time t_1 in the past when $R = 0$ and the density was infinite. If the expansion velocity were constant and equal to the present day value, then we could write

$$R_0 = \left(\frac{dR}{dt}\right)_{t=t_0} (t_0 - t_1) = H_0 R_0 (t_0 - t_1) \qquad (5.22)$$

thus obtaining for the time interval $T = t_0 - t_1$ (the so-called "age of the universe")

$$T = t_0 - t_1 = \frac{1}{H_0} = 4 \times 10^{17} \text{ sec} \sim 10^{10} \text{ YEARS} \qquad (5.23)$$

(see Fig.5.3). Obviously this result is approximate because dR/dt was greater in the past. Therefore eq.(5.22) gives only an order of magnitude. More exactly we can write

$$T = t_0 - t_1 = \frac{1}{H_0} f(\Omega) \qquad (5.24)$$

where $\Omega = \rho_0 / \rho_c$ and $f(\Omega) < 1$. The function $f(\Omega)$, which describes the dependence of the age of the universe on the matter density, may be found by solving (5.18), and its plot is given in Fig.5.4. For $\Omega > 1$, the explicit expression is

$$f = \frac{\Omega}{(\Omega - 1)^{3/2}} \left[\sin^{-1} \sqrt{\frac{\Omega - 1}{\Omega}} - \frac{1}{\Omega} \sqrt{\Omega - 1} \right] \qquad (5.25)$$

and, for $\Omega < 1$

$$f = \frac{\Omega}{(1 - \Omega)^{3/2}} \left[-\sinh^{-1} \sqrt{\frac{1 - \Omega}{\Omega}} + \frac{1}{\Omega} \sqrt{1 - \Omega} \right] \qquad (5.26)$$

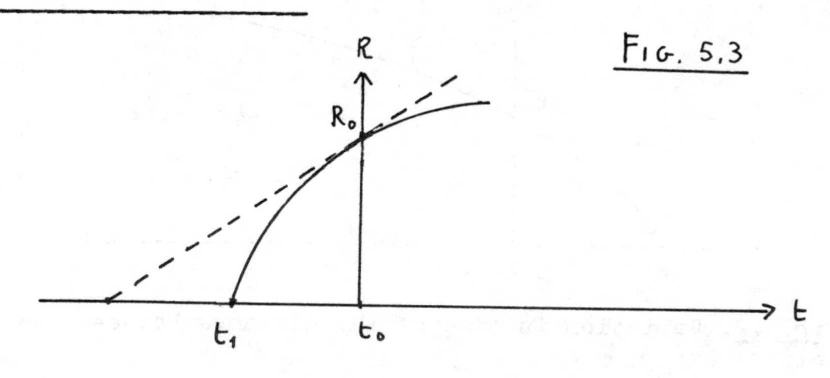

FIG. 5.3

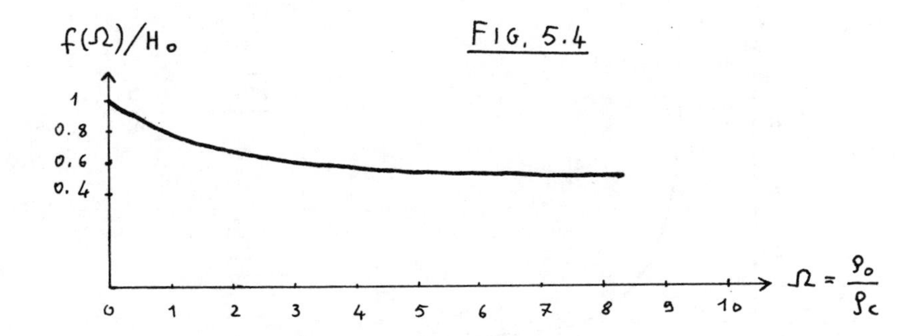

Fig.5.4. The age of the universe (in unit of $1/H_o$) plotted as a function of the adimensional matter density ρ_o/ρ_c.

In both cases, when $\Omega \to 1$ then $f \to 2/3$. Moreover $f \to 0$ when $\Omega \to \infty$, while $\Omega << 1$ we have

$$f(\Omega) \simeq 1 - \frac{\Omega}{2} \ln \frac{1}{\Omega} \tag{5.27}$$

In the case $\Omega > 1$ it is interesting to calculate the time t_M, corresponding to a minimum in the density, and the time t_2 at the end of the compression phase (see Fig.5.1). Putting

$$t_M - t_o = \frac{1}{H_o} f_M(\Omega) \tag{5.28}$$

and

$$t_2 - t_o = \frac{1}{H_o} f_2(\Omega) \tag{5.29}$$

we have, solving (5.18),

$$f_M(\Omega) = \frac{\Omega}{(\Omega-1)^{3/2}} \left[\frac{\sqrt{\Omega-1}}{\Omega} + \frac{\pi}{2} - \sin^{-1}\sqrt{\frac{\Omega-1}{\Omega}} \right] \tag{5.30}$$

$$f_2(\Omega) = \frac{\Omega}{(\Omega-1)^{3/2}} \left[\pi + \frac{\sqrt{\Omega-1}}{\Omega} - \sin^{-1}\sqrt{\frac{\Omega-1}{\Omega}} \right] \tag{5.31}$$

These two equations hold for $\Omega > 1$ only. When $\Omega \to 1$, then f_M and f_2 approach infinity like $(\Omega - 1)^{-3/2}$. The behaviour of f_M is represented in Fig.5.5.

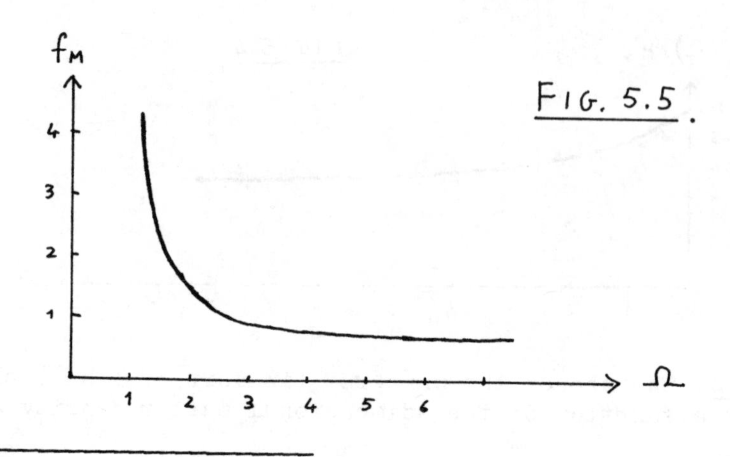

3.- Relativistic cosmology: the cosmic time

In the context of the relativistic cosmological models, the behaviour of the universe is governed by the Einstein field equations of the general relativity theory. As we have seen in the previous Section, in the framework of Newtonian cosmology, the evolution of the universe (when expansion is taken into account) is described by a function R(t) which satisfies a differential equation. Of course the Newtonian cosmology is not sufficient for a correct approach to the cosmological problem; general relativity should be used for a large scale description of the universe, because the "gravitational potential" ϕ of the matter contained inside the Hubble horizon R of an observer, measured in units of c^2, is of the order of unity, i.e. $\phi \sim G\rho_0 R^2 \sim c^2$, where R is the distance which, extrapolating the linear Hubble law (5.6), correspons to a recession with the light velocity (the Newtonian approximation is valid only for $\phi / c^2 << 1$).

However, as we will see, also in the relativistic case the main features of the universe are described by a scale function R(t) . But we will see also that a fundamental role is played by the so-called "cosmic time". Therefore we start giving a short introduction to the concept of cosmic time.

First of all consider the so-called "comoving" frame of reference, introduced first by Gauss. Consider a three dimensional hypersurface S embedded in a four-dimensional space-time with signature (+ - - -); suppose that any vector n^μ normal to S satisfies the inequality

$$n^\mu n_\mu > 0 \tag{5.32}$$

This means that S is "oriented in space" (or space-like) as n^μ, which is normal to S, is "oriented in time" (timelike). Denote with $\bar{x}^1$, $\bar{x}^2$, $\bar{x}^3$, the coordinates of a variable point $\bar{P}$ on the surface S. Through each point $\bar{P}$ of S we draw the geodesic orthogonal to S in $\bar{P}$. These geodesic will form a bundle of non-intersecting curves in some finite neighborhood $\mathcal{M}$ of S ($\mathcal{M}$ is four-dimensional), so that through each point P of $\mathcal{M}$ will pass only one of the previous geodesics (see Fig.5.6).

Now we define coordinates in $\mathcal{M}$ in the following way: consider a geodesic passing through P and the original point $\bar{P}$ in S, and denote with $\bar{P}P$ the arc length along the geodesic, and with $\bar{x}^i$ the coordinates of $\bar{P}$ (latin indices run from 1 to 3). The coordinates x^μ of P are then defined as

$$x^1 = \bar{x}^1, \; x^2 = \bar{x}^2, \; x^3 = \bar{x}^3, \; x^4 = \bar{P}P \tag{5.33}$$

In this way the three spatial coordinates x^1, x^2, x^3 remain constant along any geodesic orthogonal to S, hence along such geodesics we have

$$ds^2 = (dx^4)^2, \quad g_{44} = 1 \tag{5.34}$$

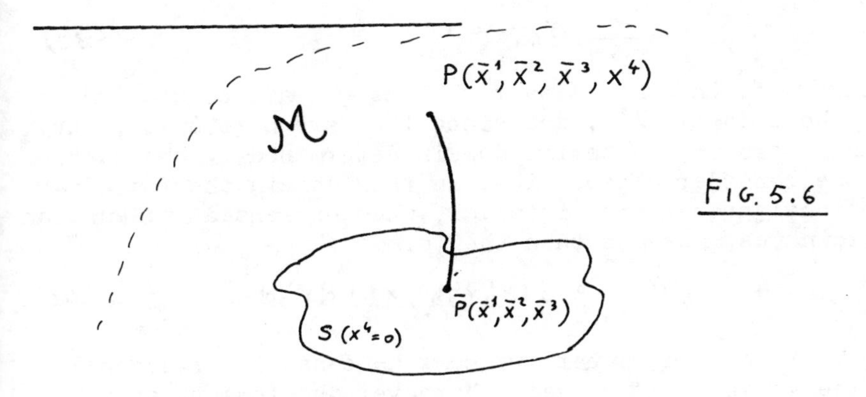

FIG. 5.6

Moreover, if we take into account the orthogonality of the geodesic with respect to the hypersurface S, it follows that any vector in S must be orthogonal to the vector (0,0,0,1) tangent to the x^4 line at the same point: then

$$g_{41} = g_{42} = g_{43} = 0 \tag{5.35}$$

in S . In fact a vector of S can be written in the form $y^\mu = (a,b,c,0)$. Putting $x^\mu = (0,0,0,1)$, from the orthogonality condition we must have $g_{\mu\nu} x^\mu y^\nu = 0$, and since the only nonvanishing products $x^\mu y^\nu$ are those for $\mu = 4$ and $\nu = 1,2,3$, we are led to the condition (5.35).

Therefore in S the line-element must assume the form

$$ds^2 = (dx^4)^2 + g_{ik}\, dx^i\, dx^k \tag{5.36}$$

This form of the line-element is valid also outside the hypersurface S, in the system of coordinates that we have constructed. In fact the geodesic equation is

$$\frac{d^2x^\alpha}{ds^2} + \Gamma_{\mu\nu}{}^{\alpha} \frac{dx^\mu}{ds} \frac{dx^\alpha}{ds} = 0 \tag{5.37}$$

and, applied to our present case in which only x^4 is variable along the geodesic, so that $ds = dx^4$, it gives $\Gamma_{44}{}^{\alpha} = 0$, and then also $\Gamma_{44\alpha} = 0$. Using the definition of the Christoffel symbols (2.60), this implies

$$2\frac{\partial}{\partial x^4} g_{4i} - \frac{\partial}{\partial x^i} g_{44} = 0 \tag{5.38}$$

and since $g_{44} = 1$ along any geodesic, we are left with

$$\frac{\partial}{\partial x^4} g_{4i} = 0 \tag{5.39}$$

along x^4 . This means that all the g_{4i} are constant on the whole domain $\mathcal{M}$, and since they are zero on S , they remain zero in the entire domain determined by the bundle of the considered geodesics. In this domain the four-dimensional proper-time interval, when expressed in Gaussian coordinates, assumes then the form

$$ds^2 = (dx^4)^2 + g_{ik}(x^1,x^2,x^3,x^4)\, dx^i\, dx^k \tag{5.40}$$

Therefore any hypersurface $x^4 =$ const is orthogonal to the geodesic x^4 lines . Moreover the lenghts of all the arcs of geodesic between two hypersurfaces S_1 and S_2 are the same, because they coincide with the unique time-interval Δx^4 between S_1 and S_2. We can say that the hypersurfaces x^4 = constant are equidistant level surface (in the case of an euclidean space they correspond to parallel planes).

The use of a system of Gaussian coordinates provides the possibility, as we shall see, of defining a cosmic time. In fact these coordinates introduce a sharp distinction between the time-like coordinate x^4 and the other three spatial coordinates, in agreement with the intuitive point of view which regards events as occurring in three-dimensional space, and describes their evolution in terms of a universal temporal parameter.

We are led to the definition of a comoving frame of reference, starting from the following three hypotheses:

a) there exist a universal time-like coordinate which coincides with the fourth component x^4 of a Gaussian coordinate system;

b) the three-dimensional hypersurfaces corresponding to various constant values of x^4 are isotropic;

c) any two points which belong to the same hypersurface (that is are characterized by the same value of x^4) are physically equivalent.

The fact that x^4 is a Gaussian coordinate of time-like character means that, at any given instant, the matter is at rest, on the average, in that particular hypersurface. Following the geodesics orthogonal to the surface, which remain always parallel to x^4, at any given time x^4 the matter remains at rest in that particular hypersurface. For this reason this Gaussian coordinate system is called comoving system.

From a physical point of view, this approach to the cosmological problem may follow from the cosmological principle and from the so-called "Weyl principle"[3], which can be expressed as follows: "The particles of the substratum (that is the galaxies) lie in a pencil of geodesics diverging from a common event in the past".

It is clear that with this postulate we can immediately introduce a Gaussian system of coordinates (remember that, following Gauss, we have considered before a bundle of geodesic forming a pencil of non-intersecting curves) and then a cosmic time. On the other hand the Weyl postulate is perfectly justified if one has in mind the

Hubble law, that is the fact the universe is expanding (and therefore the galaxies follow non-intersecting trajectories).

Combining then general relativity with the cosmological principle (homogeneity and isotropy of the universe), and with Weyl's postulate, we can introduce a cosmic time t, that is we can write the line-element in the form(see eq.5.40)

$$ds^2 = dt^2 - g_{ik}\, dx^i\, dx^k \quad , \quad i,k = 1,2,3 \qquad (5.41)$$

The cosmic time t is the proper time of an observer moving with the expanding matter that fill the universe (fundamental observer). This concept of cosmic time is very important, because when we refer to some global property of the universe, for instance when we say that the universe presents the same aspect to every fundamental observer, it means that it must be possible, for an observer A, to find a time t_A , according to his clock, at which he sees the universe in the same state as an observer B sees it at a time t_B of his clock. If the clock of one of these observers measures his proper time, then all these clocks will measure the proper time of their owners. In this way we have a universal cosmic time.

In conclusion, we observ that it is useful, in a theory which employs the concept of cosmic time, to define the notions of "world map" and "world picture".

In the former all the events are mapped at a common epoch at a time t. It represents the distribution of the physical characteristics of a model of universe on an hypersurface corresponding to a constant value t of the cosmic time.

The world picture, on the contrary, is the aspect that the universe presents to a given observer, at a given instant of time. The observer sees the distant parts of the universe at an early value of the cosmic time than the parts he sees near to him. It is the world picture that is to be compared more directly with the astronomical observations

REFERENCES

1) I.Newton, in a letter of Dec.10, 1962 , to R.Bentley

2) E.Halley: Proc.Soc.London Phil.Trans. 31, 22-26 (1720)

3) H.Weyl: Phys.Zeits.24,230(1923) ;
Phil.Mag.9,936(1930)

CHAPTER VI

RELATIVISTIC COSMOLOGICAL MODELS

1.- The Einstein static model

Before considering the cosmological models which take into account the Hubble law, and correspond to a line-element of the type given in eq.(5.41), we wish to present a short discussion of the first cosmological models formulated using the general relativistic theory, considering, first of all, the static Einstein model.

In order to obtain a homogeneous, isotropic and static cosmological solution to the field equations of general relativity, Einstein was forced to introduce the so-called cosmological constant Λ into the theory. The additional induced corrections must be small enough to be compatible with the other results of the theory which are experimentally verified. And in fact, as will be shown at the end of this Section (see eq.6.45), the cosmological constant Λ may be related to the mean density of the matter of the universe by $\Lambda = 4\pi G \rho / c^2$ and then, for a density $\rho \sim 10^{-30}$ g/cm^3 we find $\Lambda \sim 10^{-57}$ cm^{-2} (the contribution of Λ becomes relevant then only on a very large scale of distances).

Introducing the cosmological term Λ, the total action (3.57) becomes

$$S = -\frac{1}{2\chi}\int \sqrt{-g}\, d^4x \,(R + 2\Lambda) + S_m \qquad (6.1)$$

and its variation gives the following field equations

$$R_{\mu\nu} - \frac{1}{2} g_{\mu\nu} R = \chi T_{\mu\nu} + \Lambda g_{\mu\nu} \qquad (6.2)$$

Solving these equations in the case of a static and spherically symmetric field, starting from the general form (4.3) of the proper-time interval and following the same procedure used in Chapter IV, Sect.1 , we obtain the following modified Schwarzschild solution

$$ds^2 = \left(1 - \frac{2m}{r} - \frac{\Lambda}{3} r^2\right) dt^2 - \frac{dr^2}{1 - \frac{2m}{r} - \frac{\Lambda}{3} r^2} -$$

$$- r^2 \, d\theta^2 - r^2 \sin^2\theta \, d\varphi^2 \tag{6.3}$$

For $\Lambda \sim 10^{-57}$ cm $^{-2}$, the ratio between $\Lambda r^2/3$ and $2m/r$ is $\sim 10^{-19}$, if we take for r the radius of the orbit of Neptunus. Therefore the experimental tests of general relativity performed in the solar system are not affected by this value of Λ.

The cosmological constant was originally introduced by Einstein also to recover the agreement between general relativity and Mach'conception of inertia: in fact, for $T_{\mu\nu} = 0$ and $\Lambda = 0$, the equations (6.2) can be solved by $g_{\mu\nu} = \eta_{\mu\nu}$, and a test body moving in such a field (flat space) possesses all its inertia, though no matter is present to produce it. With $\Lambda \neq 0$, the flat-space solution is not allowed.

Eistein believed, though mistakenly, that with a positive Λ the field equations had no solution for $T_{\mu\nu} = 0$, that is in empty space: in this sense general relativity was then in agreement with Mach's principle.

But when De Sitter showed that eq.(6.2) may be solved for $T_{\mu\nu} = 0$ and $\Lambda \neq 0$ (it is possible to have a curved space without any matter present, in manifest contradiction with Mach's principle), Einstein rejected the cosmological term.

Quite independently from the presence or not of the cosmological term, we should note that the field $g_{\mu\nu}$ is not fully determined by the distribution of masses and energy in the universe. According to the field equations $g_{\mu\nu}$ is affected by $T_{\mu\nu}$ but, since we have differential equations, the metric is not determined without the boundary conditions at infinity. Moreover, the fact that Mach's principle is not completely incorporated into the theory may be seen directly also considering the Gödel cosmological model[1].

But let us begin to discuss the Einstein model in some detail. The starting point are the field equations (6.2) with $\Lambda \neq 0$. We assume a static and spherically symmetric line-element

$$ds^2 = e^{\nu} dt^2 - e^{\lambda} dr^2 - r^2 \left(d\theta^2 + \sin^2\theta \, d\varphi^2\right) \tag{6.4}$$

where λ and ν are functions only of r , and we suppose that the universe is filled with a perfect fluid of proper density ρ and pressure p . Then $(c=1)$

$$T^{\mu\nu} = (\rho + p)\frac{dx^\mu}{ds}\frac{dx^\nu}{ds} - p\, g^{\mu\nu} \qquad (6.5)$$

where p and ρ are the values measured by a locally inertial observer. In the static case we have $dr/ds = d\theta/ds = d\varphi/ds = 0$, and then, from (6.4), $dt/ds = \exp(-\nu/2)$. Therefore

$$\frac{dx^\mu}{ds} = u^\mu = \delta^\mu_4\, g_{44}^{-1/2} \qquad (6.6)$$

(note that the normalization condition $g_{\mu\nu}u^\mu u^\nu = 1$ is satisfied). In this case the components of the energy-momentum tensor are, from (6.5),

$$T_1{}^1 = T_2{}^2 = T_3{}^3 = -p\,, \qquad T_4{}^4 = \rho \qquad (6.7)$$

Now we follow the same procedure used in Chapter IV, to obtain the Schwarzschild exterior solution, with the only difference that we are looking now for an interior solution of the field equations, i.e. the space is filled with a continuous distribution of matter, whose energy-momentum tensor is nonvanishing and is given by eq.(6.7).

Using the explicit expression of $R_{\mu\nu}$, already computed, for the line-element (4.4), in eq.(4.18-21), we have (a prime denotes derivatives with respect to r)

$$R = R_1{}^1 + R_2{}^2 + R_3{}^3 + R_4{}^4 =$$

$$= e^{-\lambda}\left[-2\frac{\lambda'}{r} - \frac{\lambda'\nu'}{2} + \nu'' + \frac{\nu'^2}{2} + \frac{2}{r^2} + \frac{2\nu'}{r}\right] - \frac{2}{r^2} \qquad (6.8)$$

The field equations (6.2) give, for $T_4{}^4 = \rho$

$$\chi\rho = e^{-\lambda}\left(\frac{\lambda'}{r} - \frac{1}{r^2}\right) + \frac{1}{r^2} - \Lambda \qquad (6.9)$$

for $T_1{}^1 = -p$

$$\chi p = e^{-\lambda}\left(\frac{\nu'}{r} + \frac{1}{r^2}\right) - \frac{1}{r^2} + \Lambda \qquad (6.10)$$

and for $T_2{}^2 = -p$

$$\chi p = e^{-\lambda}\left[\frac{\nu''}{2} + \frac{\nu'^2}{4} - \frac{\lambda'\nu'}{4} + \frac{1}{2r}(\nu' - \lambda')\right] + \Lambda \qquad (6.11)$$

The equation for $T_3{}^3$ is equivalent to this last one, because $R_3{}^3 = R_2{}^2$. Moreover, from the continuity equation $T^{\mu\nu}{}_{;\nu} = 0$, using the explicit expressions (6.5,6) we obtain

$$T^{\mu\nu}{}_{;\nu} = -\partial^\mu p - (p+\rho)\,\partial^\mu(\ln\sqrt{g_{44}}) \qquad (6.12)$$

which in this case becomes

$$p' = -\frac{1}{2}\nu'(p+\rho) \qquad (6.13)$$

Looking for a solution describing a model which, besides being static, is also homogeneous, we must impose that the pressure p and the density ρ, as measured by a local observer, be everywhere the same, that is be independent from r.

If ρ is a constant, the integration of (6.9) is immediate and gives

$$e^{-\lambda} = 1 - \frac{1}{3}(\chi\rho + \Lambda)r^2 + \frac{A}{r} \qquad (6.14)$$

where A is an integration constant. Imposing the boundary condition that the metric reduces to the special relativistic one for small values of r, i.e. $\lambda = \nu = 0$ for r = 0, then A = 0, so that

$$e^{-\lambda} = 1 - \frac{\Lambda}{3}r^2 - \frac{1}{3}\chi\rho r^2 \qquad (6.15)$$

or

$$e^{-\lambda} = 1 - \frac{r^2}{R^2} \qquad (6.16)$$

where we have defined

$$\frac{1}{R^2} = \frac{\Lambda}{3} + \frac{1}{3}\chi\rho \qquad (6.17)$$

Moreover, if p'= 0, then (6.13) gives

$$\nu'(p+\rho) = 0 \qquad (6.18)$$

which can be satisfied in 3 ways, namely $\nu' = 0$, $\rho + p = 0$ or both equal to zero.

This last case leads, as will be shown (see 6.44), to the special relativistic Minkowski metric. The second possibility corresponds to the De Sitter universe. Finally the first case leads to the static Einstein model.

In this case, using $\nu' = 0$, eq.(6.10) for the pressure gives

$$e^{-\lambda} = 1 - (\Lambda - \chi p)r^2 \tag{6.19}$$

and defining

$$\Lambda - \chi P = \frac{1}{R^2} \tag{6.20}$$

we obtain the Einstein line-elemnt in the form

$$ds^2 = dt^2 - \frac{dr^2}{1 - r^2/R^2} - r^2(d\vartheta^2 + \sin^2\vartheta\, d\varphi^2) \tag{6.21}$$

This line-element can be written also in other different forms, by performing suitable coordinate transformations. For instance, by putting

$$r = \frac{\bar{r}}{1 + \bar{r}^2/4R^2} \tag{6.22}$$

we obtain the isotropic form

$$ds^2 = dt^2 - \frac{1}{(1 + \bar{r}^2/4R^2)}(d\bar{r}^2 + \bar{r}^2 d\vartheta^2 + \bar{r}^2 \sin^2\vartheta\, d\varphi^2) \tag{6.23}$$

Another useful form is obtained defining $r = R \sin\beta$,

$$ds^2 = dt^2 - R^2(d\beta^2 + \sin^2\beta\, d\vartheta^2 + \sin^2\beta \sin^2\vartheta\, d\varphi^2) \quad (6.$$

A still simpler expression is achieved introducing a larger number of variables as follows

$$z_1 = R\left(1 - \frac{r^2}{R^2}\right)^{1/2} ; \qquad z_2 = r \sin\vartheta \cos\vartheta$$

$$z_3 = r \sin\vartheta \sin\varphi ; \qquad z_4 = r \cos\vartheta \tag{6.25}$$

which satisfy

$$z_1^2 + z_2^2 + z_3^2 + z_4^2 = R^2 \tag{6.26}$$

The line-element becomes

$$ds^2 = dt^2 - dz_1^2 - dz_2^2 - dz_3^2 - dz_4^2 \qquad (6.27)$$

and we can see from the condition (6.26) that the spatial part of the Einstein universe can be thought as a hypersurface embedded in a higher-dimensional euclidean space. The geometry corresponding to the space-like variables is that of a spherical space of radius R, provided that the antipodal points of the sphere are considered to be distinct, while if they are identified we obtain a spatial geometry of ellyptical type. These two geometries have the same metric but different topologies (to visualize this situation consider for example a plane of finite size like this

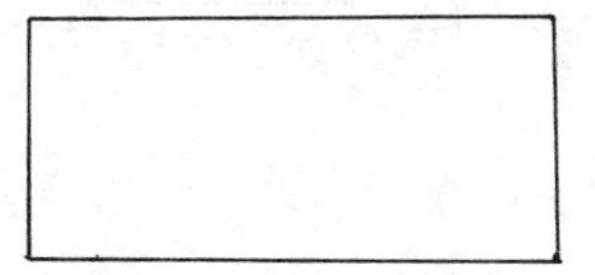

If the antipodal points are identified we obtain a cilynder)

In the case of spherical geometry we must take β variable from 0 to π : the volume of an Einstein universe is, from (6.24),

$$V = \int \sqrt{-g}\, d\beta\, d\vartheta\, d\varphi = \int_0^{\pi} d\beta \int_0^{\pi} d\vartheta \int_0^{2\pi} d\varphi\, R^3 \sin^2\beta \sin\vartheta =$$

$$= 4\pi R^3 \int_0^{\pi} \sin^2\beta\, d\beta = 2\pi^2 R^3 \qquad (6.28)$$

and the maximum distance from the origin is given by (using again 6.24)

$$\int_0^{\pi} R\, d\beta = \pi R \qquad (6.29)$$

The proper total distance around the Einstein universe is then

$$L = 2R\int_0^{\pi} d\beta = 2\pi R \tag{6.30}$$

If, on the contrary, the geometry is chosen to be ellyptical, then these values are reduced by one half. In fact, starting for example from the line-element (6.21), as

$$\sqrt{-g} = \frac{r^2 \sin\theta}{(1-r^2/R^2)^{1/2}} \tag{6.31}$$

the volume of the universe is given by

$$V = \int_0^R dr \int_0^{\pi} d\theta \int_0^{2\pi} d\varphi \, \frac{r^2 \sin\theta}{(1-r^2/R^2)^{1/2}} =$$

$$= 4\pi \int_0^R r^2 dr \, (1-r^2/R^2)^{-1/2} = \pi^2 R^3 \tag{6.32}$$

Moreover, for the maximum distance from the origin, and the total lenght of its circumference we have, respectively,

$$\int_0^R dr \, (1-r^2/R^2)^{-1/2} = R\int_0^1 \frac{d\xi}{\sqrt{1-\xi^2}} = R\left[\sin^{-1}\xi\right]_0^1 = \frac{\pi}{2} R \tag{6.33}$$

$$L = \pi R \tag{6.34}$$

As regards the total mass M of the universe we find, in the ellyptical case

$$M = \rho V = \pi^2 R^3 \rho \tag{6.35}$$

From (6.17,20) we obtain

$$\Lambda = \frac{1}{2}\chi\,(\rho + 3p) \tag{6.36}$$

$$\frac{1}{R^2} = \frac{1}{2}\chi\,(\rho + p) \tag{6.37}$$

so that, if the universe is filled with a pressureless fluid (dust matter) as originally supposed by Einstein, then $p = 0$ and

$$\Lambda = \frac{1}{R^2} = \frac{1}{2} \chi \rho \tag{6.38}$$

Eq.(6.35) gives then

$$M = \frac{2\pi^2}{\chi} R \tag{6.39}$$

Therefore, in the Einstein universe, if ρ is given, Λ and R are determined, and also the total mass M of the universe can be calculated.

If, on the other hand the model is filled only with radiation, then the equation of state is $\rho = 3p$ (this relation can be obtained noticing that for massless particles the energy-momentum tensor must be traceless, $T_\mu{}^\mu = 0$, and using eq.(6.7)), and from eq.(6.36,37) we have

$$\Lambda = 3\chi p \tag{6.40}$$

$$\frac{1}{R^2} = 2\chi p \tag{6.41}$$

that is

$$\Lambda = \frac{3}{2}\frac{1}{R^2} \tag{6.42}$$

$$\frac{1}{R^2} = \frac{2}{3}\chi\rho \tag{6.43}$$

Finally, if the model is empty, that is $\rho = p = 0$, (and obviously $\nu' = 0$) we have

$$\Lambda = \frac{1}{R^2} = 0 \tag{6.44}$$

so that the universe degenerates to the flat space-time, as noted previously.

It is interesting to consider some consequences of the Einstein model, comparing them with observations.

First of all, if we take $\rho \sim 10^{-30}$ g/cm^3 (a value which is near the current estimates) we have, from (6.38)

$$\frac{1}{R^2} = \Lambda = \frac{4\pi G}{c^2}\rho \sim 10^{-57}\ \mathrm{cm}^{-2} \tag{6.45}$$

and then the correction to the gravitational field equations due to the cosmological constant is neglegible at a planetary and interplanetary scale of dinstances (as already stressed at the beginning of this Chapter). The radius (or better the characteristic length) associated with the Einstein universe is then $R = \Lambda^{-1/2} \sim 10^{28}$ cm.

As regards light propagation, it is easy to see that no red-shift is produced in this cosmological model (also for this reason it can be said "static"). In fact, consiser the line-element (6.21), and put $ds^2 = 0$, with $d\vartheta = d\varphi = 0$, for the case of light travelling along the radial direction; we obtain

$$\frac{dr}{dt} = \pm \left(1 - \frac{r^2}{R^2}\right)^{1/2} \tag{6.46}$$

where the variable t, as we can see from the line-element, is the proper time measured by a local observer, at rest with respect to the spatial coordinates.

Suppose that an observer, placed at the origin (r = 0), receivesat a time t_2 the light emitted at a time t_1 by a galaxy placed at a point of radial coordinate r . We have then, from (6.46)

$$t_2 = t_1 + \int_0^r dr \left(1 - r^2/R^2\right)^{-1/2} = t_1 + R \sin^{-1}(r/R) \tag{6.47}$$

If the observer and the galaxy are at rest with respect to the spatial coordinates, then r = const. It follows that the interval Δt_2 between the reception of two wave crests is equal to the interval Δt_1 between their emission

$$\Delta t_2 = \Delta t_1 \tag{6.48}$$

As t is the proper time, this equation implies the equality of the proper periods of the emitted and received light, so that no shift of frequency is observed, comparing the received light with the one emitted by a source (of the same type) at rest in the observer's laboratory.

This conclusion is correct, provided that the observer and the galaxy, initially at rest at two different places of the Einstein universe, acquire no relative motion in the course of time. As regards this point, consider the geodesic equation for a test particle initially at rest in the Einstein model: from the line-element (6.21) we have, putting $dr/ds = d\vartheta/ds = d\varphi/ds = 0$,

$$\frac{d^2x^\mu}{ds^2} + \Gamma_{44}{}^{\mu}\left(\frac{dt}{ds}\right)^2 = 0 \tag{6.49}$$

An explicit computation of the Christoffel symbols, for the metric (6.21), gives $\Gamma_{44}{}^{\mu} = 0$; therefore the components of the acceleration are vanishing

$$\frac{d^2 r}{ds^2} = \frac{d^2\vartheta}{ds^2} = \frac{d^2\varphi}{ds^2} = 0 \qquad (6.50)$$

and then a particle, initially at rest, will remain permanently at rest in this model.

2.- The De Sitter universe

The cosmological model considered by De Sitter is empty, according to a classical interpretation, because it satisfy the condition $p + \rho = 0$. By adding the two equations (6.9,10) we obtain then

$$e^{-\lambda}\left(\frac{\lambda'}{r} + \frac{\nu'}{r}\right) = 0 \qquad (6.51)$$

so that

$$\nu' = -\lambda' \qquad (6.52)$$

Imposing that, at the origin, $\lambda(0) = \nu(0) = 0$, it follows that

$$\lambda = -\nu \qquad (6.53)$$

On the other hand, if ρ is constant, integrating (6.9) we obtain, like before, eq.(6.16); the line-elemnt of the De Sitter universe becomes then

$$ds^2 = \left(1 - \frac{r^2}{R^2}\right)dt^2 - \frac{dr^2}{1 - r^2/R^2} - r^2\left(d\vartheta^2 + \sin^2\vartheta\, d\varphi^2\right) \qquad (6.54)$$

where R is given by (see 6.17)

$$\frac{1}{R^2} = \frac{\Lambda}{3} + \frac{1}{3}\chi\rho \qquad (6.55)$$

We can perform a coordinate transformation, to obtain a geometrically simpler form, by putting

$$r = R \sin\eta \qquad (6.56)$$

so that eq.(6.55) becomes

$$ds^2 = \cos^2\eta \, dt^2 - R^2(d\eta^2 + \sin^2\eta \, d\vartheta^2 + \sin^2\eta \sin^2\vartheta \, d\varphi^2) \tag{6.57}$$

Another interesting form may be obtained introducing, like in the Einstein case, five variables

$$\alpha = r \sin\vartheta \cos\varphi, \quad \beta = r \sin\vartheta \sin\varphi, \quad \gamma = r\cos\vartheta$$

$$\delta + \varepsilon = R e^{t/R} (1 - r^2/R^2)^{1/2}, \quad \delta - \varepsilon = R e^{-t/R} (1 - r^2/R^2)^{1/2} \tag{6.58}$$

so that we obtain

$$ds^2 = d\varepsilon^2 - d\alpha^2 - d\beta^2 - d\gamma^2 - d\delta^2 \tag{6.59}$$

If we define

$$z_1 = i\alpha, \; z_2 = i\beta, \; z_3 = i\gamma, \; z_4 = i\delta, \; z_5 = \varepsilon \tag{6.60}$$

the line-element becomes

$$ds^2 = dz_1^2 + dz_2^2 + dz_3^2 + dz_4^2 + dz_5^2 \tag{6.61}$$

and the five new variables z are related by the condition

$$z_1^2 + z_2^2 + z_3^2 + z_4^2 + z_5^2 = -R^2 \tag{6.62}$$

which describes a four-dimensional hypersurface in a five-dimensional manifold. We can visualize the De Sitter universe as the surface of a hypersphere embedded in a five-dimensional euclidean space. It should be noted, however, that this formal simplicity is achieved loosing the distinction between space-like and time-like intervals.

An important transformation of coordinates was discovered by Lemaitre and Robertson: defining (c = 1)

$$\bar{r} = \frac{r e^{-t/R}}{(1 - r^2/R^2)^{1/2}}, \quad \bar{t} = t + \frac{1}{2} R \ln(1 - r^2/R^2) \tag{6.63}$$

we are led to line element

$$ds^2 = d\bar{t}^2 - e^{2\bar{t}/R}(d\bar{r}^2 + \bar{r}^2 d\vartheta^2 + \bar{r}^2 \sin^2\vartheta \, d\varphi^2) \tag{6.64}$$

which may be written, putting k = 1/R and dropping the bars over r and t

$$ds^2 = dt^2 - e^{2kt}\left(dr^2 + r^2 d\vartheta^2 + r^2 \sin^2\vartheta \, d\varphi^2\right) \qquad (6.65)$$

Because of the time dependence of the exponential factor, this equation seems to describe a non-static model of universe, as we have in this case that the metric $g_{\mu\nu}$ depends on the time-like coordinate t . However this situation should not be surprising, as any static line-element may be transformed into a time dependent one by a suitable choice of coordinates. As shown by Robertson, the De Sitter metric is intrinsically time-independent, because the transformation $r' = r \exp(kt_0)$, $t' = t - t_0$, corresponding to a change in the spatial scale, toghether with the choice of a new zero point for the time variable, leaves the line-element unchanged. Following Robertson, this type of universe may be called "stationary" rather than static.

One of the main differences between the De sitter and the Einstein model is that the De Sitter universe is empty (in a classical sense) . In fact, from the condition $p + \rho = 0$ one can deduce $p = \rho = 0$, because the density of matter is a positive quantity, and the pressure cannot be negative (it should be mentioned, however, that in the framework of the so-called "inflationary" cosmology [2] , the De Sitter metric is interpreted as corresponding to a vacuum background, described by the equation of state $p = - \rho$, relative to a phase during which the energy content of the universe is dominated by the vacuum contributions).

If $p = \rho = 0$, the cosmological constant is then simply related to R by (see 6.65)

$$\frac{1}{R^2} = \frac{\Lambda}{3} \qquad (6.66)$$

The geodesic equation of a test particle in this universe describes a radial motion away from the origin (or towards the origin) without return. Starting from the line-element (6.54) and following the same procedure used in Chapter IV, Sect.2, we find in fact the following equations of motion

$$\frac{dr}{ds} = \pm\left(\alpha^2 - 1 + \frac{r^2}{R^2} - \frac{h^2}{r^2} + \frac{h^2}{R^2}\right)^{1/2} \qquad (6.67)$$

$$\frac{d\varphi}{ds} = \frac{h}{r^2} \qquad (6.68)$$

$$\frac{dt}{ds} = \frac{\alpha}{1 - r^2/R^2} \qquad (6.69)$$

where α and h are integration constants. In particular h can be negative of positive according to the direction of motion. Moreover $\alpha > 0$ for $r < R$ (as the coordinate time t is increasing when the proper time s is increasing). Integrating these equations one can find that the orbit of a test particle corresponds to that of a particle subject to a repulsive force (for the case of the plus sign in eq.(6.67)).

As regards the behaviour of light rays, for a pure radial motion we have, putting ds = 0 in the metric (6.54)

$$\frac{dr}{dt} = \pm \left(1 - \frac{r^2}{R^2}\right) \qquad (6.70)$$

It is interesting to note that an infinite time is needed by light to travel the distance between r = R and the origin

$$\lim_{r_1 \to R} \int_0^{r_1} \frac{dr}{1 - r^2/R^2} = \lim_{r_1 \to R} \frac{R}{2} \left[\ln \left(\frac{1 + r/R}{1 - r/R} \right) \right]_0^{r_1} = +\infty \qquad (6.71)$$

Therefore an observer placed at the origin cannot receive informations about any event occurring beyond a radial distance R , so that we have, in this model, an " event horizon", in the sense of Rindler[3] .

Moreover, considering the light received by an observer at the origin, r = 0, at a time t_2 , and emitted by a freely moving particle (a galaxy) in r at a time t_1 , we have

$$t_2 = t_1 + \int_0^{r} \frac{dr}{1 - r^2/R^2} \qquad (6.72)$$

so that the interval Δt_2 between the reception of two successive wave crests is related to the corresponding interval of emission Δt_1 by

$$\Delta t_2 = \Delta t_1 + \frac{1}{1-r^2/R^2} \frac{dr}{dt} \Delta t_1 \qquad (6.73)$$

where dr/dt is the radial velocity of the source at the instant of emission. The proper time interval can be expressed as a function of Δ t as follows (see 6.69)

$$\Delta s = \Delta t \frac{1-r^2/R^2}{\alpha} \qquad (6.74)$$

The emitted and received wave-lengths , $\lambda_1 = c/\nu_1$ and $\lambda_2 = c/\nu_2$, are related then by (remember the definition of proper frequency, $\nu = \partial\psi/\partial\tau$, introduced in Chapter IV, Sect.4)

$$\frac{\lambda_2}{\lambda_1} = \frac{\nu_1}{\nu_2} = \frac{\Delta s_2}{\Delta s_1} = \frac{\Delta t_2}{\Delta t_1}\left(1-\frac{r^2}{R^2}\right)^{-1} \qquad (6.75)$$

Combining this result with eq.(6.73) we obtain then

$$\frac{\lambda_2}{\lambda_1} = \frac{1}{1-r^2/R^2} + \frac{1}{(1-r^2/R^2)^2}\frac{dr}{dt} \qquad (6.76)$$

The frequency shift may be toward the red, $\lambda_2 > \lambda_1$, or the violet, $\lambda_2 < \lambda_1$, depending on the motion of the source. If the emitting particle is moving away from the origin (dr/dt > 0) then we have always red-shift. If, on the contrary, dr/dt < 0, we can have red or violet shifts according to whether the first term is greater or smaller than the second in the right-hand side of eq.(6.76). We have however a prevalence of red-shifts.

Remember, at this point, the Weyl hypothesis (Chapter V, Sect.3) according to which the particles of the universe must be regarded as lying on a pencil of geodesics which diverge from a common point in the past; moreover, using the line-element (6.65) and the geodesic equations we can easily see that that (like in the Einstein model) if a particle is initially at rest, it will remain permanently at rest with respect to r, ϑ, φ . Since the coordinate time t coincides with the proper time, not only for a particle placed at the origin, but for every particle at rest in this system of coordinates, any of these particles may be chosen as origin, hence all these particles must be regarded as defining equivalent observers that will see the same aspect of the universe.

In spite of being at rest with respect to spatial coordinates, the proper distance between these observers changes in time (as $g_{\mu\nu}$ is time depending), producing then a frequency shift when the light travels from one to another observer. Using the line-element (6.65), the radial velocity of light is

$$\frac{dr}{dt} = \pm e^{-kt} \tag{6.77}$$

and for a light signal emitted at a time t_1 at r, and received at the origin at a time t_2, we have

$$\int_{t_1}^{t_2} e^{-kt}\, dt = \int_0^r dr = r = \text{const} \tag{6.78}$$

so that, differentiating, we get

$$\Delta t_2 = e^{k(t_2 - t_1)} \Delta t_1 \tag{6.79}$$

As t is also the proper time, we can conclude that the shift of wave-lengths is

$$\frac{\lambda_2}{\lambda_1} = \frac{\nu_1}{\nu_2} = \frac{\Delta t_2}{\Delta t_1} = e^{k(t_2 - t_1)} \tag{6.80}$$

For values of $r = c(t_2 - t_1)$ much smaller that the cosmological distance $R = c/k$ we obtain, in first approximation, the linear Hubble law

$$\frac{\Delta\lambda}{\lambda} = \frac{\lambda_2 - \lambda_1}{\lambda_1} = e^{k(t_2 - t_1)} - 1 \simeq$$

$$\simeq k(t_2 - t_1) = \frac{k}{c} r = \frac{r}{R} \tag{6.81}$$

If k is taken to be equal to the Hubble constant, i.e. $k/c \sim 2 \times 10^{10}$ (light years)$^{-1}$, the horizon distance is found to be

$$R = \frac{c}{k} \sim 10^{28} \text{ cm} \tag{6.82}$$

and

$$\Lambda = \frac{3}{R^2} \sim 10^{-56} \text{ cm}^{-2} \tag{6.83}$$

Concluding this Chapter, we can say that the Einstein model of the universe is not realistic as no red-shift is predicted, in disagreement with experience. On the other hand also the De Sitter model is unsatisfactory, because, although being compatible with the observed red-shift, it describes an empty universe without any matter or radiation (unless negativepressures are allowed). It is possible, however, according to the inflationary cosmology[2], that such a period of"vacuum dominance" was realized during the very early stages of the evolution of our universe, producing a De Sitter phase whichcan explain the observed large scale homogeneity and isotropy, even for regions of matter that are at present causally disconnetted.

REFERENCES

1) K.Gödel: Rev.Mod.Phys. 21,447(1949)

2) A.H.Guth: Phys.Rev.D23,347(1981)

3) W.Rindler: Mon.Not.Roy.Ast.Soc.116,663(1956)

CHAPTER VII

NON-STATIC MODELS OF UNIVERSE

1.- The Roberstson-Walker metric

According to the Hubble law, we should use an expanding model to describe the behaviour of our universe.

Starting from the cosmological principle, which implies that the universe is the same for any observer at any given instant of time and the spatial directions are all equivalent, it follows that all the points of the hypersurfaces $x^4 = t = \text{const}$ are physically equivalent, and these three-dimensional spaces must have a constant curvature (positive or negative) which is a function only of time. From the Weyl hypothesis (see Chapter V) we are led to introduce a cosmic time t, and in general the line-element can be written using comoving coordinates (see 5.41) as

$$ds^2 = dt^2 - e^{\mu}\left(dr^2 + r^2 d\vartheta^2 + r^2 \sin^2\vartheta \, d\varphi^2\right) \tag{7.1}$$

where $\mu = \mu(r,t)$, and the spatial isotropy implies

$$\mu(r,t) = f(r) + g(t) \tag{7.2}$$

In fact, the equivalence of all the points of a thredimensional hypersurface at a given time t requires that two observers, placed at two different points, must observe the same physics; therefore the ratio between proper distances, at the point $P(r_1, \vartheta_1, \varphi_1)$ and $Q(r_2, \vartheta_2, \varphi_2)$ must remain fixed in time, so that

$$e^{\mu(r_1,t)} / e^{\mu(r_2,t)}$$

must be independent of t . It follows that $\mu(r_1,t) = \mu(r_2,t) + \nu(r_1,r_2)$; fixing r_2 , we can write then $\mu(r_1,t) = \nu(r_1) + \mu(t)$ and we are led to eq.(7.2).

The function f(r) could be determined using the well known Schur's theorem : the isotropy at every point of a three-dimensional space implies its homogeneity. In other words, if the riemannian curvature of the three-space is the same in all directions, it is also the same at each point. The function f(r) is then obtained imposing a constant curvature on the three-dimensional space.

However, we prefer to determine here f(r) in another way, directly from the Einstein field equations. Following the usual procedure, starting from the metric tensor of eq.(7.1), i.e. computing the corresponding Christoffel symbols and the components of the Ricci tensor, and writing explicitly the field equations (6.2) we obtain

$$\chi T_1{}^1 = -e^{-\mu}\left(\frac{f'^2}{4} + \frac{f'}{r}\right) + \ddot{g} + \frac{3}{4}\dot{g}^2 - \Lambda$$

$$\chi T_2{}^2 = \chi T_3{}^3 = -e^{-\mu}\left(\frac{f''}{2} + \frac{f'}{2r}\right) + \ddot{g} + \frac{3}{4}\dot{g}^2 - \Lambda$$

$$\chi T_4{}^4 = -e^{-\mu}\left(f'' + \frac{f'^2}{4} + \frac{2f'}{r}\right) + \frac{3}{4}\dot{g}^2 - \Lambda \tag{7.3}$$

and $\chi\, T_\mu{}^\nu = 0$ for $\mu \neq \nu$ (a prime denotes the derivative with respect to r, and a dot with respect to t).

Now we impose the condition of local isotropy on the energy-momentum tensor $T_{\mu\nu}$. At each point we introduce a proper coordinate system $\tilde{x}^1$, $\tilde{x}^2$, $\tilde{x}^3$, for a local observer at rest with respect to r, ϑ, φ, in which the spatial line-element is proportional to

$$d\sigma^2 = (d\tilde{x}^1)^2 + (d\tilde{x}^2)^2 + (d\tilde{x}^3)^2 \tag{7.4}$$

We have then the relations

$$d\tilde{x}^1 = e^{\frac{1}{2}\tilde{\mu}}\, dr\,, \qquad d\tilde{x}^2 = e^{\frac{1}{2}\tilde{\mu}}\, r\, d\vartheta$$

$$d\tilde{x}^3 = e^{\frac{1}{2}\tilde{\mu}}\, r \sin\vartheta\, d\varphi\,, \qquad d\tilde{t} = dt \tag{7.5}$$

By using the general transformation rules of a tensor under general coordinate transformations we obtain $\tilde{T}_\mu{}^\nu = T_\mu{}^\nu$. In fact

$$\tilde{T}_1{}^1 = \frac{\partial \tilde{x}^1}{\partial x^\alpha}\frac{\partial x^\beta}{\partial \tilde{x}^1} T_\beta{}^\alpha = e^{\frac{1}{2}\tilde{\mu}} e^{-\frac{1}{2}\tilde{\mu}} T_1{}^1 = T_1{}^1 \tag{7.6}$$

and so on for the other components. This means that the components of $T_\mu{}^\nu$ given in eq.(7.3) are referred to a proper coordinate system used by a local observer at rest with respect to the cosmic matter. From the assumed spatial isotropy his measures of stress must lead to a symmetry between the x,y,z directions, so that the equality

$$T_1{}^1 = T_2{}^2 = T_3{}^3 \tag{7.7}$$

must be staisfied. We have then, from (7.3)

$$\frac{f'^2}{4} + \frac{f'}{r} = \frac{f''}{2} + \frac{f'}{2r} \tag{7.8}$$

or

$$f'' - \frac{1}{2} f'^2 - \frac{1}{r} f' = 0 \tag{7.9}$$

from which

$$f' = a r e^{\frac{1}{2} f} \tag{7.10}$$

This leads to the general solution

$$e^{f(r)} = \frac{b^2}{\left(1 - \frac{ab}{4} r^2\right)^2} \tag{7.11}$$

where a and b are two integration constants. Absorbing the constant b into the definition of the function exp [g(t)] we define a new constant (in order to agree with the usual notations) putting

$$|ab| = \frac{1}{R_0^2} \tag{7.12}$$

We obatin then the following line-element

$$ds^2 = dt^2 - \frac{e^{g(t)}}{\left(1 + \frac{k}{4} \frac{r^2}{R_0^2}\right)^2} \left(dr^2 + r^2 d\vartheta^2 + r^2 \sin^2\vartheta \, d\varphi^2\right) \tag{7.13}$$

where k = 0, +1, -1 , corresponds to ab = 0, negative and positive respectively, that is to an euclidean, spherical and pseudospherical three-dimensional space, as we will see. This is the well known Robertson-Walker metric.

Note that from eq.(7.3) we have

$$G_1{}^1 = G_2{}^2 = G_3{}^3 = -e^{-f}\left(\frac{f'^2}{4} + \frac{f'}{r}\right) e^{-g} + \ddot{g} + \frac{3}{4} \dot{g}^2 \tag{7.14}$$

and eq.(7.11) gives

$$e^{-f}\left(\frac{f'^2}{4} + \frac{f'}{r}\right) = \frac{a}{b} = \text{const} \tag{7.15}$$

Therefore the spatial part of the Einstein tensor (which is proportional to the energy-momentum tensor) is not only isotropic

$$G_1{}^1 = G_2{}^2 = G_3{}^3 \tag{7.16}$$

but also constant throughout the three-space at any given time t . We have in fact

$$G_1{}^1 = G_2{}^2 = G_3{}^3 = -e^{-g}\frac{a}{b}\cdot + \ddot{g} + \frac{3}{4}\dot{g}^2 \tag{7.17}$$

It is easy to see that the Robertson-Walker metric accounts for the red-shift of light rays emitted by distant sources. In fact, following the same arguments used for the Einstein and De Sitter models, the coordinate velocity of a light ray moving radially is, from (7.13),

$$\frac{dr}{dt} = \pm e^{-\frac{1}{2}g}\left(1 + \frac{Kr^2}{4R_0^2}\right) \tag{7.18}$$

Integrating over the time interval (t_e, t_o) needed for a ray to travel between the origin and any coordinate position r, we have

$$\int_0^r \frac{dr}{1 + Kr^2/4R_0^2} = \int_{t_e}^{t_o} e^{-\frac{1}{2}g(t)}\,dt \tag{7.19}$$

Consider two wave crests, emitted at times t_e and $t_e + \Delta t_e$, from a source at radial coordinate r , and received by an observer placed at the origin (r=0) at times t_o and $t_o + \Delta t_o$, with $t_o > t_e$. From (7.19) we have then

$$\int_0^r \frac{dr}{1 + Kr^2/4R_0^2} = \int_{t_e}^{t_o} e^{-\frac{1}{2}g}\,dt = \int_{t_e+\Delta t_e}^{t_o+\Delta t_o} e^{-\frac{1}{2}g}\,dt \tag{7.20}$$

Being $(t_e + \Delta t_e) < t_o$, the two integrals may be decomposed as follows

$$\int_{t_e}^{t_o} = \int_{t_e}^{t_e+\Delta t_e} + \int_{t_e+\Delta t_e}^{t_o} \tag{7.21}$$

$$\int_{t_e+\Delta t_e}^{t_o+\Delta t_o} = \int_{t_e+\Delta t_e}^{t_o} + \int_{t_o}^{t_o+\Delta t_o}$$

and then we obtain

$$\int_{t_e}^{t_e+\Delta t_e} e^{-\frac{1}{2}g}\,dt = \int_{t_o}^{t_o+\Delta t_o} e^{-1/2\,g}\,dt \tag{7.22}$$

If the function $g(t)$, which describes the temporal evolution of the universe, may be regarded nearly constant during the infinitesimal intervals Δt_e and Δt_o, we obtain, from (7.22)

$$\frac{\Delta t_o}{\Delta t_e} = e^{\frac{1}{2}(g_o - g_e)} \qquad (7.23)$$

where $g_o = g(t_o)$ and $g_e = g(t_e)$.

As the coordinate time intervals Δt_o and Δt_e coincide with the proper time intervals for both the observer and the source of light (we suppose that the source is at rest with respect to the comoving coordinates r, θ, φ) we obtain that the observed wave-length λ_o is related to the emitted one λ_e by

$$\frac{\lambda_o}{\lambda_e} = \frac{\nu_e}{\nu_o} = \frac{\Delta t_o}{\Delta t_e} = e^{\frac{1}{2}(g_o - g_e)} \qquad (7.24)$$

and then

$$z = \frac{\Delta\lambda}{\lambda} = \frac{\lambda_o - \lambda_e}{\lambda_e} = e^{\frac{1}{2}(g_o - g_e)} - 1 \qquad (7.25)$$

defines the red-shift factor z in terms of the temporal evolution of the Robertson-Walker metric.

In order to obtain explicitly the Hubble law from this relation, we rewrite the line-element (7.13) in an equivalent form (more used by astronomers). Instead of the radial coordinate r we introduce a dimensionless marker $u = r/R_o$, and define a new function $R^2(t) = R_o^2 \exp[g(t)]$, so that

$$ds^2 = dt^2 - \frac{R^2(t)}{(1+\frac{K}{4}u^2)^2}\left(du^2 + u^2 d\theta^2 + u^2 \sin^2\theta \, d\varphi^2\right) \qquad (7.26)$$

R(t) is also called the Robertson-Walker scale factor and may be interpreted as describing the "size" of the universe because, as we will see later, the spatial part of this model may be represented geometrically as the hypersurface of a four dimensional sphere (or pseudo sphere) with radius R(t) . We have then for the red-shift

$$z = \frac{\Delta\lambda}{\lambda} = \frac{R_o}{R_e} - 1 = \frac{\Delta R}{R} = \frac{R_o - R_e}{R_e} \qquad (7.27)$$

where $R_e = R(t_e)$ is the radius of the model at the time of light emission, and $R_o = R(t_o)$ (which should not be confused with the constant of eq.(7.12)) is the radius at the time in which the light is received. In this case the red-shift is then closely related to the expansion of the universe.

From the line-element (7.26) we have that a source of light (a galaxy), placed at a radial coordinate r, is characterized by the time independent, dimensionless distance marker ℓ such that

$$\ell = \int_0^{r/R_0} du \left(1 + \frac{K}{4} u^2\right)^{-1} \tag{7.28}$$

and related to the time of emission t_e, and observation t_0 $(t_0 > t_e)$ by

$$\ell = \int_{t_e}^{t_0} \frac{dt}{R(t)} \tag{7.29}$$

To derive the Hubble law (5.2) we expand $R^{-1}(t)$ in terms of the time of travel of the light signal (this is allowed provided this time interval is smaller than the time required by light to travel a distance of the order of R, that is $r/c = t_0 - t_e << R/c$).

$$\frac{1}{R(t)} \simeq \frac{1}{R_0} + \frac{\dot{R}_0}{R_0^2}(t_0 - t) + \frac{1}{2}\left(\frac{-\ddot{R}_0 R_0^2 + 2R_0\dot{R}_0^2}{R_0^4}\right)(t_0-t)^2 + \mathcal{O}\left(\frac{t_0-t}{R_0}\right)^3 =$$

$$= \frac{1}{R_0} + \frac{\dot{R}_0}{R_0}\frac{(t_0-t)}{R_0} + \left(\frac{\dot{R}_0^2}{R_0} - \frac{\ddot{R}_0}{2}\right)\left(\frac{t_0-t}{R_0}\right)^2 + \mathcal{O}\left(\frac{t_0-t}{R_0}\right)^3 \tag{7.30}$$

where $R_0 = R(t_0)$, $\dot{R}_0 = (dR/dt)_{t=t_0}$, and so on, and terms of order $(t_0 - t)^3$ have been neglected.

From (7.29), putting $h = (t_0 - t_e)/R_0$, we have then

$$\ell = \int_{t_e}^{t_0} \frac{dt}{R} = \int_{t_e}^{t_0}\left[\frac{1}{R_0} + \frac{\dot{R}_0}{R_0}\frac{(t_0-t)}{R_0} + \cdots\right] dt =$$

$$= \frac{t_0-t_e}{R_0} + \frac{1}{2}\frac{\dot{R}_0}{R_0}\frac{(t_0-t_e)^2}{R_0} = h + \frac{1}{2}\dot{R}_0 h^2 + \mathcal{O}(h^3) \tag{7.31}$$

and from (7.27)

$$z = \frac{R_0}{R_e} - 1 = R_0\left[\frac{1}{R(t_e)}\right] - 1 =$$

$$= R_0 \left\{ \frac{1}{R_0} + \frac{\dot{R}_0}{R_0} \frac{t_0 - t_e}{R_0} + \left(\frac{\dot{R}_0^2}{R_0} - \frac{\ddot{R}_0}{2} \right) \left(\frac{t_0 - t_e}{R_0} \right)^2 + \cdots \right\} - 1$$

$$= \dot{R}_0 h + R_0 \left(\frac{\dot{R}_0^2}{R_0} - \frac{\ddot{R}_0}{2} \right) h^2 + O(h^3) \qquad (7.32)$$

Eliminating h from (7.31,32) (in the approximation $\ell^2 \simeq h^2$) we have

$$z = \dot{R}_0 \left(\ell - \frac{1}{2} \dot{R}_0 \ell^2 \right) + \dot{R}_0^2 \ell^2 - \frac{1}{2} R_0 \ddot{R}_0 \ell^2$$

$$z = \dot{R}_0 \ell + \frac{1}{2} \left(\dot{R}_0^2 - \ddot{R}_0 R_0 \right) \ell^2 + O(\ell^3) \qquad (7.33)$$

This equation is of fundamental importance because it establishes a theoretical relation between the red-shift and the astronomical distance, both observable quantities. Consider in fact the first term on the right-hand side of eq.(7.33), noting that, in first approximation, the distance measured by astronomers d may be identified with $d = R_0 \ell$:

$$z \simeq \frac{\dot{R}_0}{R_0} (R_0 \ell) = \frac{\dot{R}_0}{R_0} d \qquad (7.34)$$

This is precisely the Hubble law (5.2), provided that the Hubble constant is identified as follows

$$H_0 = \frac{\dot{R}_0}{R_0} \qquad (7.35)$$

Moreover, from eq.(7.32), we have $z^2 \simeq \dot{R}_0^2 h^2 \simeq \dot{R}_0^2 \ell^2$, and then, putting $\ell^2 \simeq z^2 / \dot{R}_0^2$, we can solve eq.(7.33) for ℓ, obtaining

$$\ell = \frac{z}{\dot{R}_0} \left[1 + \frac{1}{2} \left(\frac{R_0 \ddot{R}_0}{\dot{R}_0^2} - 1 \right) z + \cdots\cdots \right] \qquad (7.36)$$

The term containing the second derivative of R is called the deceleration parameter q , and its value at the time t_0 is

$$q_0 = - \frac{R_0 \ddot{R}_0}{\dot{R}_0^2} = - \frac{1}{R_0} \frac{\ddot{R}_0}{H_0^2} \qquad (7.37)$$

In order to explain in which sense the scale factor R(t) may be regarded as the "radius" of the universe, we follow a procedure analogous to that used when discussing the geometry of the Einstein static model. Defining the new coordinate

$$\bar{r} = \frac{r}{1 + kr^2/4R_0^2} = \frac{u R_0}{1 + \frac{k}{4} u^2}$$

the line-element becomes, neglecting the bar over the radial coordinate,

$$ds^2 = dt^2 - e^{g(t)} \left(\frac{dr^2}{1 - kr^2/R_0^2} + r^2 d\vartheta^2 + r^2 \sin^2\vartheta \, d\varphi^2 \right) \tag{7.38}$$

Introducing an additional space-like dimension, we can define four coordinates z_i (we put k = 1 in this example for simplicity)

$$z_1 = R_0 (1 - r^2/R_0^2)^{1/2} \quad , \quad z_2 = r \sin\vartheta \cos\varphi$$

$$z_3 = r \sin\vartheta \sin\varphi \quad , \quad z_4 = r \cos\vartheta \tag{7.39}$$

such that

$$z_1^2 + z_2^2 + z_3^2 + z_4^2 = R_0^2 \tag{7.40}$$

and we have

$$ds^2 = dt^2 - e^{g(t)} (dz_1^2 + dz_2^2 + dz_3^2 + dz_4^2) \tag{7.41}$$

According to this expression, the spatial part of the universe can be regarded, at any given time t, as a spherical hypersurface embedded in a four-dimensional euclidean space and expanding in time with a radius R(t) = $R_0 \exp[g(t)/2]$. Introducing in eq.(7.26) the new variable ω, such that

$$u = 2 \operatorname{tg} \frac{\omega}{2} \tag{7.42}$$

we obtain

$$\begin{aligned} ds^2 &= dt^2 - R^2(t) \left[d\omega^2 + \sin^2\omega \, (d\vartheta^2 + \sin^2\vartheta \, d\varphi^2) \right] \\ &= dt^2 - d\sigma^2 \end{aligned} \tag{7.43}$$

where the spatial part $d\sigma^2$ describes the three-dimensional surface of a hypersphere (remember that the metric of the two-dimensional surface of a sphere of radius a is $a^2 (d\vartheta^2 + \sin^2\vartheta \, d\varphi^2)$).

As in the case of the Einstein static universe, the Robertson-Walker geometry is not completely determined without explicit hypothes on the identification of points (remember the discussion of Chapter VI, Sect.1). In the spherical case (k=1) the total volume of the three-dimensional space can be expressed, using (7.43), as

$$V = \int_0^\pi d\omega \int_0^\pi d\vartheta \int_0^{2\pi} d\varphi \, R^3 \sin^2\omega \sin\vartheta = 2\pi^2 R^3 \qquad (7.44)$$

(assuming a geometry of spherical kind we have taken $0 \leq \omega \leq \pi$, see eq.(6.28)).

In the case k=0 the line-element (7.26) gives an euclidean space whose volume is infinite. In fact, at any given instant of time t_1, putting $R(t_1) = R_1$ and $r' = R_1 u = R_1 r/R_0$ we have, from (7.26)

$$ds^2 = dt^2 - (dr'^2 + r'^2 d\vartheta^2 + r'^2 \sin^2\vartheta \, d\varphi^2) \qquad (7.45)$$

and the three-dimensional space-like hypersurfaces have an euclidean metric ,

$$d\sigma^2 = dr'^2 + r'^2 (d\vartheta^2 + \sin^2\vartheta \, d\varphi^2) \qquad (7.46)$$

The volume of the space, for $0 \leq r' \leq \bar{r}'$, at the fixed instant t_1, is that of a sphere of radius $\bar{r}'$

$$V = \frac{4}{3}\pi \bar{r}'^3 = \frac{4}{3}\pi \frac{R_1^3}{R_0^3} \bar{r}^3 \qquad (7.47)$$

which goes to infinity for $\bar{r} \to \infty$.

The last case k=-1 corresponds to a pseudo-spherical (hyperbolic) space. Consider in fact the new variable

$$r = 2 \, \mathrm{tgh} \frac{\omega}{2} \qquad (7.48)$$

so that the line-elemnt (7.26) becomes

$$ds^2 = dt^2 - R^2(t) \left[d\omega^2 + \sinh^2\omega \left(d\vartheta^2 + \sin^2\vartheta \, d\varphi^2 \right) \right] \qquad (7.4$$

The spatial part of this expression at a fixed time

$$d\sigma^2 = R^2 \left[d\omega^2 + \sinh^2\omega \, (d\vartheta^2 + \sin^2\vartheta \, d\varphi^2) \right] \qquad (7.50)$$

describes a three-dimensional hyperboloid. The volume of the universe is infinite also in this case.

All these three cases (k = 0,+1,-1) can be represented in a compact form by putting

$$ds^2 = dt^2 - R^2(t)\left[d\omega^2 + S_k^2(\omega)\left(d\theta^2 + \sin^2\theta\, d\varphi^2\right)\right] \qquad (7.51)$$

where

$$S_k(\omega) = \begin{cases} \sin\omega & \longleftrightarrow \quad k = +1 \\ \omega & \longleftrightarrow \quad k = 0 \\ \sinh\omega & \longleftrightarrow \quad k = -1 \end{cases} \qquad (7.52)$$

(see Fig.7.1)

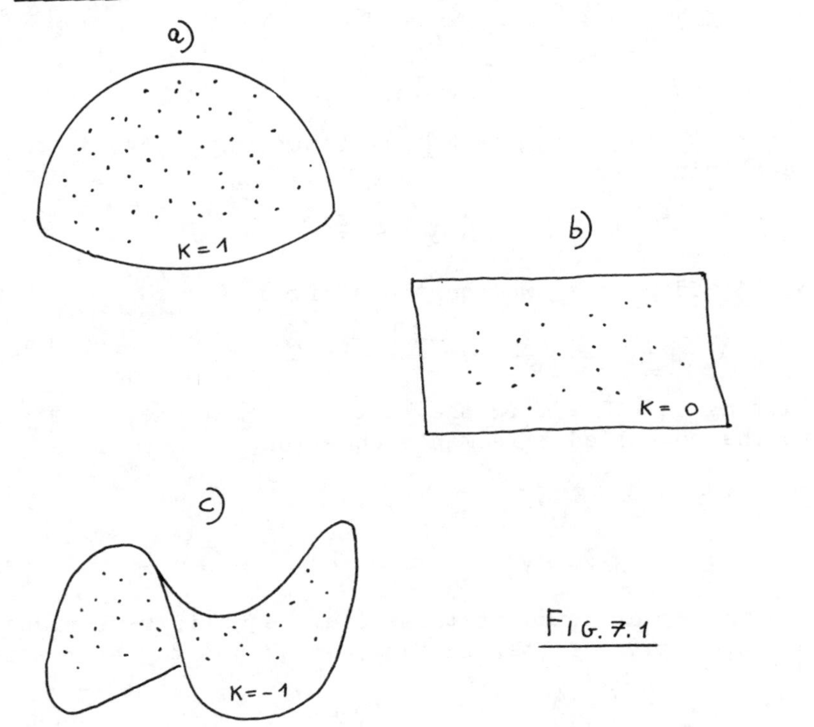

Fig.7.1. The three spatial geometries: a) spherical, b) euclidean, c) hyperbolic

After having presented the general geometrical features of the Robertson-Walker metric, now we discuss the dynamics of the scale-factor R(t) which is governed by the Einstein field equations.

The first solution of the field equations corresponding to a non-static model of universe was given by Friedman. Representing galaxies as an idealized fluid characterized by an average proper density ρ and an average internal pressure p , both depending on time but not on position in space, we are led to a very simple form of the stress energy tensor, namely $T_1{}^1 = T_2{}^2 = T_3{}^3 = -p$, and $T_4{}^4 = \rho$ in a comoving frame in which the four-velocity is given by $dx^\mu/ds = (0,0,0,1)$ (see eq.6.65).

Inserting into the equations (7.3) the metric tensor (7.26) we obtain the following two independent equations

$$\chi p = -2\frac{\ddot{R}}{R} - \frac{\dot{R}^2}{R^2} - \frac{k}{R^2} + \Lambda \tag{7.53}$$

$$\chi \rho = \frac{3}{R^2}(\dot{R}^2 + k) - \Lambda \tag{7.54}$$

(where $R = R_0 \exp[g(t)/2]$). Inserting (7.54) into (7.53) we obtain

$$2\frac{\ddot{R}}{R} + \chi\left(p + \frac{1}{3}\rho\right) = \frac{2}{3}\Lambda \tag{7.55}$$

while adding the two equations gives

$$\chi(p+\rho) = \frac{2k}{R^2} + 2\frac{\dot{R}^2}{R^2} - 2\frac{\ddot{R}}{R} \tag{7.56}$$

Putting $\Lambda = 0 = k$ in the last two relations, we are led to the so-called Friedman's equation

$$2\frac{\ddot{R}}{R} + \chi\left(p + \frac{1}{3}\rho\right) = 0 \tag{7.57}$$

$$2\frac{\ddot{R}}{R} + \chi(p+\rho) - 2\frac{\dot{R}^2}{R} = 0 \tag{7.58}$$

The combination of these field equations provides some important results. In fact, eq.(7.58) may be rewritten

$$2\frac{d}{dt}\left(\frac{\dot{R}}{R}\right) = -\chi(p+\rho) \tag{7.59}$$

and the derivative of (7.54) (with $\Lambda = k = 0$) gives

$$\chi\frac{d\rho}{dt} = 6\frac{\dot{R}}{R}\frac{d}{dt}\left(\frac{\dot{R}}{R}\right) \tag{7.60}$$

From these two equations we have

$$\chi \dot{\rho} = -3 \frac{\dot{R}}{R} \chi (\rho + p) \qquad (7.61)$$

$$\dot{\rho} + 3 \frac{\dot{R}}{R} (\rho + p) = 0 \qquad (7.62)$$

and multiplying by R^3

$$R^3 \dot{\rho} + 3 \rho R^2 \dot{R} + 3 p R^2 \dot{R} = 0 \qquad (7.63)$$

we obtain

$$\frac{d}{dt} (\rho R^3) + p \frac{dR^3}{dt} = 0 \qquad (7.64)$$

In the Friedman model the universe is filled with incoherent matter (dust), that is $p = 0$, and then

$$\rho R^3 = const \qquad (7.65)$$

This is the so-called Friedman integral, which shows that the matter density varies in time as R^{-3} , and the quantity of matter contained in a comoving volume-element is constant during the expansion.

In particular, however, we may consider two cases: 1) the matter-dominated universe, in which we can neglect pressure; and 2) the radiation-dominated universe, in which pressure and density are related by $p = \rho/3$. The first case is relevant to describe the present epoch, while the other is to be used when considering the very early universe.

In the matter-dominated case eq.(7.65) is valid, and (7.54) (with $\Lambda = 0$) becomes

$$\dot{R}^2 = \chi \rho \frac{R^2}{3} - K = \frac{\chi \rho_0 R_0^3}{3R} - K \qquad (7.66)$$

where ρ_0 and R_0 are the present values of ρ and R. For $k = 0$ the integration of this equation gives

$$R(t) = \left(\frac{3}{4} \chi \rho_0 R_0^3\right)^{1/3} t^{2/3} \qquad (7.67)$$

and this temporal evolution of R(t) corresponds to the so-called Einstein-De Sitter universe (it was found when Einstein decided to drop the cosmological constant and to allow R, which appears in his model, see 6.24, to become a function of time).

If, however, $k > 0$ in equation (7.66), R is no longer a monotonically increasing function of time: it reaches a maximum value

$$R_{MAX} = \frac{\chi \rho_0 R_0^3}{3k} \qquad (7.68)$$

corresponding to $\dot{R} = 0$. The radius of the universe is initially zero and after reaching the maximum it goes to zero again: the life and the total mass of the universe are finite, and one says that the universe is closed.

Finally, if $k < 0$ then R increases monotonically, and in the limits $t = 0$ and $t = \infty$ we have

$$R \underset{t\to 0}{\sim} t^{2/3} \quad , \quad R \underset{t\to\infty}{\sim} t \qquad (7.69)$$

If we consider the radiation-dominated model, putting $p = \rho/3$, an equation for R is obtained by subtracting (7.53) and (7.54) (with $\Lambda = k = 0$)

$$R\ddot{R} + \dot{R}^2 = \frac{1}{2}\chi\left(\frac{1}{3}\rho - p\right)R^2 = 0 \qquad (7.70)$$

and its integration gives

$$R \propto t^{1/2} \qquad (7.71)$$

From this equation we can deduce the temporal evolution of density and temperature during the era in which the energy content of the universe is radiation-dominated. Consider in fact the energy-momentum conservation law $T_\mu{}^\nu{}_{;\nu} = 0$; Putting $T_\mu{}^\nu = \text{diag}\,(-p,-p,-p,\rho)$ and using the metric (7.26) it follows that

$$\frac{\dot{\rho}}{\rho + p} = -3\,\frac{\dot{R}}{R} \qquad (7.72)$$

It is important to note that this equation is exactly equivalent to eq.(7.64) obtained directly from the field equations. This happens because, in the general relativity, the conservation laws and the equations of motion of matter are a consequence of the gravitational field equations.

Putting $p = 0$ and $\rho = \rho_m$ (matter-dominated universe) we obtain from (7.72) for the matter energy density

$$\rho_m \propto R^{-3} \qquad (7.73)$$

in agreement with (7.65). For $p = \rho/3$ and $\rho = \rho_\gamma$ (radiation), eq.(7.72) gives

$$\rho_\gamma \propto R^{-4} \tag{7.74}$$

The radiation density, therefore, evolves in time, according to eq.(7.71,74) as

$$\rho_\gamma \propto t^{-2} \tag{7.75}$$

As the energy density of the radiation is related to its absolute temperature T by the Stephan law

$$\rho_\gamma = a\, T_\gamma^4$$

(where a is a constant) it follows that its temperature varies in time as

$$T_\gamma \propto t^{-1/2} \tag{7.76}$$

This relation can be obtained also imposing on the radiation in thermal equilibrium to preserve the form of the Planck distribution during the expansion:

$$d\varepsilon = \rho_\gamma\, dV = \frac{8\pi h}{c^3}\,\frac{\nu^3\, d\nu\, dV}{e^{h\nu/KT} - 1} \tag{7.77}$$

where $d\varepsilon$ is the radiation energy contained on the volume dV . In fact, being $\varepsilon \propto \nu$, the energy evolves in time, like the frequency ν, as R^{-1} (see 7.24), while the spatial volume is proportional to R^3 . Therefore the Planck distribution is preserved if

$$T_\gamma \propto \nu \propto R^{-1} \tag{7.78}$$

and then, from (7.71), $T_\gamma \propto t^{-1/2}$.

It is also important to stress that the time development of the early universe is not influenced by the value of k . Consider in fact the field equation (7.54), which for $\Lambda = 0$ becomes

$$\dot{R}^2 = \frac{1}{3}\chi\rho R^2 - k \tag{7.79}$$

As the early universe is radiation dominated (in fact, comparing eqs.(7.73,74) we obtain that $\rho_m/\rho_\gamma \to 0$ for

$R \to 0$) we have, at a sufficiently early epoch, $\rho R^2 = \rho_\gamma R^2 \propto R^{-2} \propto T_\gamma^2$.

For $T_\gamma \sim 10^3$ °K, the ratio $\chi \rho R^2/3$ was $\gtrsim 10$, and it grows like $1/t$ for $t \to 0$. Therefore k may be certainly neglected with respect to the other term on the right-hand side of eq.(7.79), when discussing the early state of the universe, so that it will be unimportant whether space is open or closed.

Finally, remembering the deceleration parameter q which characterizes the deviation from a linear Hubble law (see 7.36,37), we note that its present values q_0 can be simply related to the total matter density in the framework of the matter-dominated Friedman model.

In fact from the Friedman eq.(7.57) we have, for $p = 0$,

$$\frac{\ddot{R}}{R} = -\frac{1}{6}\chi\rho \tag{7.80}$$

and then

$$q = -\frac{1}{R}\frac{\ddot{R}}{H^2} = \frac{\chi\rho}{6H^2} \tag{7.81}$$

(remember the definition of q, eq.(7.37), and of the Hubble constant H, eq.(7.35)). Moreover, eq.(7.66) may be rewitten as

$$H^2 = \frac{1}{3}\chi\rho - \frac{k}{R^2} \tag{7.82}$$

Therefore a precise determination of q could be used to decide whether ous universe is open or closed. In fact, combining (7.81,82) we obtain the relation

$$H^2(2q-1) = \frac{k}{R^2} \tag{7.83}$$

(valid in the case of vaninshing cosmological constant).

We have then three possibilities:

a) if $q > 1/2$, then $k > 0$ and the universe has a positive spatial curvature: it is closed and it will eventually recontract again. In this case, being $q = \rho/2\rho_c$ (see 7.81 and remember the definition of critical density ρ_c, eq.5.20), we have $\rho > \rho_c$;

b) if $q = 1/2$ it follows that $k = 0$, $\rho = \rho_c$ so that the universe has a zero spatial curvature: the three-dimensional space is euclidean and the expansion has no end;

c) if $q < 1/2$, then $k < 0$ and $\rho < \rho_c$;the universe has a negative spatial curvature, it is open and it will expand for ever.

Similar considerations can be developed also in the case of a nonvanishing cosmological constant, $\Lambda \neq 0$. Unfortunately, no precise determination of q may be obtained at present by the available experimental data.

2.- The magnitude-redshift relation as a test of the non-static models

In this Section, and in the following one, we will discuss some observational test of the Expanding models of universe based on the Robertson-Walker geometry.

To this aim we start considering the most easily determined observable parameters in cosmology, namely the redshift z and the apparent luminosity e_o (defined as the amount of energy received from a source per unit time and unit area).

These two parameters can be related in the framework of a Robertson-Walker geometrical background. Consider a source placed at the origin of a comoving coordinate system, and an observer at a radial distance u_o . According to the previous Section (see 7.27,28,29), we have

$$\ell = \int_0^{u_o} du \left(1 + \frac{k}{4} u^2\right)^{-1} \tag{7.84}$$

$$\ell = \int_{t_e}^{t_o} \frac{dt}{R(t)} \tag{7.85}$$

$$z = \frac{R_o}{R_e} - 1 \tag{7.86}$$

where the index "o" is related to the observer position and time, while the index "e" denotes the emission position and time.

Once R(t) is known, (7.85) defines t as a function of the distance marker ℓ . Moreover, as ℓ and u_o are related by (7.84), and t_e and z by (7.86), we obtain an implicit relation between z and u_o

$$A(z, u_o) = 0 \tag{7.87}$$

We must replace now the radial coordinate u_o with some observable parameter. If we suppose to know the total rate of energy emission, E_e, of a galaxy at a radial coordinate distance u from the earth, the energy flux received, e_0 can be determined for a given geometrical model. Then we can write a relation of the form

$$B\left(\frac{e_0}{E_e}, u_0\right) = 0 \qquad (7.88)$$

Eliminating u_0 between (7.87,88), then e_0 and z can be related

$$C\left(\frac{e_0}{E_e}, z\right) = 0 \qquad (7.89)$$

provided that E_e is known.

An estimate of E_e is provided by astronomical observations for some spectral types of stars and galaxies. Therefore the relation (7.89) can be tested for a group of sources of the same spectral type, which are characterized by the same value of E_e. Let us obtain then an explicit form of eq.(7.89).

First of all we note that astronomical measurements of luminosities are performed introducing a logarithmic scale and defining the apparent magnitude m_0 of a star as

$$m_0 = -2.5 \operatorname{Log} e_0 + \text{const} \qquad (7.90)$$

(where $\operatorname{Log} x = \lg_{10} x$). The zero of the scale is conventionally fixed defining the apparent magnitude of the Polar star to be $m_0 = 2.15$. Then

$$m_0 = -2.5 \operatorname{Log} \frac{e_0}{e_P} + 2.15 \qquad (7.91)$$

where e_P is the energy flux received from the Polar star. In order to calculate $C(e_0/E_e, z)$ we start deriving the relation (7.88) using the general form of the Robertson-Walker metric. Consider the total energy received per unit time on the whole surface of a sphere of radius $(t_0 - t_e)$ (remember c=1) . This energy is smaller than E_e for two reasons: the received photons have a red-shifted frequency $\nu_0/\nu_e = R_e/R_0$, and moreover if they are emitted at regular intervals Δt_e, they will be received separated by longer intervals Δt_0, such that $\Delta t_e/\Delta t_0 = R_e/R_0$. Thus the energy received per unit time on the whole sphere surrounding the galaxy placed at a coordinate distance u_0 will be

$$E_0 = E_e \left(\frac{R_e}{R_0} \right)^2 \qquad (7.92)$$

Considering the form (7.26) of the Robertson-Walker metric, the infinitesimal surface-element dA corresponding to $d\vartheta$ and $d\varphi$, at a fixed coordinate distance u_0, is given by

$$dA = \frac{R_0 u_0}{1+\frac{K}{4}u_0^2}\, d\vartheta\, \frac{R_0 u_0 \sin\vartheta}{1+\frac{K}{4}u_0^2}\, d\varphi \qquad (7.93)$$

so that the total area of the spherical surface with coordinate radius u_0 will be

$$A = \frac{4\pi R_0^2 u_0^2}{(1+\frac{K}{4}u_0^2)^2} \qquad (7.94)$$

The apparent luminosity, i.e. the total energy received per unit area and unit time on the earth (considered as a point on this sphere) will be then

$$e_0 = \frac{E_0}{A} = E_e \left(\frac{R_e}{R_0} \right)^2 \frac{(1+\frac{K}{4}u_0^2)^2}{4\pi R_0^2 u_0^2} \qquad (7.95)$$

This quantity is just the energy measured by astronomers using a bolometer (as regards this point, see the Appendix at the end of this Chapter).

The apparent magnitude of eq.(7.90) thus becomes

$$m_0 = 5 \,\mathrm{Log} \left[\frac{R_0 u_0}{R_e(1+\frac{K}{4}u_0^2)} \right] - 2.5 \,\mathrm{Log}\, E_e + \mathrm{const} \qquad (7.96)$$

where we have absorbed a Log R_0 factor in the constant, because the time of observation t_0 is the same for all the stars. The ratio R_0/R_e, and the function of u_0 appearing inside the square brackets of eq.(7.96) can be expressed in terms of z, as we will see in a moment, so we are led directly to a magnitude-red-shift relation which can be written, in general, as

$$m_0 = 5 \,\mathrm{Log}\, F(z) - 2.5 \,\mathrm{Log}\, E_e + \mathrm{const}$$

In particular, for a flat space-time geometry, one has, instead of (7.95), $e_0 = E_e/4\pi d^2$ (where d is the astronomical distance) and the Hubble law is strictly valid in the linear form $z = H d$ (see 7.34), where H is a constant. Therefore we can write

$$\ell_o = \frac{E_e H^2}{4\pi z^2} \qquad (7.97)$$

and then, from (7.90)

$$m_o = 5 \operatorname{Log} z - 2.5 \operatorname{Log} E_e + \text{const} \qquad (7.98)$$

This is the expression of the linear Hubble law in terms of the apparent magnitude, in the context of a euclidean geometry.
In order to obtain the corresponding generalized relation in a Robertson-Walker geometrical background, we rewrite (7.96) in the form

$$m_o = 5 \operatorname{Log} \frac{R_o}{R_e} + 5 \operatorname{Log} \frac{u_o}{1+\frac{K}{4}u_o^2} + \text{const} \qquad (7.99)$$

(we have included in the constant also the $\operatorname{Log} E_e$ term, so that the constant now depends also on the type of galaxy considered). Using (7.86) then

$$m_o = 5 \operatorname{Log}(1+z) + 5 \operatorname{Log} \frac{u_o}{1+\frac{K}{4}u_o^2} + \text{const} \qquad (7.100)$$

To express also the second term as a function of z, note that the integration of (7.84) gives

$$\ell = \int_0^{u_o} du \left(1+\frac{K}{4}u^2\right)^{-1} = \frac{2}{\sqrt{K}} \operatorname{arctg}\left(\frac{\sqrt{K}}{2} u_o\right) \qquad (7.101)$$

so that

$$u_o = \frac{2}{\sqrt{K}} \operatorname{tg}\left(\frac{\sqrt{K}}{2}\ell\right) \qquad (7.102)$$

from which, supposing $u_o = r_o/R_o \ll 1$, and $\ell \ll 1$, we have

$$\frac{u_o}{1+\frac{K}{4}u_o^2} = \frac{1}{\sqrt{K}} \sin(\sqrt{K}\ell) \simeq \ell + O(\ell^3) \qquad (7.103)$$

Remembering now the expansion of ℓ as a function of z, m_o can be expressed entirely in terms of z. Inserting (7.103) and (7.36) into eq.(7.100) we obtain

$$m_o = 5 \operatorname{Log}(1+z) + 5 \operatorname{Log} z + 5 \operatorname{Log}\left[1+\frac{1}{2}\left(\frac{R_o \ddot{R}_o}{\dot{R}_o^2} - 1\right)z\right] + \text{const} \qquad (7.104)$$

(we have included into the constant also the term $-5 \operatorname{Log} \dot{R}_o$)
Noting that

$$\mathrm{Log}\,(1+x) = \frac{\ln(1+x)}{\ln 10} \simeq \frac{1}{\ln 10}\left(x - \tfrac{1}{2}x^2 + \tfrac{1}{3}x^3 \cdots\right) \tag{7.105}$$

to first order in z we obtain, from (7.104)

$$m_0 = \frac{5}{\ln 10}\, z + 5\,\mathrm{Log}\, z + \frac{5}{2\ln 10}\left(\frac{\ddot{R}_0 R_0}{\dot{R}_0^2} - 1\right) z + \mathrm{const} \tag{7.106}$$

and since $5/(2 \ln 10) \simeq 1.086$

$$m_0 = 5\,\mathrm{Log}\, z + 1.086\left(1 + \frac{\ddot{R}_0 R_0}{\dot{R}_0^2}\right) z + \mathrm{const} \tag{7.107}$$

This equation, relating m_0 and z, generalizes the euclidean relation (7.98) to the case of a Robertson-Walker universe, and holds for any given scale function $R(t)$. The term in this equation depending linearly on z begins to become significant only for large distances; it is related to the deceleration parameter q, and we can write, using eq. (7.37),

$$m_0 = 5\,\mathrm{Log}\, z + 1.086\,(1 - q_0)\, z + \mathrm{const} \tag{7.108}$$

For small z, the relation becomes simply

$$m_0 = 5\,\mathrm{Log}\, z + \mathrm{const} \tag{7.109}$$

representing, in terms of the variables m_0 and z, a family of curves (one for each spectral type, as the constant depends on $\mathrm{Log}\, E_e$).

3.- Other observational tests

a) Relation between the apparent diameter and the magnitude

Another interesting information which can be extracted from the observational data is the relation between the apparent diameter of a source and its magnitude. First of all we note that, in the case of a universe with a flat space-time geometry, filled with galaxies all having the same dimensions, it is obvious that the apparent diameter of these objects (measured by the subtended angle $\Delta\theta$)

can be related to the astronomical distance d by

$$\Delta\theta = \frac{\text{const}}{d} \tag{7.110}$$

Consider now eq.(A.6) of the Appendix A at the end of this Chapter :

$$m_o - M = -2.5 \,\text{Log}\, \ell_o + 2.5 \,\text{Log}\, L \tag{7.111}$$

where M is the absolute magnitude and L the luminosity of the object at the conventional distance of 10 parsec. As

$$\frac{\ell_o}{L} = \frac{D^2}{d^2} \tag{7.112}$$

where d and D are the respective distances (D = 10 parsec) we have

$$m_o - M = 2.5 \,\text{Log}\, \frac{L}{\ell_o} = 2.5 \,\text{Log}\, \frac{d^2}{D^2} \tag{7.113}$$

that is

$$m_o - M = 5 \,\text{Log}\, d - 5 \tag{7.114}$$

$$\text{Log}\, d = 0.2\,(m_o - M) + 1 \tag{7.115}$$

From eq.(7.110)

$$\text{Log}\, \Delta\theta = \text{const} - \text{Log}\, d \tag{7.116}$$

and then apparent diameter, $\Delta\theta$, and magnitude, m_o, are related by

$$\text{Log}\, \Delta\theta = -0.2\, m_o + \text{const} \tag{7.117}$$

In the case of the Robertson-Walker universe, however, eq.(7.110) is no longer the correct relation between apparent diameter and astronomical distance, since we must take into account the red-shift effects.

To this aim we start from the Robertson-Walker metric in the form (7.38) (putting for simplicity k = 0) and we consider first of all the relation between the coordinate position and the apparent diameter of the galaxies.

Taking the diameter of interest in the direction of $d\theta$, the proper diameter, $\Delta \ell_o$, at the time t_e when the light is emitted, is, from (7.38),

$$\Delta \ell_0 = r\, e^{\frac{1}{2} g_e} \Delta \vartheta \tag{7.118}$$

where $g_e = g(t_e)$, and $\Delta\vartheta$ is the angular diameter observed at the origin.

Assuming that $\Delta \ell_0$ has the same value for different galaxies observed at the same time t_0 , i.e. $\Delta\ell_0 = \text{const} \times e^{\frac{1}{2} g_0}$, we can rewrite eq.(7.118) as

$$\Delta\vartheta = \frac{\text{const}}{r}\, e^{\frac{1}{2}(g_0 - g_e)} \tag{7.119}$$

and using (7.25)

$$\Delta\vartheta = \frac{\text{const}}{r}\left(\frac{\lambda + \Delta\lambda}{\lambda}\right) = \frac{\text{const}}{r}\,(1+z) \tag{7.120}$$

In this relation we have still a non-observable quantity, the position coordinate r . In order to eliminate it, we can rewrite (7.95) in the form

$$e_0 = E_e\, e^{(g_e - g_0)} \frac{1}{4\pi e^{g_0} r^2} \tag{7.121}$$

(remember that the radial coordinate of eq.(7.38) is related to u by $r = u R_0/(1 + \frac{k}{4} u^2)^2$) and, using (7.25)

$$e_0 = \frac{E_e}{4\pi e^{g_0} r^2}\left(\frac{\lambda}{\lambda + \Delta\lambda}\right)^2 \tag{7.122}$$

For two different galaxies G and G' of the same type ($E_e = E'_e$) (observed at the same time t_0) we have then

$$\frac{e_0}{e'_0} = \frac{r'^2}{r^2}\left(\frac{1 + \Delta\lambda'/\lambda'}{1 + \Delta\lambda/\lambda}\right)^2 \tag{7.123}$$

Therefore

$$r = \frac{\text{const}}{\sqrt{e_0}}\, \frac{\lambda}{\lambda + \Delta\lambda} = \frac{\text{const}}{\sqrt{e_0}}\, \frac{1}{1+z} \tag{7.124}$$

and using this result eq.(7.120) becomes a relation between observable quantities

$$\frac{\Delta\vartheta}{\sqrt{e_0}} = \text{const}\left(\frac{\lambda + \Delta\lambda}{\lambda}\right)^2 = \text{const}\,(1+z)^2 \tag{7.125}$$

namely apparent diameter $\Delta\vartheta$, luminosity e_0 and redshift z .

We can also relate the apparent diameter to the astronomical distance d .

In fact, starting from eq.(7.123), we have

$$\mathrm{Log}\,\frac{r}{r'} = 0.5\,\mathrm{Log}\,\frac{e_0'}{e_0} + \mathrm{Log}\left(\frac{1+\Delta\lambda'/\lambda'}{1+\Delta\lambda/\lambda}\right) \qquad (7.126)$$

that is, using (7.90)

$$\mathrm{Log}\,\frac{r}{r'} = 0.2\,(m_b - m_b') + \mathrm{Log}\left(\frac{1+\Delta\lambda'/\lambda'}{1+\Delta\lambda/\lambda}\right) \qquad (7.127)$$

where m_b and m'_b are the observed bolometric magnitudes (see the Appendix A).

Using eq.(A.4) we can write

$$\mathrm{Log}\,\frac{r}{r'} = 0.2\,(m_p - C + B.C. - m_p' + C' - B.'C.') + \mathrm{Log}\left(\frac{1+\Delta\lambda'/\lambda'}{1+\Delta\lambda/\lambda}\right) \qquad (7.128)$$

Supposing that r' is the coordinate of a galaxy at the standard distance of 10 parsec, and neglecting, at this distance, the red-shift effects, we have then

$$\mathrm{Log}\,r = 0.2\,(m_p - \Delta C + \Delta B.C. - M_p) - \mathrm{Log}\left(\frac{\lambda+\Delta\lambda}{\lambda}\right) + 1 \qquad (7.129)$$

where m'_p has been replaced with the absolute magnitude M_p . Using eqs.(A.5,7) we obtain

$$\mathrm{Log}\,r = 0.2\,(m_p - \Delta m_p - M_p) + \tfrac{1}{2}\,\mathrm{Log}\,(1+z) - \mathrm{Log}\,(1+z) + 1 \qquad (7.130)$$

$$\mathrm{Log}\,r = 0.2\,(m_p - \Delta m_p - M_p) + 1 - 0.5\,\mathrm{Log}\,(1+z) \qquad (7.131)$$

and from (A.10) we have

$$\mathrm{Log}\,r = \mathrm{Log}\,d - 0.5\,\mathrm{Log}\,(1+z) \qquad (7.132)$$

Therefore

$$d = r\,(1+z)^{1/2} = r\left(\frac{\lambda+\Delta\lambda}{\lambda}\right)^{1/2} \qquad (7.133)$$

is the relation between astronomical distance d and coordinate distance r .

Finally, from (7.133,120) we find that, in a Robertson-Walker universe, apparent diameter and astronomical distance are related by

$$\Delta\theta = \frac{const}{d}\left(\frac{\lambda+\Delta\lambda}{\lambda}\right)^{3/2} = \frac{const}{d}\,(1+z)^{3/2} \tag{7.134}$$

(where of course the red-shift z is related to the astronomical distance $d \simeq R_o \ell$ according to eq.(7.33), and its explicit expression is depending on the function R(t) provided by the model).

Comparing this result with eq.(7.110), obtained in the case of an euclidean universe, we see that the apparent diameter becomes greater when the red-shift effects are included. In fact the red-shift reduces the luminosity, so that light sources appear to be fainter. In the euclidean static universe, the fainter an object is, the more distant it is (for a fixed value of E_e).

But when a source is fainter because of the red-shift, the distance of the object is smaller than the distance implied by its luminosity, and then its apparent diameter is greater than expected.

b) Relation between coordinate position and count of galactic sources

If the space is euclidean, and the universe is filled with a uniform distribution of stationary galaxies, the number N of sources within a distance d is given simply by

$$N = const \times d^3 \tag{7.135}$$

so that, using eq.(A.9), we have

$$\mathrm{Log}\, N = 0.6\, m + const \tag{7.136}$$

where m is the limit magnitude considered.

Comparing this equation with observations, we may note, first of all, a lack of galaxies in the plane of the Milky Way: this however can be understood as due to the presence of clouds of dark material in that direction; the counts increase as one goes toward the galactic poles just in the way one could expect if the galaxies were seen through a layer of dark material. Moreover one may note irregularities in the density distribution which reveal the tendency of galaxies to assemble in clusters, but this phenomenon seems to disappear as the field of observation is extended both in width and in depth. For instance the total number of galaxies until the thirteenth magnitude in the Northern emisphere is twice the number of that found in th Southern emisphere: this is due to the presence, in the Northern emisphere, of the Virgo cluster within this observational range.

But taking into account the dark material of our galaxy and the lack of uniformity corresponding to an insufficient exploration both in width and in depth, Hubble found that the distribution of galaxies until the twentieth magnitude is well represented by the previous formula (7.136

$$\mathrm{Log}\, N = 0.6\, m - 9.12 \tag{7.137}$$

where N is the number of galaxies per squared degree and m the correct photographic apparent magnitude.

To obtain an expression for the number of galaxies within a given coordinate distance r in the Robertson-Walker model, we denote with n_o the number of galaxies per unit proper volume at a time t_o, when $g(t) = g_o = g(t_o)$.

From the metric (7.38) (putting for simplicity k=1) the proper volume-element is

$$dV_o = \sqrt{-g}\, dr\, d\vartheta\, d\varphi = \frac{e^{\frac{3}{2}g_o}\, r^2\, dr}{(1 - r^2/R_o^2)^{1/2}} \sin\vartheta\, d\vartheta\, d\varphi \tag{7.138}$$

and the number of galaxies between two spherical surfaces of radius r and r + dr is then

$$dN = \frac{4\pi n_o r^2 dr}{(1 - r^2/R_o^2)^{1/2}}\, e^{\frac{3}{2}g_o} \tag{7.139}$$

As we know that a particle (that is a galaxy in this case) initially at rest with respect to the comoving coordinates r, ϑ, φ, will remain permanently at rest (the demonstration is the same as in the case of the Einstein universe, given in Chapter VI) we can write

$$dN = \mathrm{const} \times \frac{r^2\, dr}{(1 - r^2/R_o^2)^{1/2}} \tag{7.140}$$

Remembering the relation (7.124) between coordinate position, luminosity and red-shift, this expression can be used to test the homogeneity of the galactic distribution.

Integrating (7.40) we have the total count for a given value of r; in the case $r^2/R_o^2 \ll 1$ we have simply

$$N = \text{const} \times r^3 \qquad (7.141)$$

which is sufficiently approximated for many purposes. Finally, using eqs.(7.133,124,120) we obtain

$$N = \text{const} \times d^3 \left(\frac{1}{1+z}\right)^{3/2} \qquad (7.142)$$

$$N = \frac{\text{const}}{\ell_0^{3/2}} \left(\frac{1}{1+z}\right)^3 \qquad (7.143)$$

$$N = \frac{\text{const}}{(\Delta\theta)^3} (1+z)^3 \qquad (7.144)$$

c) <u>The Log N versus Log S diagram</u>

In the euclidean case, the number N of radio sources until a distance d is given by eq.(7.135).

By expressing N as a function of the flux density S of the received electromagnetic radiation, as S decreases as d^{-2}, we should have $d^3 \propto S^{-3/2}$, and then

$$N(>S) \propto S^{-3/2} \qquad (7.145)$$

where $N(>S)$ means the number of radio sources whose received power is greater than a given value of S . Therefore the plot of Log N versus Log S should be a straight line with slope -3/2.

However, plotting the observational data relative to radio sources, we obtain a straight line, but with a smaller slope comprised between -1.8 and -2.0 .

Of course one could observe that we have used a formula relative to an euclidean and static universe, and then it should not be surprising if the observational data are not in agreement with the slope -1.5 . For instance, we must take into account the red-shift, and use eq. (7.142) instead of eq.(7.135) . However, in this way, we obtain a theoretical prediction corresponding to a line, in the plane Log N - Log S , with a slope greater than

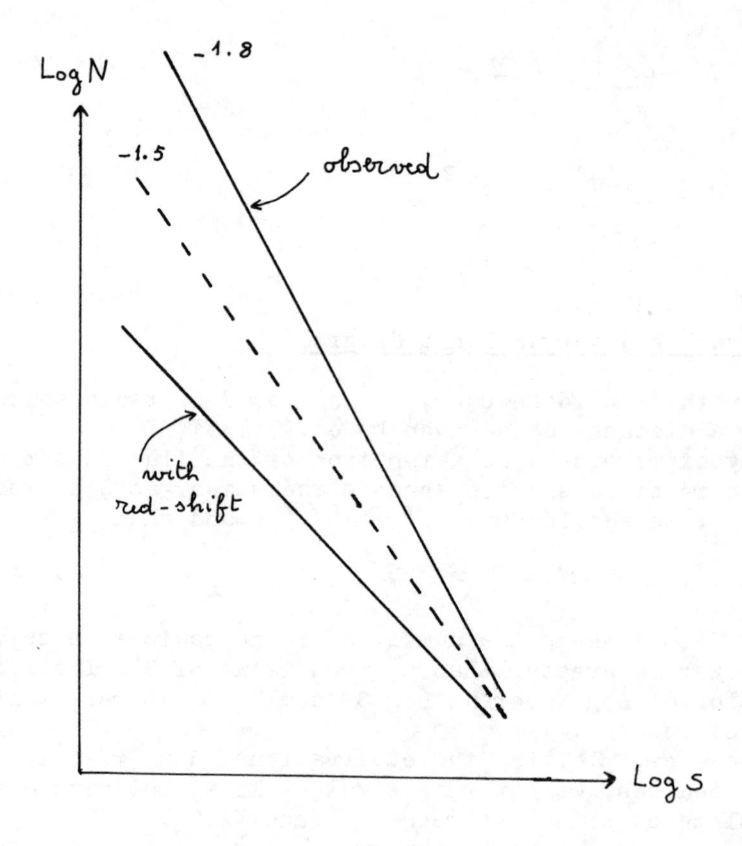

FIG. 7.2

-1.5 (see Fig.7.2).

The steeper slope of the observed line means that at a low density flux the number of sources is greater than expected.

A low density flux means that the galaxy is faint and therefore distant. We have then a great number of faint radio sources and this fact can be interpreted in two ways. One interpretation is that there is an excess of distant sources, and then the universe is not homogeneous. But there is also another interpretation. In fact eq.(7.135) is based on the hypothesis that all the galaxies have the same intrinsic luminosity. Dropping this hypothesis, we may assume that distant sources are intrinsically brighter. Remembering that when we look out into space we also look back in time, because of the finite speed of propagation of light signals, we can conclude that in the past the radio sources were more concentrated, or more luminous, or both.

In any case, the important result one can deduce from these observations is that the universe was different in the past.

This was the first proof that our universe is not only expanding, but also evolving in time, and was the first indication against the steady-state model of universe, based on the so-called perfect cosmological principle (according to which the universe appears to be the same to any observer not only at any point but also at any time).

d) The cosmic microwave background

This radiation was discovered quite accidentally in 1965 by Penzias and Wilson when calibrating a radio antenna devised to track the satellite "Echo". They found a weak radio noise of cosmic origin which was independent from the direction in which the antenna was pointing. Dicke and others immediately interpreted this result as an experimental evidence supporting the hot model of the universe.

To this regard, it must be remembered that the first theoretical estimate of the temperature of the relic radiation was performed by Gamow and Alpher in 1950. Starting from the big-bang hypothesis they found that today the temperature of the relic electromagnetic radiation should be nearly 5 °K .

The first measurements of Penzias and Wilson were performed for a wave-length of 7.35 cm. Many independent measurements at different wave-length have been performed in subsequent years, showing that its spectral shape conforms to a Plack distribution, in agreement with the standard thermodinamical predictions, and the present temperature is about 2.9 °K .

Moreover it has been found that this background radiation is also extremely isotropic around us, in good agreement with the Robertson-Walker model. The very high degree of isotropy of the radiation ($\Delta \rho / \rho < 0.1\%$) can be interpreted as evidence of the fact that the Robertson-Walker metric is a very good description of the geometry of our universe, since the time of the decoupling between matter and radiation .

e) The helium problem

The chemical comosition of cosmic matter can provide us informations on the early evolution of the universe, because the conversion of hydrogen into helium, which occurred during a hot and dense phase of the initial universe, depends both on the today observed density of matter and on the present temperature of the relic radiation.

The process of helium production, in fact, occurred at a temperature $\sim 10^9$ °K and it was dependent on the density at that epoch. Now a higher present temperature of the relic radiation implies a smaller cooling of the radiation since the epoch of helium production: this means a smaller expansion since that epoch, so that the matter density at the epoch of helium production was lower, and so also the aboundance of helium produced would have been lower.

Even without doing quantitative estimates we can then understand that the observed temperature of the relic radiation, combined with the knowledge of the matter density, can give us informations about the abundance of the primeval helium.

But why should we think that the helium we observe was produced during the first instants of the life of our universe? There at least two reasons. The first is that if we calculate the helium produced in the case of a homogeneous and isotropic universe which did expand away from a hot and dense phase with a temperature $T \geqslant 10^{12}$ °K, we find a fraction of helium over total mass

$$\text{Helium abundance} \sim 0.25 \qquad (7.146)$$

which is very near the observed helium abundance (see Table 7.1). The second reason is that the observed helium abundance in several different celestial objects is found to be nearly constant, and it does not depend on the age of the object.

These facts seem to support the idea that the observed helium was produced during the hot early universe, because if had been produced along with the heavier elements, its abundance would depend on the age and position of the celestial object, as observed for the othr elements (which are influenced, for example, by the proximity of the galactic nucleus, and by many other factors).

Table 7.1

Some observed helium abundances

Our galaxy	0.29
NGC 6822	0.27
" 4449	0.28
" 5461	0.28
" 5771	0.28
" 7679	0.29

4.- Concluding remarks

The Robertson-Walker cosmological model (also called "standard model") discussed in this chapter provides a number of predictions about the early state of the universe, near the big-bang.

First of all we note that, from eq.(7.73,74) we have

$$\frac{\rho_\gamma}{\rho_m} \propto R^{-1} \qquad (7.147)$$

and then, at an early enough time, there was more energy in radiation than in matter, and the field equation (7.79) in this case was reduced to

$$\dot{R}^2 = \frac{1}{3}\chi \rho_\gamma R^2 \qquad (7.148)$$

Using eq.(7.74) we have

$$\frac{\dot{\rho}_\gamma}{\rho_\gamma} = -4\frac{\dot{R}}{R} = -4\left(\frac{1}{3}\chi\rho_\gamma\right)^{1/2} \qquad (7.149)$$

which gives

$$t = \left(\frac{3}{32\pi G\rho_\gamma}\right)^{1/2} \qquad (7.150)$$

Using the Stephan law $\rho_\gamma = a\,T^4$ we get, in c.g.s. units, $t \simeq (T/10^{10}\ {}^\circ K)^{-2}$, or

$$T \simeq \frac{10^{10}\ {}^\circ K}{\sqrt{t}} \qquad (7.151)$$

For $t < 1$ sec the temperature is higher than the threshold for the creation of electron-positron pairs, and weak interactions can mantain matter in thermodynamical equilibrium through the reactions

$$p + \bar{\nu}_e \longleftrightarrow n + e^+ \quad , \quad p + e^- \longleftrightarrow n + \nu_e$$

which occur faster than the matter density changes due to the expansion .

Below 10^{10} °K weak interactions can no longer mantain neutrons in statistical balance with protons, because the concentration of the electronic pairs is beginning to disappear abrubtly. Then the ratio of the number of protons and neutrons is frozen, until a few hundred seconds have passed and neutron decay begins to be appreciable.

The first important nuclear reaction occurs at a temperature of about 10^9 °K (deuterium production)

$$n + p \longrightarrow D + \gamma \qquad (7.152)$$

and this is followed by a series of rapid reactions, all having the helium as their end product,

$$\begin{aligned} D + D &\longrightarrow He^3 + n \\ D + D &\longrightarrow H^3 + p \\ He^3 + n &\longrightarrow H^3 + p \\ H^3 + D &\longrightarrow He^4 + n \end{aligned} \qquad (7.153)$$

As these reaction are much faster than (7.152), the amount of helium produced depends essentially on how many deuterons are formed according to (7.152).

This explains better the dependence of the helium abundance on the actual observed temperature of the relic radiation.

In fact if the matter density at the epoch t = 100 sec is lowered (which corresponds to an increase of the temperature of the relic radiation presently measured) the reaction is reduced, and so a smaller amount of helium is formed. For higher densities at t = 100 sec, the rate of the reaction (7.152) increases, and so more helium is formed. The quantity of helium produced soon levels off, however, because when the rate of the reaction (7.152) exceedes the expansion rate of the universe at that time, essentially all the neutrons are incorporated into the helium nuclei. In this way we can qualitatively understand how to perform calculation about the aboundance of the primeval helium.

Before concluding this Chapter, there are many other points, relevant to cosmology, which should be mentioned, first of all the baryon number per photon. Experimentally this number is found to be

$$\frac{N_B}{N_\gamma} \sim 10^{-9} \qquad (7.154)$$

The very small value of this ratio implies that the contribution of the free baryons (mainly protons and neutrons) to the energy content of the universe is neglegible during the radiation dominated era.

As pointed out by Weinberg, it is interesting to stress that with this value of N_B / N_γ , the transition from a radiation dominated universe to another matter dominated occurs just at the same time in which the content of the universe become transparent to the radiation, nearly 10^5 years after the initial singularity, at a temperature $\sim$ 3000 °K . For another value of the ratio N_B / N_γ, these two independent events would have occurred at two different instants of the hystory of the universe.

Another point to explain is the isotropy of the radiation background. This indicates that at the time of the decoupling between matter and radiation (at the epoch $t \sim 10^5$ years) the universe was homogeneous and isotropic. But at that time (and earlier times) the various parts of the universe we now observe were not causally related, according to the Robertson-Walker model, as no exchange of informations among them was possible.

In fact, consider the causal horizon of a particle, that is the distance ct (where t is the cosmic time and c the light velocity) beyond which the particle cannot receive informations, because such informations would travel with a speed larger than c. Integrating the Einstein equations for the Robertson-Walker scale factor $R(t)$, we find $R(t) \propto t^{n}$, with $n = 2/3$ during the matter dominated era, and $n = 1/2$ during the radiation era (see the first Section of this Chapter): therefore in the past the spatial extension of the universe is greater than the causal horizon ct. There exist, in other words, sufficiently separated regions which do not have any reciprocal communication (see also Fig.7.3 , 7.4 and 7.5).

For example, at the present time $ct_0 \sim 10^{28}$ cm. At the time $t \sim 10^{-23}$ sec the causal horizon was of the order of the present radius of a particle, $ct \sim 10^{-13}$ cm, but the spatial size of the universe was much greater. In fact, for a given $t < t_{dec}$, where t_{dec} is the time of decoupling between matter and radiation ($t_{dec} \sim 10^{5}$ years $\sim 10^{12}$ sec) the corresponding value of $R(t)$ can be obtained rescaling the present value $R_0 = R(t_0) \sim 10^{28}$ cm down to the time t_{dec} and then to the time t , i.e.

$$R(t) = R_0 \left(\frac{R_{dec}}{R_0}\right)\left(\frac{R(t)}{R_{dec}}\right) \qquad (7.155)$$

Remembering that $RT = $ const (see eq.7.78), and that $T_0 \sim 3\ °K$, $T_{dec} \sim 3000\ °K$, we can estimate

$$\frac{R_{dec}}{R_0} = \frac{T_0}{T_{dec}} \sim 10^{-3} \qquad (7.156)$$

Moreover, for $t < t_{dec}$, the universe is radiation dominated so that eq.(7.71) can be applied, and we have

$$\frac{R(t)}{R_{dec}} = \left(\frac{t}{t_{dec}}\right)^{1/2} \qquad (7.157)$$

In the example considered, $t \sim 10^{-23}$ sec , and then $R(t)/R_{dec}$ $\sim 10^{-17}$. From (7.155) we get then $R(t) \sim 10^{-20} R_0 \sim 10^{8}$ c that is a value much greater than the causal horizon $ct \sim 10^{-13}$ cm . The universe at that time was formed by $R^3(t)/c^3 t$ $\sim 10^{63}$ mutually independent spatial volumes.

Also at the decoupling time, t_{dec} , there were regions

causally separated, the value of the scale-factor was, from (7.156), $R_{dec} \sim 10^{-3} R_o$, and these regions, as seen by us today, correspond to portions of the observed universe separated by an angular distance greater than 30 degree.

In conclusion, it must be mentioned that this problem, also known as the "horizon problem", as well as other problems of the standard Roberston-Walker model of universe (the so-called "flatness", "entropy" problems) seem to find a satisfactory solution according to the inflationary cosmological model, developed in the framework of grand unified theories of strong and electroweak interactions (the interested reader is referred to the original article by Guth [3], or to a recent review by Linde [4]).

Fig.7.3. The universe is represented by a sphere in four different instants of time . The horizon of a point P (represented by a circle on the sphere) is the distance beyond which light signals have not yet had time to reach P . The distance from P and its horizon increases proportionally to the time t, while the radius of the universe increases as $t^{1/2}$ (in the radiation dominated case). Going back in time from t_4 to t_1 the horizon surronds then a portion smaller and smaller of the universe.

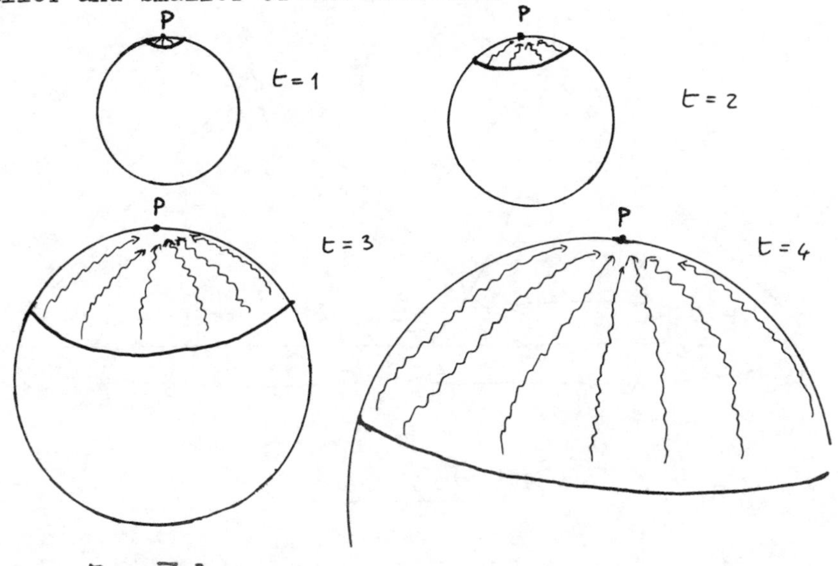

FIG. 7.3

Fig.7.4. The horizon of an observer grows in time, and includes an increasing fraction of the universe. It is a sphere centered in the observer, whose radius is equal to the distance traveled by light since the big-bang (in this Figure the horizon is the circle at the base of the cones). Because of the expansion of the universe, which initially occurs faster than the grow of the horizons (i.e. $R(t)/t \longrightarrow \infty$ for $t \longrightarrow 0$), in a early epoch there is no sufficient time for the galaxy A to see the galaxy B. As the expansion continues, after a certain time the light emitted by A can reach B, and viceversa, because the ratio between the expansion velocity and the horizon velocity, R/c, goes to zero as t goes to infinity. The edges of the cones represent the paths of the light signals, and define the limits beyond which no causal interaction can be transmitted.

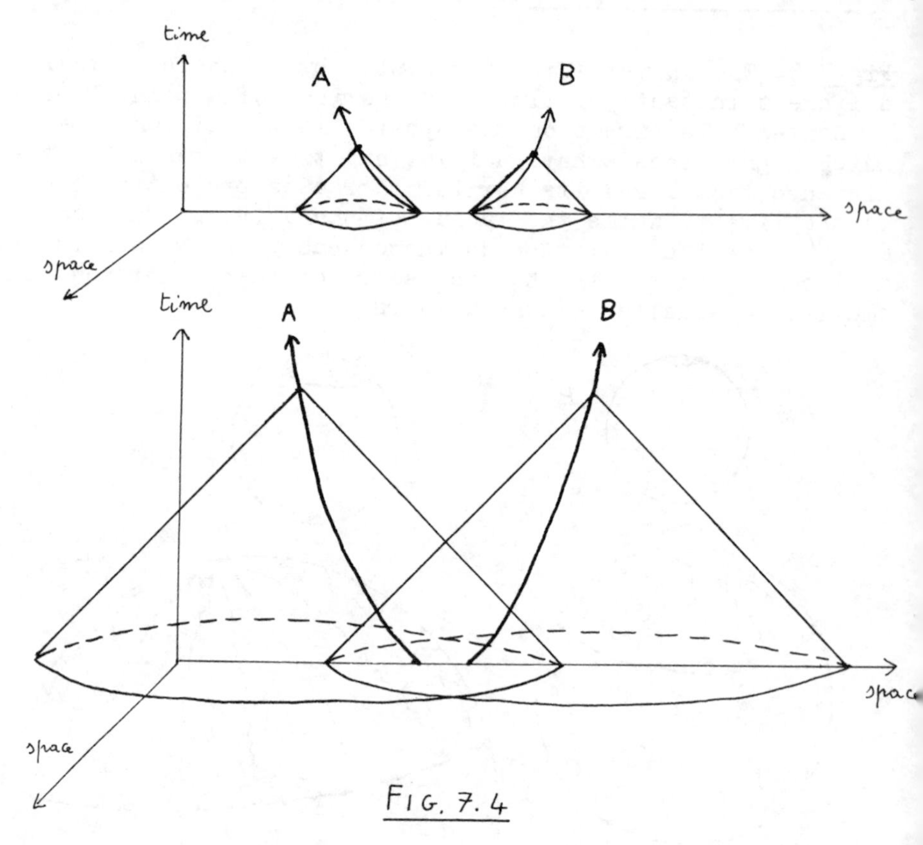

FIG. 7.4

Fig.7.5. The two observers A and B have two horizons delimited by the lines (AA', AA") and (BB', BB"). Only after a time t_1, A and B can observe common parts of the universe, and only after t_2 they enter one another's horizon.

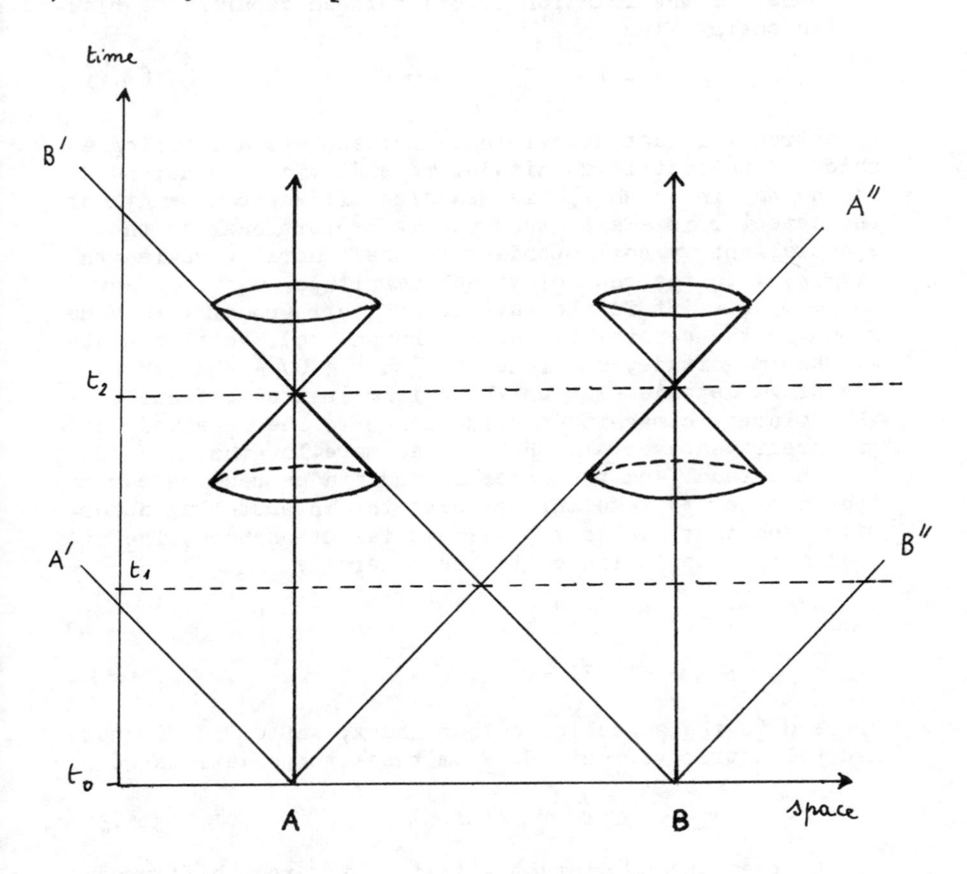

FIG. 7.5

APPENDIX A

Visual, photographic and bolometric magnitude

Consider the relation (7.86) between apparent magnitude and energy flux

$$m_0 = -2.5 \operatorname{Log} e_0 + \text{const} \tag{A.1}$$

An astronomer must discriminate between visual, photographic and bolometric magnitude. This distinction depends on the way in which e_0 is measured. If the band-width of the detector is small, then e_0 is proportional to the spectral energy corresponding to the central wave-length $\lambda = \lambda_0$. In the case of visual magnitude, m_v, we have $\lambda_0 = 5300$ Å (it is the wave-length corresponding to a maximum in the sensibility of the human eye). For the photographic magnitude, m_p, we have $\lambda_0 \simeq 4300$ Å (but the value of λ_0 depends also on which film is used). Finally, the bolometric magnitude corresponds to the received energy integrated over all the emitted wave-lengths.

The visual and photographic magnitudes must be corrected in order to take into account the interstellar absorption and that due to the terrestrial atmosphere. The visual magnitude is linked to the others by

$$m_p - m_v = C \tag{A.2}$$

$$m_b - m_v = B.C. \tag{A.3}$$

where C is the so-called colour index, while B.C. is called bolometric correction. From these two equations we have

$$m_p = C + m_b - B.C. \tag{A.4}$$

In the case of distant galaxies, the red-shift produces an increase of m_p, which modifies also the other quantities

$$\Delta m_p = \Delta C + \Delta m_b - \Delta B.C. \tag{A.5}$$

In order to calculate the red-shift effects, we must use eq.(A.1), rewritten as

$$m_o - M = -2.5 \, \mathrm{Log} \, e_o + 2.5 \, \mathrm{Log} \, L \qquad (A.6)$$

where m_o and M are respectively the observed and absolute magnitudes (i.e. the apparent magnitude at the conventional distance of 10 parsec) of a source, e_o is its observed luminosity, and L its luminosity at the distance of 10 parsec.

There are two effects, due to the red-shift, on the radiated luminosity: first, the frequency (and then the energy) associated with each individual photon received by the observer, is decreased by the factor $1/(1+z)$. Second, also the rate at which photons are received is decreased by the same ratio. But if one neglects the Doppler shift, like Hubble and Humason, the correction to the bolometric magnitude is

$$\Delta m_b = 2.5 \, \mathrm{Log} \left(\frac{\lambda + \Delta\lambda}{\lambda} \right) = 2.5 \, \mathrm{Log} \, (1+z) \qquad (A.7)$$

Assuming that the absolute magnitude is the same for all the galaxies, it follows that their apparent luminosity is proportional to the inverse of their squared astronomical distances, $e_o / L = D^2 / d^2$, where D=10 parsec. From (A.6) we have then

$$m_o - M = 2.5 \, \mathrm{Log} \, \frac{d^2}{D^2} = 5 \, \mathrm{Log} \, d - 5 \qquad (A.8)$$

Finally we obtain

$$\mathrm{Log} \, d = 0.2 \, (m_o - M) + 1 \qquad (A.9)$$

which relates the distance d (in parsec) to the apparent and absolute magnitude.

If one uses, following Hubble and Humason, the photographic magnitude, then

$$\mathrm{Log} \, d = 0.2 \, (m_p - \Delta m_p - M_p) + 1 \qquad (A.10)$$

where Δm_p is the correction for red-shift effects.

REFERENCES

1) For a more detailed discussion of the microwave radiation background, see for example A.K.Raychaudhuri : "Theoretical Cosmology" (World Scientific, Singapore, 1980), Chapter 6

2) See for example S.Weinberg:"Gravitation and Cosmology" (Wiley, New York, 1971), Chapter 15

3) A.H.Guth: Phys.Rev.D23,347 (1981)

4) A.D.Linde: Rep.Prog.Phys.47,925(1984)

CHAPTER VIII

GRAVITATIONAL WAVES

1.- The weak field approximation

One of the most interesting problems associated with the field equations of general relativity is that of the possible existence of gravitational waves.

The experimental detection of gravitational radiation is very difficult mainly because of the extreme weakness of the coupling between matter and gravitation, and up to now no independent confirmation of the pioneer experimental results, due to Weber[1], has been obtained, although many experimental apparatus for the search of gravitational waves are at present operating in the world. Moreover it must be stressed, from the theoretical point of view, that up to now no exact solution of the nonlinear Einstein field equations, representing outgoing spherical waves, has been obtained.

The possible existence of gravitational waves was first investigated by Einstein, using approximate solutions of the linearized field equations, in the so-called "weak field approximation", and he was led to the conclusion that gravitational waves should be produced by a system of accelerated masses, just like electromagnetic radiation is produced by accelerated charges.

Starting with the field equations (3.77)

$$R_{\mu\nu} - \frac{1}{2} g_{\mu\nu} R = \frac{8\pi G}{c^4} T_{\mu\nu} \tag{8.1}$$

in the weak field limit one assumes that the world metric tensor $g_{\mu\nu}$ is not very much different from the Minkowski metric $\eta_{\mu\nu} = \mathrm{diag}(-1,-1,-1,+1)$, and can be written then in the form

$$g_{\mu\nu} = \eta_{\mu\nu} + h_{\mu\nu} \tag{8.2}$$

where $h_{\mu\nu}$ represents the small corrections to the flat space-time metric $\eta_{\mu\nu}$ due to the presence of a weak gravitational field. In this approximation $|h_{\mu\nu}| \ll 1$, so that terms of order higher than the first in $h_{\mu\nu}$ can be neglected in the field equations, and we have

$$h_\mu{}^\nu = \eta^{\nu\alpha} h_{\mu\alpha} \; , \quad h = h_\mu{}^\mu = \eta^{\mu\nu} h_{\mu\nu} \qquad (8.3)$$

(i.e. the indices of $h_{\mu\nu}$ are raised and lowered with the unperturbed metric tensor $\eta_{\mu\nu}$) . The controvariant components of $g^{\mu\nu}$ are, in this approximation,

$$g^{\mu\nu} = \eta^{\mu\nu} - h^{\mu\nu} \qquad (8.4)$$

so that, to first order in h , one has

$$g_{\mu\nu} g^{\nu\alpha} = (\eta_{\mu\nu} + h_{\mu\nu})(\eta^{\nu\alpha} - h^{\nu\alpha}) \simeq$$
$$= \delta_\mu{}^\alpha - h_\mu{}^\alpha + h_\mu{}^\alpha = \delta_\mu{}^\alpha \qquad (8.5)$$

The Christoffel coefficients corresponding to the metric (8.2) are of the same order as $h_{\mu\nu}$,

$$\Gamma_{\mu\nu}{}^\alpha \simeq \frac{1}{2} \eta^{\alpha\beta} (\partial_\mu h_{\beta\nu} + \partial_\nu h_{\beta\mu} - \partial_\beta h_{\mu\nu}) \qquad (8.6)$$

then we can neglect the Γ^2 terms in the definition (3.7) of the curvature tensor, and we obtain

$$R_{\mu\nu\alpha\beta} = \frac{1}{2}\left(-\frac{\partial^2 h_{\mu\beta}}{\partial x^\nu \partial x^\alpha} - \frac{\partial^2 h_{\nu\alpha}}{\partial x^\mu \partial x^\beta} + \frac{\partial^2 h_{\nu\beta}}{\partial x^\mu \partial x^\alpha} + \frac{\partial^2 h_{\mu\alpha}}{\partial x^\nu \partial x^\beta}\right) \qquad (8.7)$$

The Ricci tensor is then (see 3.23)

$$R_{\mu\nu} = g^{\alpha\beta} R_{\alpha\mu\nu\beta} \simeq \eta^{\alpha\beta} R_{\alpha\mu\nu\beta} =$$
$$= \frac{1}{2}\left(-\eta^{\alpha\beta}\partial_\alpha \partial_\beta h_{\mu\nu} + \partial_\mu \partial_\beta h_\nu{}^\beta + \partial_\nu \partial_\alpha h_\mu{}^\alpha - \partial_\mu \partial_\nu h\right) \qquad (8.8)$$

It is convenient, in this chapter, to denote partial derivatives with a comma. This equation can be rewritten then

$$R_{\mu\nu} = -\frac{1}{2}\eta^{\alpha\beta} h_{\mu\nu,\alpha\beta} - \frac{1}{2}(h_{,\mu\nu} - h_\nu{}^\beta{}_{,\mu\beta} - h_\mu{}^\alpha{}_{,\nu\alpha}) \qquad (8.9)$$

and since

$$h_{,\mu\nu} = \frac{1}{2} h_{,\mu\nu} + \frac{1}{2} h_{,\nu\mu} =$$
$$= \frac{1}{2} \delta_\mu{}^\alpha h_{,\alpha\nu} + \frac{1}{2} \delta_\nu{}^\beta h_{,\mu\beta} \qquad (8.10)$$

we have

$$h_{,\mu\nu} - h_\nu{}^\beta{}_{,\mu\beta} - h_\mu{}^\alpha{}_{,\nu\alpha} =$$
$$= \left(\frac{1}{2} \delta_\mu{}^\alpha h - h_\mu{}^\alpha\right)_{,\nu\alpha} + \left(\frac{1}{2} \delta_\nu{}^\beta h - h_\nu{}^\beta\right)_{,\mu\beta} \qquad (8.11)$$

Defining

$$\psi_\mu{}^\nu = h_\mu{}^\nu - \frac{1}{2} \delta_\mu{}^\nu h \qquad (8.12)$$

and using the freedom in the choice of the coordinate system, we can impose the four conditions (see Appendix A)

$$\psi_\mu{}^\nu{}_{,\nu} = 0 \qquad (8.13)$$

(this choice is also called "armonic gauge"). In this case the Ricci tensor (8.9) reduces simply to

$$R_{\mu\nu} = -\frac{1}{2} \eta^{\alpha\beta} h_{\mu\nu,\alpha\beta} = -\frac{1}{2} \Box h_{\mu\nu} \qquad (8.14)$$

where

$$\Box \equiv \eta^{\alpha\beta} \partial_\alpha \partial_\beta = -\nabla^2 + \frac{1}{c^2}\frac{\partial^2}{\partial t^2} \qquad (8.15)$$

and the curvature scalar becomes

$$R \simeq \eta^{\mu\nu} R_{\mu\nu} = -\frac{1}{2} \Box h \qquad (8.16)$$

The field equations (8.1) become then

$$-\frac{1}{2} \Box h_{\mu\nu} + \frac{1}{4} \eta_{\mu\nu} \Box h =$$
$$= -\frac{1}{2} \Box \left(h_{\mu\nu} - \frac{1}{2} \eta_{\mu\nu} h\right) = \frac{8\pi G}{c^4} T_{\mu\nu} \qquad (8.17)$$

where $T_{\mu\nu}$ is the energy-momentum tensor of matter sources, evaluated to first order in h . Taking into account (8.12), we obtain that the Einstein equations, in the weak field approximation, may be written in the form

$$\Box \psi_{\mu\nu} = -\frac{16\pi G}{c^4} T_{\mu\nu} \qquad (8.18)$$

The solution of these linear equations can be obtained easily, using for example the Green's function and the method of Fourier's transforms. The solution is exactly analogous to the retarded potentials of the electromagnetic theory, and can be written

$$\psi_{\mu\nu}(\vec{x}, t) = -\frac{4G}{c^4} \int \frac{d^3x'}{|\vec{x}-\vec{x}'|} T_{\mu\nu}(\vec{x}', t') \qquad (8.19)$$

where the gravitational sources $T_{\mu\nu}$ are considered at the retarded time $t' = t - |\vec{x} - \vec{x}'|/c$.

In the limit of a static mass distribution, the field equations (8.18) lead to the relation between the scalar Newtonian potential $\phi(\vec{x})$ and the g_{44} component of the metric tensor already found in Chapter III.

In fact in the static case the temporal derivatives of $\psi_{\mu\nu}$ are vanishing, and the only nonzero component of the energy-momentum tensor is the rest energy density, $T_{44} = \rho c^2$. From (8.18) we have

$$\nabla^2 \psi_{44} = \frac{16\pi G \rho}{c^2} \qquad (8.20)$$

and the general solution of this equation is

$$\psi_{44} = -\frac{4G}{c^2} \int \frac{d^3x'}{|\vec{x}-\vec{x}'|} \rho(\vec{x}') = \frac{4}{c^2}\phi \qquad (8.21)$$

where

$$\phi(\vec{x}) = -G \int \frac{d^3x'}{|\vec{x}-\vec{x}'|} \rho(\vec{x}') \qquad (8.22)$$

is the Newton gravitational potential. From (8.12) we have

$$\psi = \eta^{\mu\nu}\psi_{\mu\nu} = -h = \eta^{\mu\nu} h_{\mu\nu} \qquad (8.23)$$

and then in this approximation $h = -\psi_{44}$, since $\psi_{\mu\nu} = 0$ for $\mu \neq 4$, $\nu \neq 4$. Therefore

$$\psi_{44} = h_{44} - \frac{1}{2}h = h_{44} + \frac{1}{2}\psi_{44} \qquad (8.24)$$

or

$$\psi_{44} = 2h_{44} \qquad (8.25)$$

Using (8.21), we can then relate the metric tensor to the Newtonian potential as follows

$$g_{44} = \eta_{44} + h_{44} = 1 + \frac{1}{2}\psi_{44} = 1 + \frac{2}{c^2}\phi \qquad (8.26)$$

in agreement with eq.(3.48) obtained previously. The Newtonian approximation is then obtained as the static limit of the weak field approximation.

Finally, we note that in vacuum ($T_{\mu\nu} = 0$) the linearized equations (8.18) become

$$\Box \psi_{\mu\nu} = 0 \qquad (8.27)$$

or , equivalently,

$$\Box h_{\mu\nu} = 0 \qquad (8.28)$$

(because h = 0) and they describe then a weak gravitational perturbation, propagating in vacuum with the velocity of light.

2.- Plane gravitational waves

The linearized equations (8.28) for the gravitational field in vacuum, toghether with the conditions (8.13) and $h_{\mu\nu} = h_{\nu\mu}$, $h_\mu{}^\mu = 0$, describe, physically, a massless spin-two particle, as pointed out first by Pauli and Fierz[2] .

The solution of eq.(8.28) can be written, in general, as

$$h_{\mu\nu} = \varepsilon_{\mu\nu}\, e^{i k_\alpha x^\alpha} + \varepsilon^*_{\mu\nu}\, e^{-i k_\alpha x^\alpha} \qquad (8.29)$$

(a star denotes complex conjugation) where $k_\mu k^\mu = 0$, and the condition (8.13) implies

$$k_\nu \left(\varepsilon_\mu{}^\nu - \frac{1}{2}\delta_\mu{}^\nu \varepsilon \right) = 0 \qquad (8.30)$$

The polarization tensor $\varepsilon_{\mu\nu} = \varepsilon_{\nu\mu}$, being symmetric, has ten components, but only two are independent and correspon to physical degree of freedom. In fact the ten components of $\varepsilon_{\mu\nu}$ are subject to the four constraints (8.30); moreover, under the infinitesimal change of coordinates $x'^\mu = x^\mu + \xi^\mu$, the polarization tensor becomes (see

Appendix A, eq.A.10),

$$\varepsilon'_{\mu\nu} = \varepsilon_{\mu\nu} - \partial_\nu \xi_\mu - \partial_\mu \xi_\nu \qquad (8.31)$$

and $\varepsilon'_{\mu\nu}$ represents the same physical situation for arbitrary values of the four parameters $\xi^\mu(x)$. The number of independent components of $\varepsilon_{\mu\nu}$ is then 10 - 4 - 4 = 2, and then there are only two polarization states, as appropriate for a massless partcle.

If the solution (8.29) describes, for example, a field propagating along the x^3 direction, the only nonvanishing two components of $\varepsilon_{\mu\nu}$ are, as will be shown in a moment, $\varepsilon_{11} = -\varepsilon_{22}$, and $\varepsilon_{12} = \varepsilon_{21}$. Performing a rotation around the x^3 axis, i.e.

$$\varepsilon'_{\mu\nu} = U_\mu{}^\alpha U_\nu{}^\beta \varepsilon_{\alpha\beta} \qquad (8.32)$$

where

$$U_\alpha{}^\beta = \begin{pmatrix} \cos\vartheta & \sin\vartheta & 0 & 0 \\ -\sin\vartheta & \cos\vartheta & 0 & 0 \\ 0 & 0 & 1 & 0 \\ 0 & 0 & 0 & 1 \end{pmatrix} \qquad (8.33)$$

we obtain

$$\varepsilon'_\pm = e^{\pm 2i\vartheta} \varepsilon_\pm \qquad (8.34)$$

where

$$\varepsilon_\pm = \varepsilon_{11} \mp i\,\varepsilon_{12} \qquad (8.35)$$

And since a plane wave ψ, with helicity H, under a rotation of an angle ϑ around the direction of propagation transforms as

$$\psi' = e^{\pm iH\vartheta} \psi \qquad (8.36)$$

we can conclude, from (8.34), that the gravitational wave (8.29) has helicity H = 2, i.e. it describes a spin-two particle.

It follows that, in the theory of general relativity, the quanta of the gravitational radiation, the so-called "gravitons", arinsing from the quantization of the classical gravitational waves (just like photons are introduced by quantizing the electromagnetic waves), should be (if they exist) massless and spin-two particles.

The massless property reflects the fact that the range of the classical gravitational interaction is infinite, like that of the electromagnetic interaction, while the value two of the helicity corresponds to the fact that the gravitational field is described by a second rank, symmetric tensor $h_{\mu\nu}$, unlike the electromagnetic field, represented by a potential vector A_μ .

Let us consider now in some detail a plane gravitational wave propagating along the axis $x^1 = x$. In this case (8.28) becomes

$$\left(\frac{\partial^2}{\partial x^2} - \frac{1}{c^2}\frac{\partial^2}{\partial t^2}\right) h_\mu{}^\nu = 0 \tag{8.37}$$

as the field is variable only along the direction of propagation. From the condition (8.13) we have

$$\psi_\mu{}^4{}_{,4} + \psi_\mu{}^1{}_{,1} = 0 \tag{8.38}$$

and if we consider a plane wave propagating along the positive direction of the x axis, then $h_{\mu\nu}$ is a function of $t - x/c$, and we have

$$\frac{\partial \psi}{\partial x} = -\frac{1}{c}\frac{\partial \psi}{\partial t} \tag{8.39}$$

The condition (8.38) becomes then

$$(\psi_\mu{}^4 - \psi_\mu{}^1)_{,4} = 0 \tag{8.40}$$

from which we obtain

$$\psi_1{}^4 = \psi_1{}^1 , \quad \psi_2{}^4 = \psi_2{}^1 , \quad \psi_3{}^4 = \psi_3{}^1 , \quad \psi_4{}^4 = \psi_4{}^1 \tag{8.41}$$

(in the integration of (8.40) we have neglected the integration constants, as only the variable part of the field is physically interesting).

As already pointed out, the gravitational potentials $\psi_{\mu\nu}$ can be constrained also by four additional conditions, obtained performing a coordinate transformation of the form $x'^\mu = x^\mu + \xi^\mu$, where $\Box \xi^\mu = 0$ (see the Appendix A) and these additional conditions can be used, with a suitable choice of the four vectors ξ^μ, to impose

$$\psi'_1{}^4 = \psi'_2{}^4 = \psi'_3{}^4 = \psi'_2{}^2 + \psi'_3{}^3 = 0 \qquad (8.42)$$

so that, from (8.41),

$$\psi'_1{}^1 = \psi'_2{}^1 = \psi'_3{}^1 = \psi'_4{}^4 = 0 \qquad (8.43)$$

It is important to note that $\psi_2{}^3$ and $\psi_2{}^2 - \psi_3{}^3$ cannot be made vanishing by any choice of the coordinate system, as, under a coordinate transformation (see Appendix A) we have

$$h'_{\mu\nu} = h_{\mu\nu} - \xi_{\mu,\nu} - \xi_{\nu,\mu}$$

and since $\xi^\mu = \xi^\mu(t-x/c)$, for μ, ν = 2,3 we have $h'_{\mu\nu} = h_{\mu\nu}$.

The only nonvanishing components of the gravitational field in this case are then (neglecting primes) $\psi_2{}^3$ and $\psi_2{}^2 = -\psi_3{}^3$; moreover, from eq.(8.42,43) we have also $\psi = \psi_1{}^1 + \psi_2{}^2 + \psi_3{}^3 + \psi_4{}^4 = 0$, so that, from (8.12), $\psi_{\mu\nu} = h_{\mu\nu}$.

We can say then that the polarization of a plane gravitational wave, propagating along the x^1 axis, is described, in the (x^2, x^3) plane, by a traceless, symmetric tensor determined only by the two quantities h_{23} and $h_{22} = -h_{33}$,

$$\varepsilon_{\mu\nu} = \begin{pmatrix} h_{22} & h_{23} \\ h_{32} & h_{33} \end{pmatrix} \qquad (8.44)$$

The corresponding Riemann tensor in this case is given by

$$R_{4343} = -R_{4242} = -R_{1212} = R_{4212} = R_{4331} =$$

$$= R_{3131} = \frac{1}{2} h_{33,44} = -\frac{1}{2} h_{22,44}$$

$$R_{4243} = -R_{1231} = -R_{4312} = R_{4231} = \frac{1}{2} h_{23,44} \qquad (8.45)$$

all the other components being vanishing.

3.- Emission of gravitational radiation

In the presence of matter, one must apply the linearized equations (8.18), where $T_{\mu\nu}$ represents, to first order in $h_{\mu\nu}$, the energy-momentum tensor of the sources in the case of weak gravitational fields.

From the condition $\psi_\mu{}^\nu{}_{,\nu} = 0$ it follows that

$$T_\mu{}^\nu{}_{,\nu} = 0 \tag{8.46}$$

that is the linearized energy-momentum tensor satisfies the usual conservation equation, instead of the covariant one. Using this condition, we can express the solution (8.19) of the field equations in terms of $T_4{}^4$.

In fact, separating the spatial and temporal components and lowering indices with $\eta_{\mu\nu}$ we get, from (8.46) (remember that $T_\mu{}^4 = T_{\mu 4}$, $T_\mu{}^i = -T_{\mu i}$)

$$T_{ik,k} - T_{i4,4} = 0 \tag{8.47}$$

$$T_{4k,k} - T_{44,4} = 0 \tag{8.48}$$

(latin indices run from 1 to 3). Multiplying (8.47) by x_a, and integrating over a three-dimensional space-like hypersurface we have

$$\frac{\partial}{\partial x^4}\int T_{i4}x_a\, d^3x = \int \partial_k T_{ik}x_a\, d^3x =$$

$$= \int d^3x\, \partial_k(T_{ik}x_a) - \int d^3x\, T_{ia} = -\int d^3x\, T_{ia} \tag{8.49}$$

(we have used the Gauss theorem and the fact that $T_{ik} = 0$ at spatial infinity). Therefore, since T_{ia} is a symmetric tensor,

$$\int d^3x\, T_{ia} = -\frac{1}{2}\frac{\partial}{\partial x^4}\int (T_{i4}x_a + T_{a4}x_i)\, d^3x \tag{8.50}$$

Multiplying (8.48) by $x_i x_a$, integrating and using again the Gauss theorem we have

$$\frac{\partial}{\partial x^4}\int T_{44}x_i x_a\, d^3x = \int d^3x\, T_{4k,k}x_i x_a =$$

$$= \int d^3x\, \partial_k(T_{4k}x_i x_a) - \int d^3x\, T_{4i}x_a - \int d^3x\, T_{4a}x_i =$$

$$= -\int d^3x \, (\tau_{i4} x_a + \tau_{a4} x_i) \qquad (8.51)$$

Combining these last two equations we get finally

$$\int d^3x \, \tau_{ik} = \frac{1}{2c^2} \frac{\partial^2}{\partial t^2} \int \tau_{44} x_i x_k \, d^3x \qquad (8.52)$$

The general expression for $\psi_{\mu\nu}$, given in eq.(8.19), can be rewritten now in terms of τ_{44} only, which, in this approximation, is given by $\tau_{44} = \rho c^2$. At a great distance from the sources, i.e. $|\vec{x}| >> |\vec{x}'|$ (see Fig.8.1), we consider gravitational radiation with a wave-lenght λ much greater than the size of the source, $\lambda >> |\vec{x}'|$ (like in the case of the dipolar approximation in classical electrodynamics). We have then $|\vec{x} - \vec{x}'| \simeq |\vec{x}|$, and the general solution (8.19) becomes

$$\psi_{\mu\nu}(\vec{x}, t) = -\frac{4G}{c^4 R_0} \int d^3x' \, T_{\mu\nu}(\vec{x}', t') \qquad (8.53)$$

where $R_0 = |\vec{x}|$ is the distance from the point P where the field is calculated from the origin O of the coordinate system, placed inside the matter distribution (see Fig.8.1 and the retarded time t' is given simply by $t' = t - R_0/c$.

Using (8.52) we obtain then

$$\psi_{ik} = -\frac{2G}{c^4 R_0} \frac{\partial^2}{\partial t^2} \int d^3x' \, \rho(\vec{x}', t') \, x'_i x'_k \qquad (8.54)$$

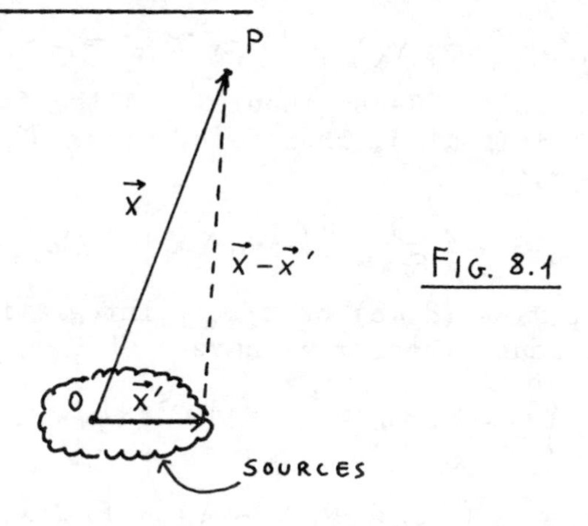

FIG. 8.1

In order to evaluate the total energy radiated away by the system, we can approximate, at a large distance from the sources, the gravitational wave as a plane wave, and the flux of energy along the direction x^1 (for example), will depend then only on h_{22}, h_{33} and h_{23}.

The energy flux of the gravitational radiation can be expressed through the components of the energy-momentum pseudotensor $t^{\mu\nu}$ (see the Appendix B), and, in the case of a plane wave propagating along the x^1 axis, the flux of energy is (eq.B.30)

$$c\sqrt{-g}\, t^{41} = \frac{c^3}{16\pi G}\left[\dot{h}_{23}^2 + \frac{1}{4}\left(\dot{h}_{22} - \dot{h}_{33}\right)^2\right] \qquad (8.55)$$

(a dot denotes the derivative with respect to t).

Introducing the quadrupole momentum tensor

$$D_{ik}(\vec{x}) = \int \rho(\vec{x}')\,(3x'_i x'_k - r'^2\delta_{ik})\, d^3x' \qquad (8,56)$$

where $r'^2 = \delta_{ik}x'^i x'^k$, we obtain from (8.54)

$$\psi_{23} = -\frac{2G}{3c^4 R_0}\ddot{D}_{23} = h_{23} \qquad (8,57)$$

$$\psi_{22} - \psi_{33} = -\frac{2G}{3c^4 R_0}\left(\ddot{D}_{22} - \ddot{D}_{33}\right) = h_{22} - h_{33} \qquad (8,58)$$

(remember that, for a plane gravitational wave, $\psi_{\mu\nu} = h_{\mu\nu}$, as shown in the previous Section, and then (8.55) can be rewritten

$$c\sqrt{-g}\, t^{41} = \frac{G}{36\pi c^5 R_0^2}\left[\dddot{D}_{23}^2 + \frac{1}{4}\left(\dddot{D}_{22} - \dddot{D}_{33}\right)^2\right] \qquad (3.59)$$

The intensity of radiation, dI, within the solid angle $d\Omega$, is defined as the density of energy flux through the element of spherical surface $R_0^2\, d\Omega$, centered at the origin and with radius R_0.

Generalizing the expression (8.59) for the density of energy flux of a plane wave propagating along an arbitrary direction, characterized by the unit vector $\vec{n} = (n_1, n_2, n_3)$, $|\vec{n}| = 1$ (see the Appendix C, eq.C.9), we obtain

$$dI = \frac{G}{36\pi c^5 R_0^2}\left[\frac{1}{4}\left(\dddot{D}_{ik} n_i n_k\right)^2 + \frac{1}{2}\left(\dddot{D}_{ik}\right)^2 - \dddot{D}_{ik}\dddot{D}_{ij} n_k n_j\right] R_0^2\, d\Omega \qquad (3,60)$$

The total energy radiated in all directions, i.e. the energy loss by the sources in the unit of time, $-dE/dt$, can be obtained by integrating (8.60) over the solid angle $d\Omega$, and putting $n_1 = \sin\vartheta\cos\varphi$, $n_2 = \sin\vartheta\sin\varphi$, $n_3 = \cos\vartheta$, $d\Omega = \sin\vartheta\, d\vartheta\, d\varphi$. The final result (for an explicit computation see the Appendix C) is

$$-\frac{dE}{dt} = \frac{G}{45c^5}\left(\dddot{D}_{ik}\right)^2 \tag{8.61}$$

This formula will be applied in the following Section.

4.- Possible sources of radiation

a) Linear quadrupole oscillator

Consider a system formed by two bodies, each with a mass m, oscillating linearly along the x^3 axis, which can be also represented as a particle with reduced mass $\mu = m/2$, and coordinates given by

$$x^1 = x^2 = 0 \quad , \quad x^3 = l\cos\omega t \tag{8.62}$$

The gravitational energy radiated away in this case can be calculated starting from the conservation of the total energy-momentum, for the physical system which includes both the matter sources and the gravitational field (see the Appendix B). From eq.B15

$$(\sqrt{-g}\, T_\mu{}^\nu)_{,\nu} + (\sqrt{-g}\, t_\mu{}^\nu)_{,\nu} = 0 \tag{8.63}$$

we have

$$-c\frac{\partial}{\partial x^4}\int\sqrt{-g}\, T_4{}^4 d^3x - c\int(\sqrt{-g}\, T_4{}^k)_{,k}\, d^3x =$$

$$= c\frac{\partial}{\partial x^4}\int\sqrt{-g}\, t_4{}^4 d^3x + c\int(\sqrt{-g}\, t_4{}^k)_{,k}\, d^3x \tag{8.64}$$

and using the Gauss theorem one obtains

$$\int(\sqrt{-g}\, T_4{}^k)_{,k}\, d^3x = \int_F \sqrt{-g}\, T_4{}^k df_k = 0 \tag{8.65}$$

because the surface integral in df_K is performed over a closed bidimensional surface F surrounding the matter sources at a large distance in the radiation zone, and $T_{\mu\nu}$ is vanishing outside the sources. The first integral in eq. (8.64) represents the total energy E of the sources, while the third integral, for the oscillating system we are considering, is a periodical function of time which gives a vanishing contribution if averaged over a period.

Averaging in time eq.(8.64) over a period $T = 2\pi/\omega$ we obtain then

$$-\left\langle \frac{dE}{dt} \right\rangle = \left\langle \int_F c\sqrt{-g}\, t_4{}^{\kappa}\, df_{\kappa} \right\rangle \tag{8.66}$$

The total radiated power (time-averaged) by the system, is given then by the flux of the energy-momentum pseudotensor over the bidimensional surface F, averaged in time.

The density of energy flux, $c\sqrt{-g}\, t_4{}^{\kappa}$, can be expressed using the gravitational potential $\psi_{\mu\nu}$ (see the Appendix B) which is obtained from eqs.(8.54,62), and applying the condition $\psi_{\mu}{}^{\nu}{}_{,\nu} = 0$.

We can also directly apply, however, the expression for the energy loss in all directions, that is eq.(8.61),

$$-\left\langle \frac{dE}{dt} \right\rangle = \frac{G}{45c^5} \left\langle (\dddot{D}_{ik})^2 \right\rangle \tag{8.67}$$

For an oscillating pointlike particle of reduced mass $\mu = m/2$ one has, from (8.62), $r^2 = x_3^2 = \ell^2\cos^2\omega t$. The only nonvanishing components of the quadrupole tensor are then, putting $\rho(x') = \mu\,\delta(x_1')\,\delta(x_2')\,\delta(x_3'-x_3)$ in eq.(8.56)

$$\begin{aligned} D_{11} &= -\mu r^2 = -\mu x_3^2 = -\mu \ell^2 \cos^2\omega t \\ D_{22} &= -\mu r^2 = -\mu x_3^2 = -\mu \ell^2 \cos^2\omega t \\ D_{33} &= \mu(3x_3^2 - r^2) = 2\mu x_3^2 = 2\mu \ell^2 \cos^2\omega t \end{aligned} \tag{8.68}$$

from which

$$\begin{aligned} \dddot{D}_{11} &= -8\mu \ell^2 \omega^3 \sin\omega t \cos\omega t = \dddot{D}_{22} \\ \dddot{D}_{33} &= 16\mu \ell^2 \omega^2 \sin\omega t \cos\omega t \end{aligned} \tag{8.69}$$

Therefore

$$\begin{aligned} (\dddot{D}_{ik})^2 &= \dddot{D}_{11}^2 + \dddot{D}_{22}^2 + \dddot{D}_{33}^2 = \\ &= 384\mu^2 \ell^4 \omega^6 \sin^2\omega t \cos^2\omega t \end{aligned} \tag{8.70}$$

and since, averaging over a period $T = 2\pi/\omega$,

$$\langle \sin^2\omega t \cos^2\omega t \rangle =$$

$$= \frac{1}{T}\int_0^T dt \, \frac{\sin^2 2\omega t}{4} = \frac{1}{8} \tag{8.71}$$

we obtain, from (8.67)

$$\langle \frac{dE}{dt} \rangle = -\frac{G}{45c^5} \frac{384}{8} \mu^2 \ell^4 \omega^6 \tag{8.72}$$

Defining the moment of inertia with respect to the center of mass, $I = 4\mu \ell^2 = 2\,m\,\ell^2$, the (time-averaged) power radiated by a linear quadrupole, oscillating with a frequency $\nu = \omega/2\pi$, and with a moment of inertia I, is given then by

$$\langle \frac{dE}{dt} \rangle = -\frac{G}{45c^5} \frac{384}{8} \frac{I^2}{16} \omega^6 =$$

$$= -\frac{G I^2 \omega^6}{15 c^5} \tag{8.73}$$

b) Rotating masses

Consider two bodies, with masses m_1 and m_2 , rotating in the plane (x,y) around their center of mass. The system can be equivalently described as a body with reduced mass $\mu = m_1 m_2/(m_1 + m_2)$ rotating with angular velocity ω along an orbit which, for simplicity, we suppose circular with radius a . We have then

$$x_1 = a\cos\omega t \, , \quad x_2 = a\sin\omega t \, , \quad x_3 = 0$$

$$r^2 = a^2 \tag{8.74}$$

and the nonvanishing components of the quadrupole tensor are

$$D_{11} = \mu a^2 (3\cos^2\omega t - 1) = \mu a^2 (2\cos^2\omega t - \sin^2\omega t)$$

$$D_{22} = \mu a^2 (3\sin^2\omega t - 1) = \mu a^2 (2\sin^2\omega t - \cos^2\omega t)$$

$$D_{33} = -\mu a^2$$

$$D_{21} = \mu a^2 3 \sin\omega t \cos\omega t = D_{12} \tag{8.75}$$

It is easy to obtain then

$$\dddot{D}_{11} = 24\mu a^2\omega^3 \sin\omega t\cos\omega t = -\dddot{D}_{22}$$

$$\dddot{D}_{21} = -12\mu a^2\omega^3(\cos^2\omega t - \sin^2\omega t) \qquad (8.76)$$

Averaging over a period we have $\langle \sin^2\omega t \cos^2\omega t\rangle = 1/8$, and

$$\langle(\cos^2\omega t - \sin^2\omega t)^2\rangle = \langle\cos^2 2\omega t\rangle =$$

$$= \frac{1}{T}\int_0^T dt\cos^2 2\omega t = \frac{1}{2} \qquad (8.77)$$

The radiated power is then

$$\left\langle\frac{dE}{dt}\right\rangle = -\frac{G}{45c^5}\langle \dddot{D}_{11}^2 + \dddot{D}_{22}^2 + 2\dddot{D}_{12}^2\rangle =$$

$$= -\frac{32GI^2\omega^6}{5c^5} \qquad (8.78)$$

where $I = \mu a^2$ is the moment of inertia of the system relative to the center of mass. Note that, because of the factor G/c^5 in eqs.(8.73,78) the emission of gravitational radiation is always extremely weak.

c) Astronomical sources

The first astronomical objects considered as possible sources of gravitational radiation were the binary stars, because of the presence of a quadrupole moment in those systems. If they lose energy emitting gravitational radiation, then the parameters of the stellar orbits must change in time and such variations could be recorded by astronomical observations.

Considering two stars with masses m_1 and m_2, the moment of inertia relative to their center of mass is

$$I = \mu r^2 = \frac{m_1 m_2}{m_1+m_2} r^2 \qquad (8.79)$$

where μ is the reduced mass of the system, and

$$\vec{r} = -\frac{m_1+m_2}{m_2}\vec{r}_1 = \frac{m_1+m_2}{m_1}\vec{r}_2 \qquad (8.80)$$

where $\vec{r}_1$ and $\vec{r}_2$ are the position vectors of the two stars with respect to their center of mass. Moreover, the angular velocity ω, in the simplified case of a circular orbit, is, according to Kepler's third law

$$\omega^2 = \frac{G(m_1 + m_2)}{r^3} \tag{8.81}$$

We have then, from (8.78),

$$\left\langle \frac{dE}{dt} \right\rangle = -\frac{32}{5} \frac{G^4}{c^5} m_1^2 m_2^2 \frac{m_1 + m_2}{r^5} \tag{8.82}$$

or

$$\left\langle \frac{dE}{dt} \right\rangle = -\frac{32}{5} \frac{G^{7/3}}{c^5} m_1^2 m_2^2 \frac{\omega^{10/3}}{(m_1 + m_2)^{2/3}} \tag{8.83}$$

Supposing, for simplicity, $m_1 = m_1 = M$, then $|\vec{r}_1| = |\vec{r}_2| = |\vec{r}| / 2$, and putting $|\vec{r}_1| = R$ we have, from (8.82),

$$-\left\langle \frac{dE}{dt} \right\rangle = \frac{2G^4}{5c^5} \frac{M^5}{R^5} \tag{8.84}$$

We can express then the radiated power in terms of the masses and their separation (or alternatively in terms of the angular velocity).

Another parameter characterizing a binary system in the case of energy losses due to the emission of gravitational radiation, is the decay time. In fact, if the binary stars are radiating energy away, they cannot mantain their orbits, but must circle closer and closer, emitting gravitational radiation of shorter and shorter wavelength. If the two masses are equal and the orbit is circular, they will fall toghether in the finite time

$$t = \frac{5c^5}{32G^3} \frac{R^4}{M^3} \tag{8.85}$$

In fact their total energy E is given by

$$E = \frac{1}{2} \frac{m_1 m_2}{m_1 + m_2} \omega^2 r^2 - G \frac{m_1 m_2}{r} \tag{8.86}$$

Using (8.81), and putting $m_1 = m_2 = M$, and $r = 2R$, we get

$$E = -\frac{GM^2}{4R} \quad , \quad dE = \frac{GM^2}{4R^2} dR$$

so that, from (8.84),

$$dt = -\frac{5c^5 R^3}{8G^3 M^3} dR \tag{8.87}$$

which, integrating in dt from zero and t, and in dR from

R and zero, gives eq.(8.85). In the last few minutes the system should emit a pulse of high energy radiation, constantly increasing in frequency and power. This will radiate away the mass quadrupole portion of the angular momentum of the system, leaving only a rapidly spinning and spherically symmetric mass.

To estimate the flux which could be received on the earth we report, in Table I, the parameters relative to six binary stars, according to Braginsky[3]. T is the orbital period, L the distance from the earth, $F = (4\pi L^2)^{-2} dE/dt$ is the received flux of energy, and A is a numerical factor which takes into account the orientation of the plane of the orbit with respect to the earth.

Table I

Stars	T (days)	$m_1/m_\odot$	$m_2/m_\odot$	L (cm)	dE/dt (erg/sec)	A F (erg/cm^{2}s)
UVLeo	0.60	1.36	1.25	2.1×10^{20}	1.8×10^{31}	3.5×10^{-12}
V Pup	1.45	1.66	0.8	1.2×10^{21}	4×10^{31}	2.3×10^{-12}
i Boo	0.268	1.35	0.68	3.8×10^{19}	1.9×10^{30}	1.1×10^{-10}
YYEri	0.321	0.76	0.50	1.3×10^{20}	2.6×10^{29}	1.3×10^{-12}
SVLac	0.321	0.97	0.83	2.3×10^{20}	1.1×10^{30}	1.7×10^{-12}
WZSge	81 min	0.6	0.03	3×10^{20}	3.5×10^{29}	3×10^{-13}

An estimate of the gravitational flux for the binary systems of the WUMA type, which have a narrow orbit and a small orbital period, has been given by Mironowski[4], who found a received flux $\sim 10^{-9}$ erg/cm^2 sec.

Another indirect method to reveal the existence of gravitational radiation is to observe, in a binary system the variation of the angular velocity, $\Delta\omega$, in a given time interval Δt.

For a system of two stars with different masses, starting from the expression (8.68) of their total energy, using (8.81,82) and following the same procedure as before we obtain that the decay time is given by

$$t = \frac{5}{256}\frac{c^5}{G^3}\frac{r^4}{m_1 m_2 (m_1+m_2)} \tag{8.88}$$

or, using (8.81) to eliminate r,

$$t = \frac{5}{256}\frac{c^5}{G^{5/3}}\frac{(m_1+m_2)^{1/3}}{m_1 m_2\, \omega^{8/3}} \tag{8.89}$$

In order to obtain the temporal variation of the orbital frequency we note that, using (8.81), the total energy (8.8 can be written as

$$E = -\frac{1}{2} G \frac{m_1 m_2}{r} = -\frac{1}{2} G^{2/3} \frac{m_1 m_2}{(m_1+m_2)^{1/3}} \omega^{2/3} \tag{8.90}$$

Differentiating with respect to ω , and using (8.83) to eliminate dE , we find

$$dt = \frac{5c^5}{96\, G^{5/3}} \frac{(m_1+m_2)^{1/3}}{m_1 m_2} \frac{d\omega}{\omega^{11/3}}$$

that is

$$\frac{d\omega}{\omega} = \frac{96}{5} \frac{G^{5/3} m_1 m_2\, \omega^{8/3}}{c^5 (m_1+m_2)^{1/3}}\, dt \tag{8.91}$$

For the stars of the Table I one has, in the time interval Δt = 10 years ($\sim 3\times10^8$ sec), the increments of angular velocities given in Table II

Table II

Stars		$\Delta\omega/\omega$
UVLeo	....	3.2×10^{-10}
V Pup	....	1.4×10^{-10}
i Boo	....	1.9×10^{-9}
YYEri	....	5.1×10^{-10}
SVLac	....	9.5×10^{-10}
WZSge	....	4×10^{-9}

The orbital frequency for the majority of the binary systems is known in general up to the eight decimal figure, and then a variation of the orbital period due to the emission of gravitational radiation can be observed only in some particular system.

Recently such a system has been found: it is the binary pulsar PSR1913+16, discovered by Hulse and Taylor[5], and observed for more than six years.

This pulsar is a very accurate clock moving with high speed in the strong gravitational field of its invisible companion, and the increase of its orbital frequency is in agreement with the calculated increase due to the loss of energy by emission of gravitational radiation.

This fact may be interpreted as an indirect proof of the existence of gravitational waves.

Concluding this Section, it should be mentioned that there are also many other possible sources of gravitational radiation which could be considered in astrophysics. For example the matter falling into the gravitational field of a black-hole of mass M (as regards the possible existence of black-holes see Chapter IX, Sect.3), considered by Zeldovich and Novikov[6], and giving a pulse of duration $\Delta t \sim 2GM/c^3$, or a body undergoing asymmetric collapse, as suggested by Shklovsky and Kardashev[6].

Moreover gravitons could be produced in the interactions with electromagnetic fields, as we shall see in the last Section of this Chapter, and also in many other microscopic processes, which however can occur in the framework of a quantum theory of gravity, and are then outside the scope of this book.

In order to estimate the energy density of a possible background of gravitational radiation, we can note that at the present epoch gravitational waves should be produced mainlyby the systems of binary stars, pulsars, supernovae explosions, and probably also by galactic nuclei and quasars. In all cases (with the possible exception of a collision of two black-holes) the emission of energy in the form of electromagnetic radiation is always greater than the corresponding gravitational radiation. This means that the energy density of the produced gravitational waves is smaller than the electromagnetic (non-cosmological) energy density, that is $\lesssim 10^{-13}$ erg/cm^3 (corrsponding to a mass density $\sim 10^{-34}$ g/cm^3).

The most effective sources of gravitational radiation

seem to be the binary stars systems, whose radiated power was estimated by Mironowski to be $\sim 10^{38}$ erg/sec inside each galaxy. This gives, in the whole universe, a total mean energy density of gravitons $\sim 10^{-19}$ erg/cm^3 .

Considering the possibility of black-holes collisions, supposing that black-holes contribute to the 10% of the total matter of our universe, and supposing also that 10% of their masses is transformed into gravitational radiation, one obtains an associated rest energy density $\sim 10^{-32}$ g/cm^3. Another possibility is to speculate on the production of gravitational radiation during the early stages of the universe, when it was very dense and hot. In the case of a primordial anisotropic expansion, the energy density of the produced gravitational radiation, as pointed out by Zeldovich, could be of the same order as that of the electromagnetic black body radiation background, i.e. $\sim 10^{-14}$ erg/cm^3 .

Even in this case, however, this possible gravitational background is out of the possibilities of the present tecniques of detection, and moreover it is not enough to reach the critical density and to produce a closed universe.

5.- Detection of gravitational waves

In appendix D we have deduced the so-called equation of geodesic deviation

$$\frac{D^2\eta^\mu}{Ds^2} + R_{\nu\alpha\beta}{}^{\mu}\eta^\nu u^\alpha u^\beta = 0 \qquad (8.92)$$

where η^μ is an infinitesimal space-like vector connecting two nearby geodesics, u^μ is the unitary, time-like tangent vector to the geodesic, s the proper time along it and $u^\mu\eta_\mu = \text{const} = 0$.

If we are considering a system of interacting masses, eq.(8.92) becomes (see Addendix D)

$$\frac{D^2\eta^\mu}{Ds^2} + R_{\nu\alpha\beta}{}^{\mu}\eta^\nu u^\alpha u^\beta = \frac{DF^\mu}{m\,Dv}\,dv \qquad (8.93)$$

where F^μ/m is the relative asceleration induced by the interaction force between two nearby masses, and v is a parameter such that the world line of each mass corresponds to a given value of v.

Consider, for example, two masses interacting through a spring, and forming a damped harmonic oscillator. We put

$$\eta^\mu = r^\mu + \xi^\mu \qquad (8,94)$$

where r^μ satisfies $Dr^\mu/Ds = 0$, and ξ^μ is the relative displacement of the two masses; $\eta^\mu \to r^\mu$ in the limit of large internal damping and $R_{\mu\nu\alpha\beta} = 0$. We have then

$$\frac{D^2 \xi^\mu}{Ds^2} + R_{\nu\alpha\beta}{}^{\mu} u^\alpha u^\beta (r^\nu + \xi^\nu) = \frac{f^\mu}{mc^2} \qquad (8,95)$$

where f^μ denotes the difference between the (non-gravitational) forces acting on the two masses, which we assume to contain a damping term $-c\,\gamma^\mu{}_\nu D\xi^\nu/Ds$ and a restoring term $-K^\mu{}_\nu \xi^\nu$ (where $K^\mu{}_\nu$ and $\gamma^\mu{}_\nu$ are constant parameters depending on the characteristics of the spring). Therefore

$$\frac{D^2 \xi^\mu}{Ds^2} + \frac{\gamma^\mu{}_\nu}{mc^2}\frac{D\xi^\nu}{Ds} + \frac{K^\mu{}_\nu}{mc^2}\xi^\nu = -R_{\nu\alpha\beta}{}^{\mu}(r^\nu + \xi^\nu) u^\alpha u^\beta \qquad (8,96)$$

Choosing a system of reference in which the time flows tangently to the world line of the center of mass (i.e. in which the oscillator is free falling), and using the geodesic coordinates in which all the Christoffel symbols are vanishing and $u^\mu = (0,0,0,1)$, we have $\xi^4 = 0$, as $\xi^\mu u_\mu = 0$, and, for $\xi^\mu << r^\mu$ (Latin indices run from 1 to 3)

$$\frac{d^2 \xi^i}{dt^2} + \frac{\gamma^i{}_k}{m}\frac{d\xi^k}{dt} + \frac{K^i{}_j}{m}\xi^j = c^2 R_{4k4}{}^{i} r^k \qquad (8,97)$$

This equation describes a damped harmonic oscillator subject to an external force due to the Riemann tensor: a measurement of the oscillation amplitude may lead then directly to the evaluation of some components of $R_{\mu\nu\alpha\beta}$.

It should be noted that, to detect in this way a gravitational wave, at least two particles are needed. Suppose in fact to have only one test particle which starts oscillating under the action of the wave, and suppose also we are observing this interaction at the same position of the particle: in this case we also oscillate in the same way (according to the equivalence principle) so that no effect can be detected. A local observer cannot reveal any motion of a single test particle, being free falling with the same acceleration. Gravitational forces are given by the first derivatives of the metric tensor, and they are locally vanishing in a free falling frame.

However, if we have two test particles (or, more generally, an extended macroscopical body whose microscopic components can be regarded each as a test particle), the gravitational forces can be made vanishing only at one given point, (for example at the position of the center of mass), and a strain can be evidenced between different points of the body, which produces a relative geodesic deviation. Considering two particles placed at different positions, we have that, because of phase retardation effects, one of the particles will move relatively to the other, and besides the acceleration induced by the wave, we have also that the variation of the metric affects the relative separation of the two particles.

It is important to stress, at this point, that when we are dealing with gravitational radiation, we must consider quadrupole oscillations, both for the sources and for the detectors.

In fact the dipolar radiation involves the second temporal derivatives of the total dipole moment of the system of masses forming the source (or the detector), $\frac{d^2}{dt^2} \Sigma_i m_i \vec{r}_i$. But this object is proportional to the first temporal derivative of the total momentum of the system, $\Sigma_i \vec{P}_i$, which must be vanishing for an isolated system. Therefore dipolar gravitational radiation cannot be emitted, neither received, by an isolated system. A detector of gravitational waves must consist then of a mass quadrupole oscillator, which is an extended body whose components particles are put in relative motion by the passage of the wave. Moreover, in order to detect and measure such oscillations, some mechanism converting the energy extracted by the wave into some other form of energy (acoustic vibrations, thermal energy and so on) is needed.

Suppose now that four particles are arranged in a plane perpendicular to the direction of the wave, like in Fig.8.2.

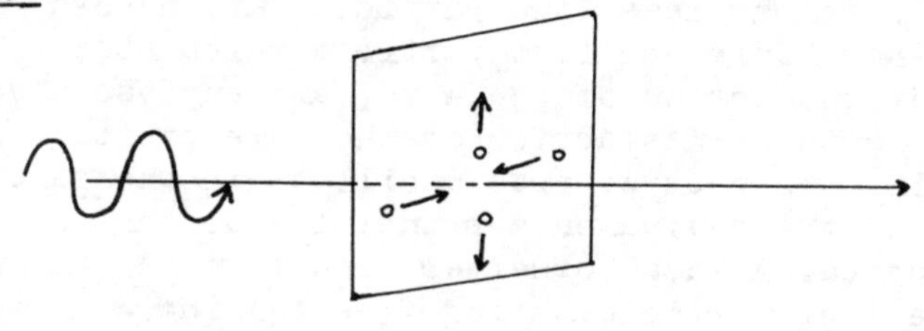

FIG. 8.2

Under the action of the wave, two of the particles will move apart, and two will get near. This is characteristic of quadrupole oscillations (see Appendix E). In the case of electromagnetic waves, for each direction of propagation there are two polarization states for a plane wave, and one may be obtained from the other by a 90 degrees rotation. In the gravitational case the two independent states of polarization are related by a 45 degrees rotation (see Fig.8.3 and also Appendix E for a discussion of the polarization).

Consider a monochromatic plane wave of frequency ω, and suppose the oscillator oriented along the x^1 axis; then, putting $K^1{}_1 = K$ and $\gamma^1{}_1 = \gamma$, and introducing the Fourier transform of $\xi^i(t)$

$$\xi^i(t) = \frac{1}{\sqrt{2\pi}} \int e^{-i\omega t} \xi^i(\omega)\, d\omega \qquad (8.98)$$

eq.(8.97) gives

$$\xi^i(\omega) = \frac{mc^2 R_{414}{}^i(\omega)\, \tau^1}{k - m\omega^2 - i\gamma\omega} \qquad (8.99)$$

This expression has a maximum of resonance at $\omega^2 m - k = 0$. The dissipation factor γ is related to the so-called quality factor Q of the system by $\gamma = \omega m/Q$, and then at the resonance we have

$$\xi^1(\omega) = i\,\frac{c^2 Q}{\omega^2}\, R_{414}{}^1(\omega)\, \tau^1 \qquad (8.100)$$

Since the enrgy flux S of a gravitational wave can be ex-

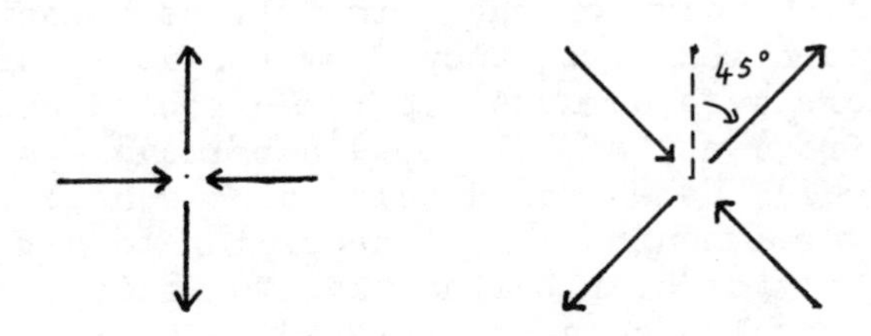

FIG. 8.3

pressed as follows[7]

$$S = \frac{c^7}{4\pi |r|^2 \omega^2 G} \left[(R_{4k4}{}^{i} z^k)^2 \right]_{Av} \qquad (8.101)$$

(where Av denotes average over all the possible orientations of the receiver) then the strain $\mathcal{E}(\omega)$, i.e. the relative displacement $|\xi(\omega)|/r$, can be related to the energy flux of the incident wave as

$$\mathcal{E}(\omega) \propto \frac{c^2 Q^2}{\omega^2} \left(\frac{4\pi^2}{c^7} \omega^2 G S \right)^{1/2} \qquad (8.102)$$

(for a more detailed discussion od this point., and of other aspects of gravitational detectors, the interested reader is referred to Weber[7]).

The first laboratory detector of gravitational radiation was built up by Weber at the University of Maryland. It was an aluminum cylinder (weight 1.5 tons) more than half a meter in diameter and about 1.5 meters high; it was suspended in a large vacuum chamber by a wire wropped around it at the center, and around the wire the cylinder was sheated in a piezoelectric quartz envelope connected to a sensitive voltmeter. Aluminum was chosen for the antenna because of its high Q-value, and the first longitudinal mode of vibration had a frequency $\omega \sim 10^4$ Hz.

The sensitivity of the mass detector was such as to enable measurements of displacements of the cylinder ends, that is expansions and contractions due to quadrupole mode excited by the incident gravitational radiation, of the order of 10^{-14} cm . In a successive refinement Weber developed a system which works on the principle of coincidence of signals of identical frequency from two detectors.

The detectors were adjusted to the expected gravitational radiation frequency of collapsed supernovas in our galaxy ($\sim$ 1660 Hz). Weber found some coincidences during a period of several months which, according to him, preclude the possibility of attributing them to random coincidences, and these results were interpreted as a proof of the existence of powerful gravitational waves in the galaxy. However until now no confirmation of this experiment has been found by other similar apparatus used by many observers in the world, and moreover the Weber interpretation is far from being indisputable and leads to contradictions with other astrophysical observations.

In conclusion, to discuss the sensibility of the detec

tor, we may note that the effects produced by the waves are proportional to the mass of the detector (see 8.99). The largest mass available is the earth itself. In fact, in spite of its shape (approximately spherical) some vibrational modes have a mass quadrupole moment which can couple to the gravitational gradients having Fourier components at the natural frequency of that mode.

This possibility seems to be attractive, since the quadrupole moment of the earth is many orders higher than the ones of the laboratory detectors, and the frequency of earth's oscillations ($\sim 10^{-3}$ Hz)˙ should make possible to record gravitational waves coming from pulsars (rapidly spinning neutron stars, see Chapter IX, Sect.2).

However this method is limited by the high noise temperature of the terrestrial core which, as pointed out by Weber, would require measurements of effective displacements $\sim 10^{-17}$ cm.

A more effective approach may be the use of individual seismically insulated inhomogeneities of the earth surface, capable of absorbing gravitational radiation in the frequency band around 1 Hz. However, according to Dyson[8], which proposed this investigation, the intensity of seismical signals due to gravitational radiation from a theoretical model of pulsar at 1 Hz frequency is five orders lower than the noise level.

The possibility of seismic recording of gravitational radiation from pulsars cannot be regarded as definitively closed however, as pointed out by De Sabbata[8], who suggested the use of local inhomogeneities of the Moon surface to detect gravitational waves of the same frequqncy.

6.- Interactions between gravitational waves and electromagnetic fields

The interaction between gravitational and electromagnetic fields suggests (at least in principle) a possible mechanism to generate high frequency gravitational radiation in a terrestrial laboratory, and to detect the gravitational waves through the elctromagnetic radiation generated by them[9].

In fact an electromagnetic (e.m.) wave travelling through a static e.m. field produces a gravitational wave of the same frequency which, propagating through another static e.m. field, creates an e.m.wave.

The ratio of the intensity of the produced e.m. wave to that of the incident wave shows that in the production (or detection) of gravitational radiation by a static e.m. field, the efficiency of the process depends on the square of the e.m. field and of its linear dimensions.

Consider a static e.m. field which is nonvanishing in the region between the two planes $x = -\ell$ and $x = 0$, and constant inside them (see Fig.8.4)

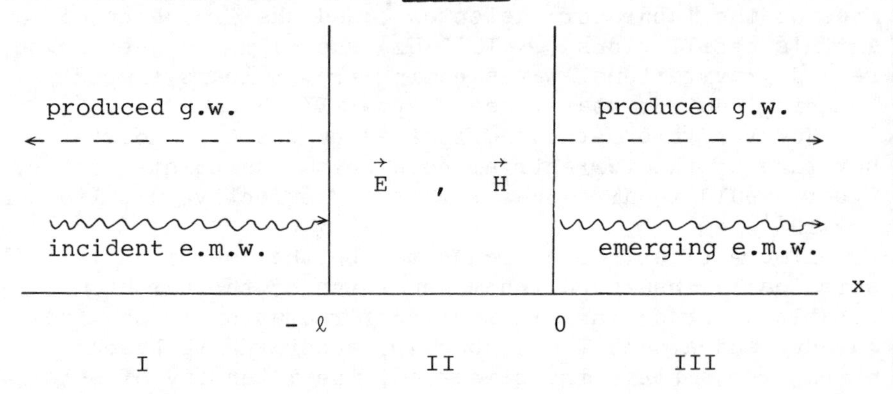

Fig. 8.4

A monochromatic plane e.m. wave, coming from $-\infty$, propagates along the axis $x^1 = x$. The Einstein equations, in the weak field approximation, are

$$\Box \psi_\mu{}^\nu = -\frac{16\pi}{c^4} G T_\mu{}^\nu \tag{8.103}$$

and since the gravitational sources are in this case the e.m. fields, then (remember 3.97)

$$T^\mu{}_\nu = \frac{1}{4\pi}\left(F^{\mu\alpha} F_{\nu\alpha} - \frac{1}{4}\delta^\mu{}_\nu F^{\alpha\beta} F_{\alpha\beta}\right) \tag{8.104}$$

$F_{\mu\nu}$ contains both the static part of the field and the field of the electromagnetic wave, whose nonvanishing components we assume to be

$$\begin{aligned} H_z(\text{WAVE}) &= a \exp\left[ik(x-ct)\right] \\ E_y(\text{WAVE}) &= a \exp\left[ik(x-ct)\right] \end{aligned} \tag{8.105}$$

(it describes a linearly polarized plane wave with frequency $\omega = ck$). Denoting with $\vec{E}$, $\vec{H}$, the constant e.m. fields of the region II, we have then in this region

$$F^{12} = H_z + a \exp[ik(x-ct)] = -F^{21}$$

$$F^{13} = -H_y = -F^{31}$$

$$F^{23} = H_x = -F^{32}$$

$$F^{14} = -E_x = -F^{41}$$

$$F^{24} = -E_y - a \exp[ik(x-ct)] = -F^{42}$$

$$F^{34} = -E_z = -F^{43} \tag{8.106}$$

An explicit computation of $\tau_{\mu\nu}$, using (8.104,106), in the three regions, shows that this tensor contains three types of terms: there are terms quadratic in H and E, terms quadratic in the field of the wave, and mixed terms: only the last ones contribute to the production of an oscillating gravitational field.

The mixed term have the following form (considering for example the $\tau_1{}^1$ component)

$$4\pi \tau_1{}^1 = a(H_z + E_y) \exp[ik(x-ct)] \tag{8.107}$$

and it is easy to verify that the condition $\partial_\nu \tau^{\mu\nu} = 0$ is always satisfied.

The field equations (8.103) become then

$$\Box \psi_\mu{}^\nu = 0$$

in the regions I, III, and

$$\Box \psi_\mu{}^\nu = \lambda_\mu{}^\nu \exp[ik(x-ct)] \tag{8.108}$$

in the region II, where $\lambda_\mu{}^\nu$ denote the constant coefficients of thecomponents of $\tau_\mu{}^\nu$, for example $\lambda_1{}^1 = a(H_z+E_y)$ (see 8.107).

Looking for solutions representing outgoing waves in regions I and III, one can write, in general[9) ,

$$\psi_{I\mu}{}^\nu = \lambda_\mu{}^\nu C(\mu,\nu) \exp[-ik(x+ct)]$$

$$\psi_{II\mu}{}^{\nu} = \lambda_\mu{}^\nu \left\{ \frac{x+\ell}{2ik} \exp[ik(x-ct)] + A(\mu,\nu)\exp[ik(x-ct)] + B(\mu,\nu)\exp[-ik(x+ct)] \right\}$$

$$\psi_{III\mu}{}^{\nu} = \lambda_\mu{}^\nu D(\mu,\nu) \exp[ik(x-ct)] \quad (8.109)$$

where A,B,C,D are arbitrary constants which, for any given value of μ and ν, can be determined imposing the continuity of ψ and $\partial\psi/\partial x$ at $x = 0$ and $x = -\ell$. We find, in the region III,

$$\psi_{III\mu}{}^{\nu} = \frac{2iG\ell a}{kc^4} \alpha_\mu{}^\nu \exp[ik(x-ct)] \quad (8.110)$$

where $\alpha_\mu{}^\nu$ are coefficients depending on the components of the static e.m. field.

The energy flux of the gravitational wave, according to (8.55), in the region III is ($\psi_{\mu\nu} = h_{\mu\nu}$)

$$t^{41} = \frac{c^2}{16\pi G}\left[\dot{\psi}_{23}^2 + \frac{1}{4}(\dot{\psi}_{22} - \dot{\psi}_{33})^2\right] = \frac{G}{4\pi c^4}\ell^2 a^2 \left[(H_y + E_z)^2 + (E_y + H_z)^2\right] \quad (8.111)$$

The energy flux is nonvanishing only if the constant e.m. field is not aligned along the direction of propagation of the e.m. wave (the x axis in this case).

It should be noted that the quadratic dependence on ℓ follows from the fact that the velocity of propagation of e.m. and gravitational waves is the same, however the presence of a medium with a refraction index different from unity can modify, in general, this dependence.

Following the same procedure, one can calculate the e.m. waves produced by a gravitational wave travelling through a static e.m. field (see Fig.8.5)

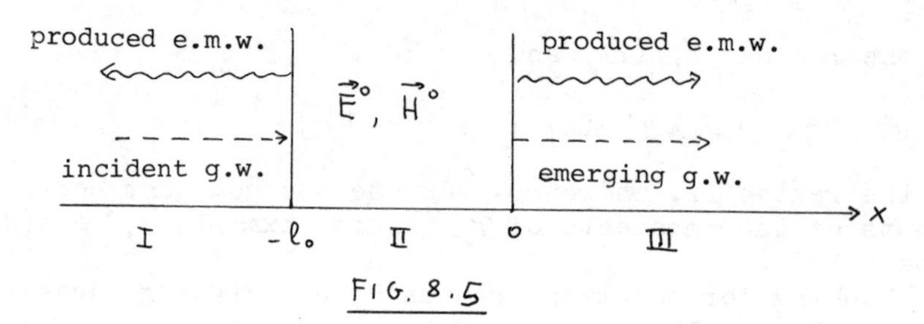

FIG. 8.5

In the region II, between 0 and $-\ell_o$, we have again a constant e.m. field $\vec{E}_o$, $\vec{H}_o$, however in the other regions the role of the e.m. and gravitational fields are exchanged, and we must solve, instead of the Einstein equations with e.m. sources, the Maxwell equations in a curved space-time.

An explicit computation[9] gives the following results in the region III :

$$E_y = -\frac{A\ell_o}{2ik}\exp[ik(x-ct)] = H_z$$
$$E_z = -\frac{B\ell_o}{2ik}\exp[ik(x-ct)] = H_y \qquad (8.112)$$

where

$$A = -k^2\left[(E_y^o - H_z^o)\alpha_{22} + (E_z^o + H_y^o)\alpha_{23}\right]$$
$$B = -k^2\left[(E_y^o - H_z^o)\alpha_{23} + (E_z^o + H_y^o)\alpha_{33}\right] \qquad (8.113)$$

and $\alpha_{\mu\nu}$ are the coefficients appearing in eq.(8.110). If A = B = 0 we obtain an e.m. wave propagating backward in the negative x direction.

Combining these two examples, i.e. considering an e.m. wave which produces a gravitational wave which is reconverted into an e.m. wave (see Fig.8.6, where G is the generator and D the detector of lineraly polarized e.m.waves, and the central screen S is opaque to e.m. waves), the computation of the enrgy density W_f of the emerging e.m. wave (being $W_i = a^2/4\pi$ the energy density of the incident wave) gives

$$W_f = \frac{a^2}{4\pi}\frac{G^2\ell^2\ell_o^2}{c^8}\left\{\left[(E_y^o - H_z^o)\alpha_{22} + (E_z^o + H_y^o)\alpha_{23}\right]^2 + \left[(E_y^o - H_z^o)\alpha_{23} + (E_z^o + H_y^o)\alpha_{33}\right]^2\right\} \qquad (8.114)$$

where the α coefficients depend on the static fields E and H .

If only static magnetic fields are present, in particular, we obtain

$$W_f = W_i\frac{G^2\ell^2\ell_o^2}{c^8}\left\{(H_z H_z^o + H_y H_y^o)^2 + (H_z^o H_y - H_y^o H_z)^2\right\} \qquad (8.115)$$

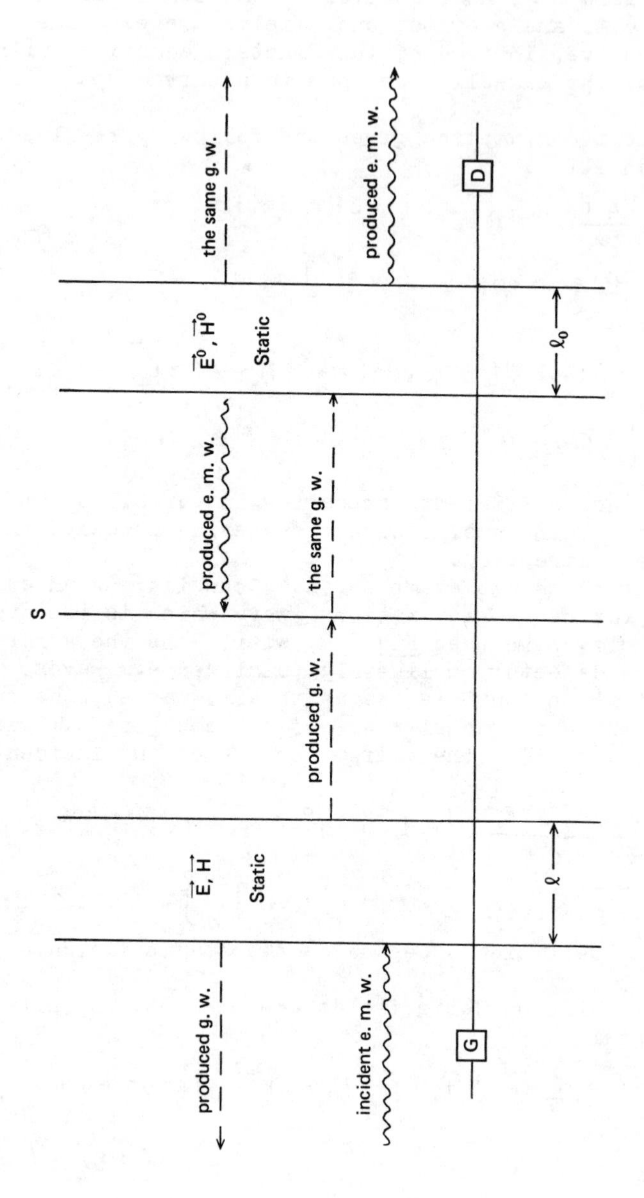
produced g. w.
incident e. m. w.
G
$\vec{E}, \vec{H}$
Static
ℓ
produced g. w.
S
the same g. w.
produced e. m. w.
$\vec{E}^0, \vec{H}^0$
Static
ℓ_0
the same g. w.
produced e. m. w.
D

FIG. 8.6

The energy density, very small because of the factor G^2/c^8, increases with ℓ^4, and this may be interesting in some astrophysical situations. It should be noted also that the produced power does not depend on the frequency of the e.m. wave, and then a suitable choice of the frequency, relative to the characteristic parameters of the detector, is allowed. The possibility of strong magnetic fields over long distances is however outside the capabilities of the present day technology for a laboratory experiment.

A more realistic process is the emission of e.m. waves when a gravitational wave is incident on a strong magnetic field of dipolar type, like the one of a neutron star. The mechanism of emission is the same as in the previous cases: solving the Maxwell equations, written in the space-time curved because of the presence of the gravitational field of the wave, one obtains that the energy density of the outgoing e.m. radiation is proportional to the energy flux S of the gravitational wave, and to the square of the frequency ω and of the magnetic dipole moment μ of the celestial body:

$$W_{e.m.} \simeq 10^{-70}\, \omega^2 \mu^2 S \qquad (8.116)$$

In the case of the magnetic field of the earth, assuming as incident flux the one recorded by Weber ($\sim 10^6 \div 10^7$ erg/cm^2 sec), we obtain an emission of e.m. waves with power $\sim 10^{-4} \div 10^{-3}$ erg/sec per pulse (at a frequency $\sim 10^3$ Hz).

APPENDIX A

Infinitesimal coordinate transformations

Consider the infinitesimal transformation of coordinates

$$x'^{\mu} = x^{\mu} + \delta x^{\mu} \tag{A.1}$$

where $\delta x^{\mu} = \xi^{\mu}(x)$ are four infinitesimal local parameters. Under this transformation, the infinitesimal variation of a second rank tensor has been given at the end of Section 2, Chapter II: for the metric tensor we have, in particular,

$$\delta g^{\mu\nu} = -\xi^{\alpha}\partial_{\alpha} g^{\mu\nu} + g^{\mu\alpha}\partial_{\alpha}\xi^{\nu} + g^{\alpha\nu}\partial_{\alpha}\xi^{\mu} \tag{A.2}$$

and

$$\delta g_{\mu\nu} = -\xi^{\alpha}\partial_{\alpha} g_{\mu\nu} - g_{\alpha\nu}\partial_{\mu}\xi^{\alpha} - g_{\mu\alpha}\partial_{\nu}\xi^{\alpha} \tag{A.3}$$

Noting that

$$\xi^{\mu;\nu} + \xi^{\nu;\mu} = g^{\alpha\nu}\xi^{\mu}{}_{;\alpha} + g^{\mu\alpha}\xi^{\nu}{}_{;\alpha} =$$

$$= g^{\alpha\nu}\partial_{\alpha}\xi^{\mu} + g^{\alpha\nu}\Gamma_{\alpha\beta}{}^{\mu}\xi^{\beta} + g^{\mu\alpha}\partial_{\alpha}\xi^{\nu} + g^{\mu\alpha}\Gamma_{\alpha\beta}{}^{\nu}\xi^{\beta} \tag{A.4}$$

and using the metricity condition (2.56)

$$\partial_{\alpha} g^{\mu\nu} + \Gamma_{\alpha\beta}{}^{\mu} g^{\beta\nu} + \Gamma_{\alpha\beta}{}^{\nu} g^{\mu\beta} = 0 \tag{A.5}$$

we obtain

$$\xi^{\mu;\nu} + \xi^{\nu;\mu} = g^{\alpha\nu}\partial_{\alpha}\xi^{\mu} + g^{\mu\alpha}\partial_{\alpha}\xi^{\nu} - \xi^{\beta}\partial_{\beta} g^{\mu\nu} \tag{A.6}$$

and, with the same procedure,

$$\xi_{\mu;\nu} + \xi_{\nu;\mu} = g_{\alpha\nu}\partial_{\mu}\xi^{\alpha} + g_{\mu\alpha}\partial_{\nu}\xi^{\alpha} + \xi^{\beta}\partial_{\beta} g_{\mu\nu} \tag{A.7}$$

The infinitesimal variations (A.2,3) can also be written then

$$\delta g^{\mu\nu} = \xi^{\mu;\nu} + \xi^{\nu;\mu} \tag{A.8a}$$

$$\delta g_{\mu\nu} = -\xi_{\mu;\nu} - \xi_{\nu;\mu} \tag{A.8b}$$

In the weak field approximation these equations can be linearized as follows

$$\delta g^{\mu\nu} = \xi^{\mu,\nu} + \xi^{\nu,\mu}$$
$$\delta g_{\mu\nu} = -\xi_{\mu,\nu} - \xi_{\nu,\mu} \tag{A.9}$$

and since $g_{\mu\nu} = \eta_{\mu\nu} + h_{\mu\nu}$ we have, in particular,

$$h'_{\mu\nu} = h_{\mu\nu} + \delta h_{\mu\nu} =$$
$$= h_{\mu\nu} - \xi_{\mu,\nu} - \xi_{\nu,\mu} =$$
$$= h_{\mu\nu} - \eta_{\alpha\mu}\xi^{\alpha}{}_{,\nu} - \eta_{\alpha\nu}\xi^{\alpha}{}_{,\mu} \tag{A.10}$$

Defining

$$\psi_{\mu}{}^{\nu} = h_{\mu}{}^{\nu} - \frac{1}{2}\delta_{\mu}{}^{\nu} h \tag{A.11}$$

we have then

$$\psi'_{\mu}{}^{\nu} = h'_{\mu}{}^{\nu} - \frac{1}{2}\delta_{\mu}{}^{\nu} h' =$$
$$= h_{\mu}{}^{\nu} - \eta_{\alpha\mu}\xi^{\alpha,\nu} - \delta_{\alpha}{}^{\nu}\xi^{\alpha}{}_{,\mu} -$$
$$- \frac{1}{2}\delta_{\mu}{}^{\nu}\left[\eta^{\sigma\lambda}(h_{\sigma\lambda} - \eta_{\alpha\lambda}\xi^{\alpha}{}_{,\sigma} - \eta_{\alpha\sigma}\xi^{\alpha}{}_{,\lambda})\right] =$$
$$= \psi_{\mu}{}^{\nu} - \eta_{\alpha\mu}\xi^{\alpha,\nu} - \delta_{\alpha}{}^{\nu}\xi^{\alpha}{}_{,\mu} + \delta_{\mu}{}^{\nu}\xi^{\sigma}{}_{,\sigma} =$$
$$= \psi_{\mu}{}^{\nu} - \eta^{\beta\nu}\xi_{\mu,\beta} - \xi^{\nu}{}_{,\mu} + \delta_{\mu}{}^{\nu}\xi^{\sigma}{}_{,\sigma} \tag{A.12}$$

By taking the divergence of this equation it follows that

$$\psi'_{\mu}{}^{\nu}{}_{,\nu} = \psi_{\mu}{}^{\nu}{}_{,\nu} - \eta^{\beta\nu}\xi_{\mu,\beta\nu} - \xi^{\nu}{}_{,\mu\nu} + \xi^{\sigma}{}_{,\sigma\mu} =$$
$$= \psi_{\mu}{}^{\nu}{}_{,\nu} - \eta^{\beta\nu}\xi_{\mu,\beta\nu} \tag{A.13}$$

Therefore, if in a given system of coordinates $\psi_{\mu}{}^{\nu}{}_{,\nu} \neq 0$, we can always perform a coordinate transformation generated by four infinitesimal functions ξ^{μ} which satisfy

$$\eta^{\beta\nu}\xi_{,\beta\nu} = \square\xi_{\mu} = \psi_{\mu}{}^{\nu}{}_{,\nu} \qquad (A.14)$$

and we obtain, in the new system, the condition (8.13)

$$\psi'_{\mu}{}^{\nu}{}_{,\nu} = 0 \qquad (A.15)$$

which simplifies the expression (8.91) for the Ricci tensor.

Moreover, the condition (8.13) does not determine univocally the choice of the coordinate system: in fact, if $\psi_{\mu}{}^{\nu}{}_{,\nu} = 0$ is satisfied in our system, we can always perform another coordinate transformation, whose generators ξ_{μ} satisfies the condition

$$\square\xi_{\mu} = 0 \qquad (A.16)$$

so that from (A.13) we obtain, also in the new system, $\psi'_{\mu}{}^{\nu}{}_{,\nu} = 0$.

To conclude this appendix, we give a table summarizing some formal analogies which can be established between the electromagnetic theory and general relativity in the weak field approximation. It must be emphasized that these analogies can be used as a mnemonic device, but cannot be pushed too far without avoiding erroneous deductions.

In fact one should remember always that electromagnetism is a vector-like interaction (corresponding to a spin one field, the photon)while gravity is tensor-like (the graviton is a spin-two field); moreover, there are two types of electric charges (positive and negative), but only one type of gravitational mass, and two charges of the same type repel each other, while two masses attract each other. With these fundamental differences clear in mind, we can consider the following analogies:

GRAVITATION	ELECTROMAGNETISM
Linearized Einstein eqs.	Maxwell eq. for the potential
$\Box \psi_\mu{}^\nu = -\frac{16\pi G}{c^4} T_\mu{}^\nu$	$\Box A_\mu = \frac{4\pi}{c} J_\mu$
armonic gauge	Lorentz condition
$\partial_\nu \psi_\mu{}^\nu = 0$	$\partial_\mu A^\mu = 0$
Infinitesimal coord. tran.	Infinitesimal gauge trans.
$\delta x_\mu = \xi_\mu(x)$	$\delta A_\mu = \partial_\mu \Lambda(x)$
leads to	leads to
$\partial_\nu \psi'_\mu{}^\nu = 0$	$\partial_\mu A'^\mu = 0$
if	if
$\partial_\nu \psi_\mu{}^\nu = \Box \xi_\mu$	$\partial_\mu A^\mu = \Box \Lambda$
Static grav. potential	Static scalar potential
$\phi = \frac{c^2}{4} \psi_{44} = -\frac{1}{4\pi\gamma} \int d^3x \frac{\rho}{r}$	$\varphi = \frac{1}{4\pi\varepsilon} \int d^3x \frac{\rho_e}{r}$
$\gamma = (4\pi G)^{-1}$ is a sort of "gravitational capacity" of the vacuum.	ρ_e is the charge density, ε is the dielectric const.
"Grav. vector potential"	Electr. vector potential
$\vec{K} = -\frac{\eta}{4\pi} \int d^3x \frac{\rho \vec{v}}{r}$	$\vec{A} = \frac{\mu}{4\pi c} \int d^3x \frac{\rho_e \vec{v}}{r}$
$K_i = \psi_{4i}$ and $\eta = 16\pi G/c^3$ is a sort of "grav. permeability"	μ is the magnetic permeability

The expressions for ϕ and $\vec{K}$ have been obtained from (8.19) putting $T_{44} = \rho c^2$ and $T_{4i} = \rho c v_i$ (to first order in v/c).

It should be noted that a qualitative analogy holds also in the radiation case: for example eq.(8.61) is similar to the expression for the quadrupole radiated power in the electromagnetic theory,

$$\frac{dE}{dt} = \frac{1}{180c^5} \left(\dddot{D}_{ik} \dddot{D}^{ik} \right)$$

and in many cases one can obtain the gravitational analogous of an electromagnetic expression (modulo numerical factors) with the simple replacement $\rho_e \longrightarrow \sqrt{G}\,\rho$.

Finally we observe, comparing the expression (8.97) for the gravitational force $F_g \sim mRr$, with the correspondin expression for an electric type force, $F_e \sim q\,\nabla Er$, that the Riemann tensor can be interpreted as the equivalent of the gradient of the field intensity.

APPENDIX B

The energy-momentum pseudotensor

In the absence of the gravitational interaction, the energy-momentum conservation law is expressed by the special relativistic equation

$$T_\mu{}^\nu{}_{,\nu} = 0 \qquad (B.1)$$

which implies the conservation of the four momentum vector

$$P_\mu = \frac{1}{c}\int T_\mu{}^\nu\, dS_\nu = \frac{1}{c}\int T_\mu{}^4\, d^3x \qquad (B.2)$$

(dS_ν denotes the integration over a three-dimensional space-like hypersurface). In fact, choosing two different hypersurfaces, Σ_1 and Σ_2 , at two different values of time t_1 and t_2 , we have (see also Chapter III, Appendix B)

$$c\,\Delta P_\mu = \int_{\Sigma_2} T_\mu{}^\nu dS_\nu - \int_{\Sigma_1} T_\mu{}^\nu dS_\nu = \oint_\Sigma T_\mu{}^\nu dS_\nu \qquad (B.3)$$

where Σ is a closed hypersurface obtained connecting Σ_1 and Σ_2 with a time-like three-dimensional hypersurface at spatial infinity, where $T_{\mu\nu}$ is vanishing . Using the Gauss theorem, the flux integral (B.3) becomes the integral of a divergence over the four volume Ω delimited by Σ , givin

$$c\,\Delta P_\mu = \int_\Omega d^4x\, \partial_\nu T_\mu{}^\nu = 0 \qquad (B.4)$$

so that the definition of P_μ is independent of the choice of the hypersurface.

In the presence of gravity, the energy-momentum tensor of the matter sources is covariantly conserved, i.e. it satisfies the generalized conservation law

$$T_\mu{}^\nu{}_{;\nu} = 0 \qquad (B.5)$$

which follows, as shown in Sect.3 of Chapter III, both from the invariance of the matter Lagrangian under local infinitesimal translations, $\delta x^\mu = \xi^\mu(x)$, and also from the Einstein field equations

$$G_\mu{}^\nu = \chi T_\mu{}^\nu \qquad (B.6)$$

because of the contracted Bianchi identity

$$G_\mu{}^\nu{}_{;\nu} = 0 \qquad (B.7)$$

where $G_\mu{}^\nu = R_\mu{}^\nu - \frac{1}{2}\delta_\mu{}^\nu R$ is the Einstein tensor. The four momentum vector (B.2) of the matter fields is no longer conserved in this case. In fact (B.5) implies

$$T_\mu{}^\nu{}_{;\nu} = \partial_\nu T_\mu{}^\nu + \Gamma_{\nu\alpha}{}^\nu T_\mu{}^\alpha - \Gamma_{\nu\mu}{}^\alpha T_\alpha{}^\nu = 0 \qquad (B.8)$$

and since (see 3.29 and 2.59)

$$\Gamma_{\alpha\nu}{}^\nu = \partial_\alpha (\ln\sqrt{-g}) = \frac{1}{\sqrt{-g}}\partial_\alpha(\sqrt{-g}) \qquad (B.9)$$

$$\Gamma_{\nu\mu}{}^\alpha T_\alpha{}^\nu = \Gamma_{\nu\mu\alpha} T^{\alpha\nu} = \frac{1}{2}\partial_\mu g_{\nu\alpha} T^{\alpha\nu} \qquad (B.10)$$

the generalized conservation law (13.5) can also be written as follows

$$T_\mu{}^\nu{}_{;\nu} = \frac{1}{\sqrt{-g}}\partial_\nu(\sqrt{-g}\,T_\mu{}^\nu) - \frac{1}{2}(\partial_\mu g_{\nu\alpha})T^{\alpha\nu} = 0 \qquad (B.11)$$

or

$$\partial_\nu(\sqrt{-g}\,T_\mu{}^\nu) = \frac{1}{2}\sqrt{-g}\,T^{\alpha\nu}\partial_\mu g_{\nu\alpha} \qquad (B.12)$$

It follows that

$$c\Delta P_\mu = \int_{\Sigma_2}\sqrt{-g}\,T_\mu{}^\nu dS_\nu - \int_{\Sigma_1}\sqrt{-g}\,T_\mu{}^\nu dS_\nu =$$

$$= \oint_{\Sigma} \sqrt{-g}\, T_{\mu}{}^{\nu} dS_{\nu} = \int_{\Omega} \partial_{\nu} (\sqrt{-g}\, T_{\mu}{}^{\nu})\, d^4x \neq 0 \qquad (B.13)$$

because of the additional terms appearing in eq.(B.12), so that it is not possible to obtain the conservation of $T_{\mu\nu}$ in general.

This result can be physically interpreted arguing that, in the presence of gravity, it is not the current of matter alone which must be conserved, but rather the total energy, including the contributions of gravity. And since in the tensor $T_{\mu\nu}$ are included all the fields which contribute to the energy content of the space-time, i.e. all the sources of curvature, the gravitational field excepted, we are led to introduce another object, $t_{\mu}{}^{\nu}$, which should describe the gravitational energy-momentum distribution, allowing then the definition of a conserved total four momentum vector.

To this aim, we must rewrite the right-hand side of (B.12) as an ordinary divergence; putting

$$(\sqrt{-g}\, t_{\mu}{}^{\nu})_{,\nu} = -\frac{1}{2} \sqrt{-g}\, T^{\alpha\nu} \partial_{\mu} g_{\alpha\nu} \qquad (B.14)$$

the condition (B.12) becomes

$$\partial_{\nu} \left[\sqrt{-g} (T_{\mu}{}^{\nu} + t_{\mu}{}^{\nu}) \right] = 0 \qquad (B.15)$$

We can introduce then the four vector

$$P_{\mu} = \frac{1}{c} \int \sqrt{-g}\, (T_{\mu}{}^{\nu} + t_{\mu}{}^{\nu})\, dS_{\nu} \qquad (B.16)$$

which is conserved,

$$c\, \Delta P_{\mu} = \oint_{\Sigma} \sqrt{-g}\, (T_{\mu}{}^{\nu} + t_{\mu}{}^{\nu})\, dS_{\nu} =$$

$$= \int_{\Omega} d^4x\, \partial_{\nu} \left[\sqrt{-g} (T_{\mu}{}^{\nu} + t_{\mu}{}^{\nu}) \right] = 0 \qquad (B.17)$$

and may be interpreted therefore as the total energy-momentum vector of matter plus gravity.

It is important to note, from the definition (B.14), that $t_{\mu\nu}$ is not a tensor, and for this reason it is called the energy-momentum "pseudo-tensor" of the gravitational field. Owing to the non-tensorial character of $t_{\mu\nu}$, it is always possible to find a system of coordinates in which all the components of $t_{\mu\nu}$ are locally vanishing, without entailing $t_\mu{}^\nu = 0$ everywhere, and this property of $t_\mu{}^\nu$ is in agreement with the principle of equivalence, as the energy of the gravitational field cannot be localized (in general relativity it is a nonsense to speak of "localization" of the gravitational energy).

The energy-momentum tensor $T_\mu{}^\nu$ of any other field determines univocally its energy density in a given element of volume; moreover it has "weight" and curves space-time (i.e. it appears as a source in the right-hand side of Einstein's equations (B.6)), and produces the relative geodesic deviation (see Appendix D) of two world lines passing through the considered portion of space.

None of these properties characterizes the energy-momentum gravitational pseudotensor $t_\mu{}^\nu$: it is not univocally defined, it is not the source of curvature according to the Einstein field equations and it does not produce any geodesic deviation. According to the equivalence principle, one can always find a frame of reference in which the gravitational field locally disappears, i.e. all the Christoffel symbols $\Gamma_{\mu\nu}{}^\alpha$ are vanishing. In the local absence of gravitational field, also the local gravitational energy-momentum is vanishing: the principle of equivalence , therefore, does not forbid the existence of an energy density associated to the gravitational field, but only its localization at a given point of space and time.

The explicit expression for the gravitational pseudotensor $t_\mu{}^\nu$ can be obtained directly from the definition (B.14), which in the weak field approximation becomes

$$(t_\mu{}^\nu \sqrt{-g})_{,\nu} = -\frac{1}{2}\sqrt{-g}\, g_{\alpha\beta,\mu} T^{\alpha\beta} \simeq -\frac{1}{2} h_{\alpha\beta,\mu} T^{\alpha\beta} \qquad (B.18)$$

From the linearized field equations (8.18)

$$\Box \psi_{\alpha\beta} = \Box \left(h_{\alpha\beta} - \frac{1}{2}\eta_{\alpha\beta} h \right) = -2\chi T_{\alpha\beta} \qquad (B.19)$$

(where $\chi = 8\pi G/c^4$) we have

$$-\frac{1}{2} h_{\alpha\beta,\mu} T^{\alpha\beta} = \frac{1}{4\chi} h^{\alpha\beta}{}_{,\mu} \left(h_{\alpha\beta,\nu}{}^{,\nu} - \frac{1}{2} h_{,\nu}{}^{,\nu} \eta_{\alpha\beta} \right) \qquad (B.20)$$

Therefore

$$(\sqrt{-g}\, t_\mu{}^\nu)_{,\nu} = \frac{1}{4\chi} \left[h^{\alpha\beta}{}_{,\mu} \left(h_{\alpha\beta,\nu}{}^{,\nu} - \frac{1}{2}\eta_{\alpha\beta} h_{,\nu}{}^{,\nu} \right) \right] =$$

$$= \frac{1}{4\chi}\left[h^{\alpha\beta}{}_{,\mu}\, h_{\alpha\beta,\nu}{}^{,\nu} - \frac{1}{2} h_{,\mu}\, h_{,\nu}{}^{,\nu}\right] =$$

$$= \frac{1}{4\chi}\left[\left(h^{\alpha\beta}{}_{,\mu}\, h_{\alpha\beta}{}^{,\nu}\right)_{,\nu} - h^{\alpha\beta}{}_{,\mu\nu}\, h_{\alpha\beta}{}^{,\nu} - \right.$$

$$\left. - \frac{1}{2}\left(h_{,\mu} h^{,\nu}\right)_{,\nu} + \frac{1}{2} h_{,\mu\nu}\, h^{,\nu}\right] \qquad (B.21)$$

Noting that

$$- h^{\alpha\beta}{}_{,\mu\nu}\, h_{\alpha\beta}{}^{,\nu} + \frac{1}{2} h_{,\mu\nu}\, h^{,\nu} =$$

$$= - h^{\alpha\beta}{}_{,\nu\rho}\, h_{\alpha\beta,}{}^{\rho}\, \delta^{\nu}_{\mu} + \frac{1}{2}\delta_{\mu}{}^{\nu} h_{,\nu\rho}\, h^{,\rho} =$$

$$= -\frac{1}{2}\delta_{\mu}{}^{\nu}\left(h^{\alpha\beta}{}_{,\rho}\, h_{\alpha\beta}{}^{,\rho}\right)_{,\nu} + \frac{1}{4}\delta_{\mu}{}^{\nu}\left(h_{,\rho} h^{,\rho}\right)_{,\nu} \qquad (B.22)$$

eq.(B.21) becomes

$$\left(\sqrt{-g}\, t_{\mu}{}^{\nu}\right)_{,\nu} = \frac{1}{4\chi}\left(h^{\alpha\beta}{}_{,\mu}\, h_{\alpha\beta}{}^{,\nu} - \frac{1}{2} h_{,\mu}\, h^{,\nu} - \right.$$

$$\left. - \frac{1}{2}\delta^{\nu}_{\mu}\, h^{\alpha\beta}{}_{,\rho}\, h_{\alpha\beta}{}^{,\rho} + \frac{1}{4}\delta_{\mu}{}^{\nu} h_{,\rho}\, h^{,\rho}\right)_{,\nu} \qquad (B.23)$$

and the explicit expression for $t_{\mu}{}^{\nu}$ is then

$$\sqrt{-g}\, t_{\mu}{}^{\nu} = \frac{1}{4\chi}\left[h^{\alpha\beta}{}_{,\mu}\, h_{\alpha\beta}{}^{,\nu} - \frac{1}{2} h_{,\mu}\, h^{,\nu} - \right.$$

$$\left. - \frac{1}{2}\delta_{\mu}{}^{\nu}\left(h^{\alpha\beta}{}_{,\rho}\, h_{\alpha\beta}{}^{,\rho} - \frac{1}{2} h_{,\rho}\, h^{,\rho}\right)\right] \qquad (B.24)$$

In the case of a plane gravitational wave propagating along the x^1 axis, the only nonvanishing components of $h_{\mu\nu}$ are $h_{22} = - h_{33}$ and $h_{23} = h_{32}$, which are functions of x^1 and $x^4 = ct$ only (see the Sect.2 of this Chapter). The energy flux density in the x^1 direction is given then by ($t_4{}^2$ and $t_4{}^3$ are vanishing)

$$c\sqrt{-g}\, t_4{}^1 = \frac{c}{4\chi}\left[h^{22}{}_{,4}\, h_{22}{}^{,1} + h^{33}{}_{,4}\, h_{33}{}^{,1} + 2 h^{23}{}_{,4}\, h_{23}{}^{,1}\right] \qquad (B.25)$$

(remember that $h = h_{\mu}{}^{\mu} = h_{22} + h_{33} = 0$), that is

$$c\sqrt{-g}\, t_4{}^1 = -\frac{1}{4\chi}\left[\frac{\partial h_{22}}{\partial t}\frac{\partial h_{22}}{\partial x} + \frac{\partial h_{33}}{\partial t}\frac{\partial h_{33}}{\partial x} + \right.$$

$$\left. + 2\frac{\partial h_{23}}{\partial t}\frac{\partial h_{23}}{\partial x}\right] \qquad (B.26)$$

(the negative sign comes from our metric conventions, as $\partial h_{ik}/\partial x = \partial_1 h_{ik} = \eta_{1\mu} h_{ik}{}^{,\mu} = \eta_{11}\, h_{ik}{}^{,1} = -h_{ik}{}^{,1}$).

For a plane wave, h_{ik} is a function of (x - ct), we have

$$\frac{\partial h_{ik}}{\partial x} = -\frac{1}{c}\frac{\partial h_{ik}}{\partial t} \qquad (B.27)$$

and then, denoting with a dot the derivative with respect to t , (B.26) becomes

$$c\sqrt{-g}\, t_4{}^1 = \frac{1}{4\chi c}\left[\dot{h}_{22}^2 + \dot{h}_{33}^2 + 2\dot{h}_{23}^2\right] \qquad (B.28)$$

Finally, as $\dot{h}_{22} = -\dot{h}_{33}$, we have

$$(\dot{h}_{22} - \dot{h}_{33})^2 = \dot{h}_{22}^2 + \dot{h}_{33}^2 - 2\dot{h}_{22}\dot{h}_{33} =$$

$$= 4\dot{h}_{22}^2 = 2\dot{h}_{22}^2 + 2\dot{h}_{33}^2 \qquad (B.29)$$

and we are led to the formula (8.55)

$$c\sqrt{-g}\, t_4{}^1 = \frac{1}{4\chi c}\left[\frac{2}{4}(\dot{h}_{22} + \dot{h}_{33})^2 + 2\dot{h}_{23}^2\right] =$$

$$= \frac{c^3}{16\pi G}\left[\dot{h}_{23}^2 + \frac{1}{4}(\dot{h}_{22} - \dot{h}_{33})^2\right] \qquad (B.30)$$

In conclusion, we wish to note that, from (B.24), we find the following expression for the energy density of the gravitational wave

$$\sqrt{-g}\, t_4{}^4 = \frac{1}{4\chi c^2}\left[\dot{h}_{22}^2 + \dot{h}_{33}^2 + 2\dot{h}_{23}^2\right] =$$

$$= \sqrt{-g}\, t_4{}^1 \qquad (B.31)$$

(see B.28). This is in agreement with the fact that a gravitational wave in vacuum travels with the speed of light, and then it can be associated to the propagation of massless particles (the gravitons), like an electromagnetic wave is associated to the propagation of photons.

In fact, the four momentum vector we can associate to the gravitational wave is

$$\Pi^\mu = \frac{1}{c}\int\sqrt{-g}\, t_\nu{}^\mu\, dS^\nu = \frac{1}{c}\int\sqrt{-g}\, t_4{}^\mu\, d^3x \qquad (B.32)$$

and its squared modulus, for a propagation along the x^1 axis, is vanishing

$$\Pi_\mu \Pi^\mu = \eta_{\mu\nu} \Pi^\mu \Pi^\nu = \Pi^4 \Pi^4 - \Pi^1 \Pi^1 =$$

$$= \frac{1}{c^2} \left(\int d^3x \sqrt{-g}\, t_4{}^4 \right)^2 - \frac{1}{c^2} \left(\int d^3x \sqrt{-g}\, t_4{}^1 \right)^2 = 0 \qquad (B.33)$$

(we have used eq.B.31). The four momentum Π^μ describes the a massless particle, since $\Pi^2 = \Pi_\mu \Pi^\mu = E^2 - |\vec{p}|^2 = m^2 = 0$

APPENDIX C

Quadrupole radiation

The density of energy flux, in the case of a plane gravitational wave propagating along the x^1 axis, is given by eq.(8.59)

$$c\sqrt{-g}\, t_4{}^1 = \frac{G}{36\pi c^5 R_0^2} \left[\dddot{D}_{23}^2 + \frac{1}{4} \left(\dddot{D}_{22} - \dddot{D}_{33} \right)^2 \right] \qquad (C.1)$$

where D_{ik} is the quadrupole momentum tensor of the source, defined in eq.(8.56), and a dot denotes derivative with respect to time.

The intensity of energy radiated in an arbitrary direction, characterized by the unit vector $\vec{n} = (n_1, n_2, n_3)$, $|\vec{n}| = 1$, is defined by

$$dI = c\sqrt{-g}\, t_4{}^k \, d\sigma_k \qquad (C.2)$$

where $d\sigma_k$ is the infinitesimal element of a bidimensional surface, perpendicular to the direction of $\vec{n}$. This intensity can be calculated, starting from D_{ik} and n_i , by constructing a scalar expression, quadratic in $\dddot{D}_{ik}$, which must reduce to the expression enclosed in squared brackets in eq.(C.1) when the considered surface is perpendicular to the x^1 axis, i.e. when $\vec{n} = (1,0,0)$.

The most general scalar object, quadratic in $\dddot{D}_{ik}$, one can write using n_i and $\dddot{D}_{ik}$, is the following

$$a \left(\dddot{D}_{ik} n_i n_k \right)^2 + b \left(\dddot{D}_{ik} \dddot{D}_{ik} \right) +$$

$$+ c\, \dddot{D}_{ji} \dddot{D}_{jk} n_i n_k \qquad (C.3)$$

(Latin indices run from 1 to 3, and scalar products are

performed with the metric δ_{ik}= diag(1,1,1) : therefore $n_i = n^i$, $D_{ik} = D^{ik}$). The constants a, b, c, are to be determined by imposing that, for $n_1 = 1$, $n_2 = 0 = n_3$, this expression reduces to the one of eq.(C.1), that is

$$a\dddot{D}_{11}^2 + b\left(\dddot{D}_{11}^2 + \dddot{D}_{22}^2 + \dddot{D}_{33}^2 + 2\dddot{D}_{12}^2 + 2\dddot{D}_{13}^2 + 2\dddot{D}_{23}^2\right) + c\left(\dddot{D}_{11}^2 + \dddot{D}_{12}^2 + \dddot{D}_{13}^2\right) =$$

$$= \dddot{D}_{23}^2 + \frac{1}{4}\dddot{D}_{22}^2 + \frac{1}{4}\dddot{D}_{33}^2 - \frac{1}{2}\dddot{D}_{22}\dddot{D}_{33} \qquad (C.4)$$

Remembering that the quadrupole tensor is traceless (see the definition 8.56)

$$\delta_{ik} D_{ik} = D_{11} + D_{22} + D_{33} = 0 \qquad (C.5)$$

we obtain

$$a\left(\dddot{D}_{22}^2 + \dddot{D}_{33}^2 + 2\dddot{D}_{22}\dddot{D}_{33}\right) + b\left(2\dddot{D}_{22}^2 + 2\dddot{D}_{33}^2 + 2\dddot{D}_{22}\dddot{D}_{33} + 2\dddot{D}_{12}^2 + 2\dddot{D}_{13}^2 + 2\dddot{D}_{23}^2\right) +$$

$$+ c\left(\dddot{D}_{22}^2 + \dddot{D}_{33}^2 + 2\dddot{D}_{22}\dddot{D}_{33} + \dddot{D}_{12}^2 + \dddot{D}_{13}^2\right) =$$

$$= \dddot{D}_{23}^2 + \frac{1}{4}\dddot{D}_{22}^2 + \frac{1}{4}\dddot{D}_{33}^2 - \frac{1}{2}\dddot{D}_{22}\dddot{D}_{33} \qquad (C.6)$$

This gives

$$2b = 1 \quad , \quad a + 2b + c = \frac{1}{4}$$

$$2a + 2b + 2c = -\frac{1}{2} \qquad 2b + c = 0 \qquad (C.7)$$

that is

$$b = \frac{1}{2} \quad , \quad c = -1 , \qquad a = \frac{1}{4} \qquad (C.8)$$

The intensity of energy dI radiated within the solid angle $d\Omega$, in the direction $\vec{n}$, i.e. the flux of energy through the element of spherical surface $R_0^2\, d\Omega$, centered at the origin and of radius R_0, normal to $\vec{n}$, can be obtained then using (C.1,2,3,8), and it is given by

$$dI = \frac{G}{36\pi c^5}\left[\frac{1}{4}\left(\dddot{D}_{ik} n_i n_k\right)^2 + \frac{1}{2}\dddot{D}_{ik}\dddot{D}_{ik} - \right.$$
$$\left. - \dddot{D}_{ji}\dddot{D}_{jk} n_i n_k\right] d\Omega \qquad (C.9)$$

The total power radiated in all directions is defined as

$$I = -\frac{dE}{dt} = c\int \sqrt{-g}\, t_4{}^k\, d\sigma_k \qquad (C.10)$$

and can be obtained integrating eq.(C.9) over the solid angle $d\Omega = \sin\vartheta\, d\vartheta\, d\varphi$. Putting

$$n_1 = \sin\vartheta\cos\varphi$$
$$n_2 = \sin\vartheta\sin\varphi \qquad (C.11)$$
$$n_3 = \cos\vartheta$$

we have

$$\int d\Omega\, n_1^2 = \int_0^{2\pi} d\varphi \int_0^{\pi} \sin\vartheta\, d\vartheta\, \sin^2\vartheta\cos^2\varphi =$$
$$= -\int_0^{2\pi} d\varphi\, \frac{1}{2}(1+\cos 2\varphi)\int_{+1}^{-1} d(\cos\vartheta)(1-\cos^2\vartheta) = \frac{4\pi}{3} \qquad (C.12)$$

An explicit computation leads also to

$$\int d\Omega\, n_2^2 = \int d\Omega\, n_3^2 = \frac{4}{3}\pi$$
$$\int d\Omega\, n_i n_k = 0 \quad , \ i \neq k \qquad (C.13)$$

Therefore

$$\int d\Omega\, n_i n_k = \frac{4\pi}{3}\delta_{ik} \qquad (C.14)$$

Moreover, putting

$$\int_0^{2\pi} d\varphi \cos^2\varphi\sin^2\varphi = A \qquad (C.15)$$

we have

$$A = \int_0^{2\pi} d\varphi(1-\sin^2\varphi)\sin^2\varphi = \int_0^{2\pi} d\varphi\,\frac{1}{2}(1-\cos 2\varphi) -$$
$$- \int_0^{2\pi} d\varphi \sin^4\varphi =$$

$$= \pi - \int_0^{2\pi} d\varphi \sin\varphi \, (1 - \cos^2\varphi)^{3/2} =$$

$$= \pi + \left[\cos\varphi \, (1 - \cos^2\varphi)^{3/2} \right]_0^{2\pi} - \int_0^{2\pi} d\varphi \, 3 \cos^2\varphi \sin^2\varphi =$$

$$= \pi - 3A \tag{C.16}$$

Therefore

$$A = \int_0^{2\pi} d\varphi \cos^2\varphi \sin^2\varphi = \frac{\pi}{4} \tag{C.17}$$

and

$$\int_0^{\pi} d\vartheta \sin\vartheta \sin^4\vartheta = \int_{-1}^{1} dx \, (1-x^2)^2 = \frac{16}{15} \tag{C.18}$$

Using the results (C.17,18) we obtain

$$\int d\Omega \, n_1 n_1 n_2 n_2 = \int_0^{2\pi} d\varphi \int_0^{\pi} d\vartheta \sin\vartheta \sin^4\vartheta \cos^2\varphi \sin^2\varphi =$$

$$= \frac{4\pi}{15} \tag{C.19}$$

and, with the same procedure, we have also

$$\int d\Omega \, n_1^2 n_3^2 = \int d\Omega \, n_2^2 n_3^2 = \frac{4\pi}{15} \tag{C.20}$$

while the integral

$$\int d\Omega \, n_i n_j n_k n_l \tag{C.21}$$

is vanishing if three or more indices of $\vec{n}$ are different. These results can be summarized by putting

$$\int d\Omega \, n_i n_j n_k n_l = \frac{4\pi}{15} \left(\delta_{ij}\delta_{kl} + \delta_{ik}\delta_{jl} + \delta_{il}\delta_{jk} \right) \tag{C.22}$$

Using eqs.(C.14,22) we can finally integrate eq.(C.9) to obtain the total power radiated by the gravitating system in all directions (and in the quadrupole approximation)

$$-\frac{dE}{dt} = \int dI = \frac{G}{36\pi c^5} \left\{ \frac{1}{4} \int d\Omega \, \dddot{D}_{ij} \dddot{D}_{kl} n_i n_j n_k n_l + \right.$$

$$\left. + \frac{1}{2} \int d\Omega \, \dddot{D}_{ik} \dddot{D}_{ik} - \int d\Omega \, \dddot{D}_{ji} \dddot{D}_{jk} n_i n_k \right\} =$$

$$= \frac{G}{36\pi c^5} \left\{ \frac{1}{4} \dddot{D}_{ij} \dddot{D}_{k\ell} \int d\Omega \, n_i n_j n_k n_\ell \right. +$$

$$\left. + \frac{1}{2} \dddot{D}_{ik}^{\,2} \int d\Omega - \dddot{D}_{ji} \dddot{D}_{jk} \int d\Omega \, n_i n_k \right\} =$$

$$= \frac{G}{36\pi c^5} \left\{ \frac{1}{4} \dddot{D}_{ij} \dddot{D}_{k\ell} \frac{4\pi}{15} (\delta_{ij}\delta_{k\ell} + \delta_{ik}\delta_{j\ell} + \delta_{i\ell}\delta_{kj}) \right.$$

$$\left. + \frac{1}{2} \dddot{D}_{ik}^{\,2} \, 4\pi - \dddot{D}_{ji} \dddot{D}_{jk} \frac{4}{3}\pi \, \delta_{ik} \right\} =$$

$$= \frac{4\pi G}{36\pi c^5} \left\{ \frac{1}{30} + \frac{1}{2} - \frac{1}{3} \right\} (\dddot{D}_{ik})^2 =$$

$$= \frac{G}{45 c^5} (\dddot{D}_{ik})^2 \qquad (C.23)$$

APPENDIX D

Geodesic deviation

The propagation of a gravitational wave through space-time produces a relative deviation of two nearby geodesics (see Fig.D.1), and this deviation can be measured in terms of the vector η^μ connecting them. The figure a) shows two nearby geodesics whose separation is measured by light signals (the dashed lines). The proper time intervals of emission and reception along the geodesics are related to the separation vector η^μ. The time dependence of η^μ is governed by the equation of geodesic deviation:

$$\frac{D^2\eta^\mu}{Ds^2} + R_{\nu\alpha\beta}{}^{\mu} \eta^\nu u^\alpha u^\beta = 0 \qquad (D.1)$$

where D denotes the covariant derivative, s is the proper time along one of the geodesics and u^μ is the tangent four velocity vector ($u^\mu \eta_\mu = 0$).

The separation of two nearby free-falling masses is governed by this equation, and by measuring η^μ (for example observing the shift of interference fringes with a Michelson interferometer) one can determine certain components of the curvature tensor $R_{\mu\nu\alpha\beta}$.

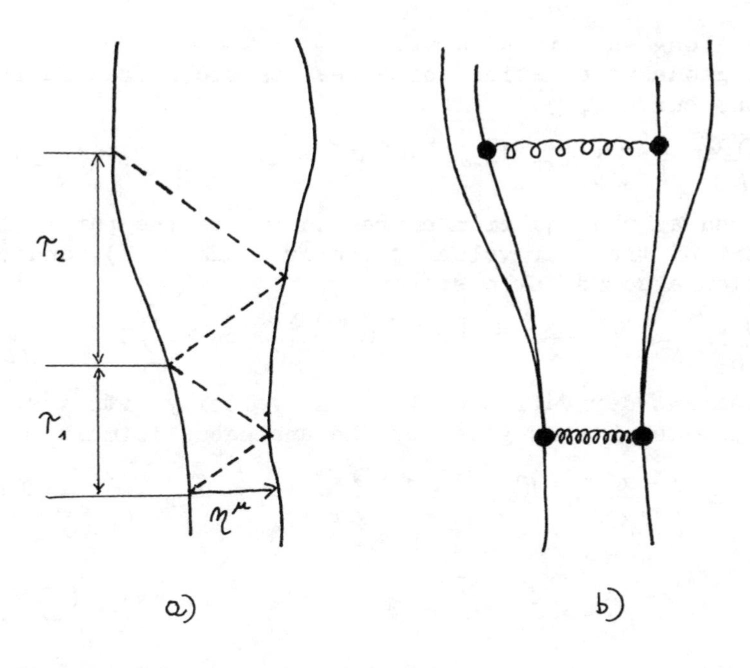

FIG. D.1

If we connect the two masses with a spring, as in figure b), then the equation (D.1) is modified by the addition of a term representing the elastic force, and becomes the equation describing a harmonic oscillator subject to the action of an external force

$$\frac{D^2\eta^\mu}{Ds^2} + K\eta^\mu + R_{\nu\alpha\beta}{}^{\mu}\eta^\nu u^\alpha u^\beta = 0 \qquad (D.2)$$

In order to deduce these two equations, consider a family of geodesics, $\gamma_v(s)$, characterized by a parameter v which is constant along any curve; the unit tangent vector for a given geodesic $\gamma_v(s)$ is the four velocity

$$u^\mu = \frac{\partial x^\mu}{\partial s} \qquad (D.3)$$

where s is an affine parameter (for example the proper time) varying along the given curve.

The geodesic equation for a test particle free falling along the curve $\gamma_v(s)$

$$\frac{Du^\mu}{Ds} = \frac{du^\mu}{ds} + \Gamma_{\nu\alpha}{}^\mu u^\alpha u^\nu = 0 \qquad (D.4)$$

is defined as the equation corresponding to the parallel transport of the four velocity vector along $\gamma(s)$, and can be written also as follows:

$$\frac{Du^\mu}{Ds} = \left(\frac{\partial u^\mu}{\partial x^\nu} + \Gamma_{\nu\alpha}{}^\mu u^\alpha\right)\frac{\partial x^\nu}{\partial s} = u^\mu{}_{;\nu} u^\nu \qquad (D.5)$$

The covariant derivative of the four velocity with respect to the parameter v is given by the analogous formula

$$\frac{Du^\mu}{Dv} = \left(\frac{\partial u^\mu}{\partial x^\nu} + \Gamma_{\nu\alpha}{}^\mu u^\alpha\right)\frac{\partial x^\nu}{\partial v} = u^\mu{}_{;\nu} v^\nu \qquad (D.6)$$

where

$$v^\nu = \frac{\partial x^\nu}{\partial v} \qquad (D.7)$$

Considering the commutator of the two covariant derivatives we have

$$\frac{D^2 u^\mu}{Dv\,Ds} - \frac{D^2 u^\mu}{Ds\,Dv} =$$

$$= (u^\mu{}_{;\nu} u^\nu)_{;\alpha} v^\alpha - (u^\mu{}_{;\nu} v^\nu)_{;\alpha} u^\alpha =$$

$$= (u^\mu{}_{;\nu;\alpha} u^\nu + u^\mu{}_{;\nu} u^\nu{}_{;\alpha}) v^\alpha -$$

$$- (u^\mu{}_{;\nu;\alpha} v^\nu + u^\mu{}_{;\nu} v^\nu{}_{;\alpha}) u^\alpha =$$

$$= u^\mu{}_{;\nu;\alpha} (u^\nu v^\alpha - v^\nu u^\alpha) +$$

$$+ u^\mu{}_{;\nu} (u^\nu{}_{;\alpha} v^\alpha - v^\nu{}_{;\alpha} u^\alpha) \qquad (D.8)$$

Noting that (as $\Gamma_{\alpha\beta}{}^\mu = \Gamma_{\beta\alpha}{}^\mu$)

$$\frac{Du^\nu}{Dv} - \frac{Dv^\nu}{Ds} = u^\nu{}_{;\alpha} v^\alpha - v^\nu{}_{;\alpha} u^\alpha =$$

$$= (\partial_\alpha u^\nu + \Gamma_{\alpha\beta}{}^\nu u^\beta) v^\alpha - (\partial_\alpha v^\nu + \Gamma_{\alpha\beta}{}^\nu v^\beta) u^\alpha =$$

$$= v^\alpha \partial_\alpha u^\nu - u^\alpha \partial_\alpha v^\nu =$$

$$= \frac{\partial x^\alpha}{\partial v} \partial_\alpha u^\nu - \frac{\partial x^\alpha}{\partial s} \partial_\alpha v^\nu = \frac{\partial u^\nu}{\partial v} - \frac{\partial v^\nu}{\partial s} =$$

$$= \frac{\partial^2 x^\nu}{\partial v \partial s} - \frac{\partial^2 x^\nu}{\partial s \partial v} = 0 \qquad (D.9)$$

from (D.8) we obtain then

$$\frac{D^2 u^\mu}{Dv Ds} - \frac{D^2 u^\mu}{Ds Dv} = (u^\mu{}_{;\nu;\alpha} - u^\mu{}_{;\alpha;\nu}) u^\nu v^\alpha \qquad (D.10)$$

and expressing the commutator of two covariant derivatives in terms of the curvature tensor (see eq.3.10) we have

$$\frac{D^2 u^\mu}{Dv Ds} - \frac{D^2 u^\mu}{Ds Dv} = - R_{\nu\alpha\beta}{}^\mu u^\beta u^\nu v^\alpha \qquad (D.11)$$

For a free falling particle along a geodesic world line we have $Du^\mu/Ds = 0$; moreover, from (D.9) we have also

$$\frac{Du^\mu}{Dv} = \frac{Dv^\mu}{Ds} \qquad (D.12)$$

The relation (D.11) therefore reduces to

$$\frac{D^2 v^\mu}{Ds^2} + R_{\alpha\nu\beta}{}^\mu v^\alpha u^\nu u^\beta \qquad (D.13)$$

Finally, we introduce the following infinitesimal four vector

$$\eta^\mu = \frac{\partial x^\mu}{\partial v} dv = v^\mu dv \qquad (D.14)$$

which is called "deviation vector", and represents physically the position of a particle, falling along the geodesic C', with respect to another particle whose world line coincides with the geodesic C (see Fig.D.2).

Multiplying (D.13) by dv we obtain then the equation of "geodesic deviation"

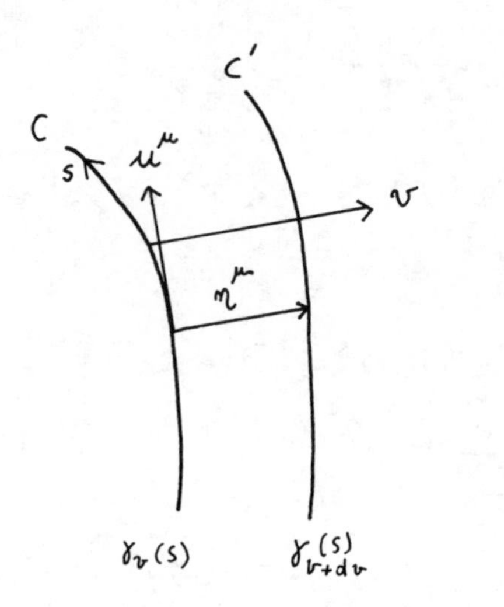

FIG. D.2

$$\frac{D^2\eta^\mu}{Ds^2} + R_{\nu\alpha\beta}{}^\mu \eta^\nu u^\alpha u^\beta = 0 \qquad (D.15)$$

which relates the second covariant derivative of η^μ to the Riemann tensor (in flat space $D^2\eta^\mu/Ds^2 = 0$). The presence of a gravitational field therefore can be revealed by the relative displacement of two nearby particles.

The equation (D.15) can be immediately generalized to the case in which an external force is present, in additio to gravity. Suppose in fact that the geodesic equation (D.4) is modified as follows

$$\frac{Du^\mu}{Ds} = \frac{F^\mu}{m} \qquad (D.16)$$

From (D.11) we get then

$$\frac{D^2 v^\mu}{Ds^2} + R_{\nu\alpha\beta}{}^\mu v^\nu u^\alpha u^\beta = \frac{DF^\mu}{mDv} \qquad (D.17)$$

and, multiplying by dv, the generalized equation is

$$\frac{D^2\eta^\mu}{Ds^2} + R_{\nu\alpha\beta}{}^\mu \eta^\nu u^\alpha u^\beta = \Phi^\mu \qquad (D.18)$$

where

$$\Phi^{\mu} = \frac{DF^{\mu}}{m\,Dv}\,dv \qquad (D.19)$$

At last, we note that the orthogonality condition $\eta^{\mu} u_{\mu} = 0$ is preserved along a geodesic. In fact

$$\frac{d}{ds}\left(v^{\mu} u_{\mu}\right) = \frac{D}{Ds}\left(v^{\mu} u_{\mu}\right) = u_{\mu}\frac{Dv^{\mu}}{Ds} \qquad (D.20)$$

(since $Du^{\mu}/Ds = 0$). Using (D.12)

$$u_{\mu}\frac{Dv^{\mu}}{Ds} = u_{\mu}\frac{Du^{\mu}}{Dv} = \frac{1}{2}\frac{D}{Dv}\left(u^{\mu}u_{\mu}\right) = \frac{1}{2}\frac{d}{dv}\left(u^{\mu}u_{\mu}\right) = 0 \qquad (D.21)$$

(since $u^{\mu}u_{\mu} = 1 = \text{const}$) . Therefore $u_{\mu}v^{\mu} = \text{const}$ along the geodesic, and the projection of $\eta^{\mu} = v^{\mu}dv$ on the geodesic, $\eta^{\mu} u_{\mu}$, is constantly vanishing, if initially $\eta^{\mu} u_{\mu} = 0$.

APPENDIX E

On the polarization of a plane wave

We have seen, in Section 2 of this Chapter, that a plane gravitational wave has only two independent components of $h_{\mu\nu}$ which are nonzero. For the propagation along the x^1 axis, for example, we have $h_{22} = -h_{33} \neq 0$ and $h_{23} = h_{32} \neq 0$. Putting

$$A_{\mu\nu} = \begin{pmatrix} 0 & 0 & 0 & 0 \\ 0 & 1 & 0 & 0 \\ 0 & 0 & -1 & 0 \\ 0 & 0 & 0 & 0 \end{pmatrix} , \quad B_{\mu\nu} = \begin{pmatrix} 0 & 0 & 0 & 0 \\ 0 & 0 & 1 & 0 \\ 0 & 1 & 0 & 0 \\ 0 & 0 & 0 & 0 \end{pmatrix} \qquad (E.1)$$

the weak gravitational field of the wave can be written then as follows

$$h_{\mu\nu} = \left(a A_{\mu\nu} + b B_{\mu\nu}\right) e^{-i\omega(t - x/c)} \qquad (E.2)$$

that is as a linear combination of two independent states of polarization.The tensor $A_{\mu\nu}$ describes a state of linear polarization corresponding to a tension along y and a compression along z , while $B_{\mu\nu}$ corresponds to shear forces in the (y,z) plane.

These two polarization states are linearly independent, since

$$A_{\mu\nu} B^{\mu\nu} = 0 \tag{E.3}$$

and they are related by a rotation of an angle $\pi/4$. Consider in fact the matrix U corresponding to a rotation of an angle ϑ in the (y,z) plane:

$$U_\mu{}^\nu = \begin{pmatrix} 1 & 0 & 0 & 0 \\ 0 & \cos\vartheta & \sin\vartheta & 0 \\ 0 & -\sin\vartheta & \cos\vartheta & 0 \\ 0 & 0 & 0 & 1 \end{pmatrix} \tag{E.4}$$

Performing a rotation of the $A_{\mu\nu}$ tensor we have

$$A'_{\mu\nu} = U_\mu{}^\alpha U_\nu{}^\beta A_{\alpha\beta} =$$

$$= \begin{pmatrix} 0 & 0 & 0 & 0 \\ 0 & \cos 2\vartheta & \sin 2\vartheta & 0 \\ 0 & \sin 2\vartheta & -\cos 2\vartheta & 0 \\ 0 & 0 & 0 & 0 \end{pmatrix} \tag{E.5}$$

and then, for $\vartheta = \pi/4$, we have $A_{\mu\nu} \rightarrow A'_{\mu\nu} = B_{\mu\nu}$ (note that $A_{\mu\nu}$ is left invariant by a rotation of an angle $\vartheta = n\pi/2$, with n = 0,1,2,....., modulo an unimportant phase factor).

Like in the electromagnetic case, two orthogonal states of circular polarization can be defined by the linear combination of the two states describing linear polarization. We have then

$$\tau_{\mu\nu} = A_{\mu\nu} + i B_{\mu\nu} = \begin{pmatrix} 0 & 0 & 0 & 0 \\ 0 & 1 & i & 0 \\ 0 & i & -1 & 0 \\ 0 & 0 & 0 & 0 \end{pmatrix} \tag{E.6}$$

for a right-handed wave, and

$$l_{\mu\nu} = A_{\mu\nu} - i B_{\mu\nu} = \begin{pmatrix} 0 & 0 & 0 & 0 \\ 0 & 1 & -i & 0 \\ 0 & -i & -1 & 0 \\ 0 & 0 & 0 & 0 \end{pmatrix} \qquad (E.7)$$

for a left-handed circularly polarized gravitational wave. These new tensors are orthogonal,

$$r_{\mu\nu} l^{*\mu\nu} = 0 \qquad (E.8)$$

(the star denotes complex conjugation) and, under a rotation of an angle ϑ in the (y,z) plane, perpendicular to the direction of wave propagation, they tranforms as follows

$$r'_{\mu\nu} = U_\mu{}^\alpha U_\nu{}^\beta r_{\alpha\beta} =$$

$$= \begin{pmatrix} 0 & 0 & 0 & 0 \\ 0 & \cos 2\vartheta + i\sin 2\vartheta & i\cos 2\vartheta - \sin 2\vartheta & 0 \\ 0 & i\cos 2\vartheta - \sin 2\vartheta & -\cos 2\vartheta - i\sin 2\vartheta & 0 \\ 0 & 0 & 0 & 0 \end{pmatrix} =$$

$$= \begin{pmatrix} 0 & 0 & 0 & 0 \\ 0 & 1 & i & 0 \\ 0 & i & -1 & 0 \\ 0 & 0 & 0 & 0 \end{pmatrix} e^{2i\vartheta} = r_{\mu\nu} e^{2i\vartheta} \qquad (E.9)$$

and likewise

$$l'_{\mu\nu} = U_\mu{}^\alpha U_\nu{}^\beta l_{\alpha\beta} = l_{\mu\nu} e^{-2i\vartheta} \qquad (E.10)$$

As previously noted, a phase shift 2ϑ for a rotation of an angle ϑ means that two unities of angular momentum are associated to the gravitational wave, and then its quanta (the gravitons) must be massless spin-two particles, with the spin parallel (or antiparallel) to the direction of propagation.

The polarization state of a gravitational wave in general is a linear combination of the possible states. For example, the gravitational radiation from a system of binary stars is circularly polarized along the rotation axis, and linearly polarized in the rotation plane, see Fig.E.1.

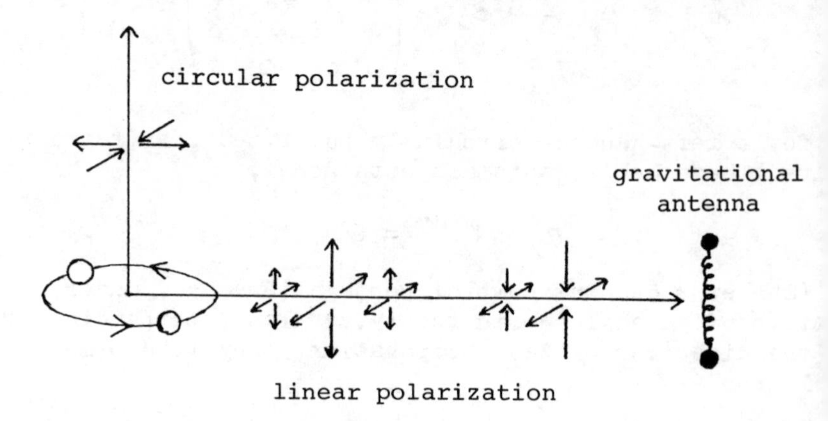

FIG. E.1

A simple anntenna for detecting gravitational waves can be represented by two masses connected by a spring and vibrating along a longitudinal direction. The response of the antenna is maximum if the detector is placed at a right angle with the direction of the incident radiation and if its axis of vibration is parallel to the polarization vector. If these conditions are not satisfied, then the antenna will respond only to the components of the Riemann tensor along the antenna axis (see Fig.E.2)

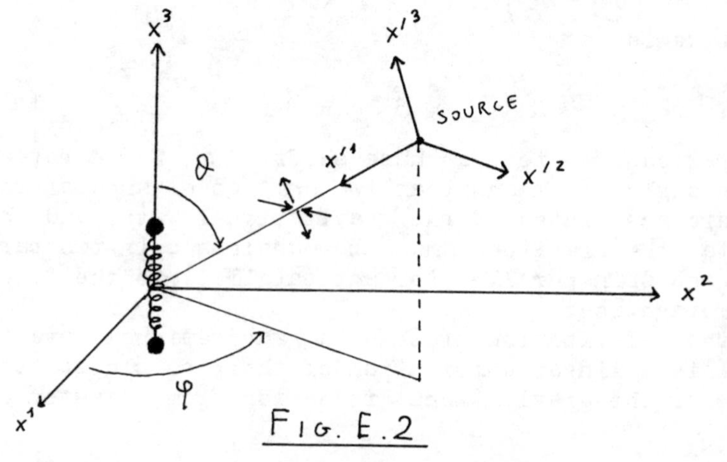

FIG. E.2

In the case considered in <u>Fig.E.2</u>, the axis of the antenna is the x^3 axis, and the response will be proportional to (see eq.8.100)

$$R_{3434} = \frac{\partial x'^i}{\partial x^3}\frac{\partial x'^4}{\partial x^4}\frac{\partial x'^k}{\partial x^3}\frac{\partial x'^4}{\partial x^4} R'_{i4k4} \qquad (E.11)$$

Considering time intervals far longer than the period of the radiation, the directional response of the antenna becomes insensitive to the polarization angle of the radiation. Assuming then for simplicity that one axis of the polarization tensor is in the (x^1,x^2) plane, and the axis x'^2 of the source is parallel to the (x^1, x^2) plane of the detector, we have

$$R_{3434} = \frac{\partial x'^3}{\partial x^3}\frac{\partial x'^4}{\partial x^4}\frac{\partial x'^3}{\partial x^3}\frac{\partial x'^4}{\partial x^4} R'_{3434} \qquad (E.12)$$

(the components along x'^2 do not contribute to the elongation of the spring, which is aligned along x^3).

The response of the antenna is then independent of the angle φ and we can assume that the source is placed in the (x^1,x^3) plane. The two coordinate systems of the source and of the detector are then related by a rotation around the x^2 axis and an inversion, i.e.

$$\begin{aligned} x^4 &= x'^4 \\ x^1 &= -x'^1 \sin\vartheta - x'^3 \cos\vartheta \\ x^2 &= -x'^2 \\ x^3 &= -x'^1 \cos\vartheta + x'^3 \sin\vartheta \end{aligned} \qquad (E.13)$$

from which we obtain

$$\frac{\partial x'^4}{\partial x^4} = 1 \quad , \quad \frac{\partial x'^3}{\partial x^3} = \sin\vartheta \qquad (E.14)$$

and then

$$R_{3434} = \sin^2\vartheta \, R'_{3434} \qquad (E.15)$$

The directional response of the antenna to the gravitational radiation varies then as $\sin^2\vartheta$, where ϑ is the angle between the axis of the antenna and the direction of the incoming radiation. In the electromagnetic case, on the contrary, the directional response of a dipolar antenna varies simply as $\sin\vartheta$.

REFERENCES

1) J.Weber: Phys.Rev.Lett.22,1320(1969); 24,276(1970); 25,180(1970)

2) W.Pauli and M.Fierz: Proc.R.Soc.A173,211(1939)

3) V.B.Braginskii: Usp.Fiz.Nauk.86,433(1965) (Sov.Phys. Usp.8,513(1966))

4) V.N.Mironowskii: Astr.Zurn.42,977(1965) (Sov.Astr.A.J. 9,752(1966))

5) R.H.Hulse and J.H.Taylor: Astr.J.(Lett.)195,L53(1975)

6) see for example Y.B.Zeldovich and I.D.Novikov: "Relativistic Astrophysics", vol.I (Chigago Press, Chigago 1971)

7) J.Weber: "General relativity and gravitational waves" (Interscience Pub., New York 1961)

8) see for example V.D.Zakharov: "Gravitational waves in Einstein's theory" (Halsted Press, Jerusalem, 1973)

9) V.De Sabbata, in "Topics in theoretical and experimental gravitation physics", ed. by V.De Sabbata and J. Weber (Plenum Press, New York 1977) p.69

CHAPTER IX

DENSE AND COLLAPSED MATTER

1.- White dwarfs

The first evidence of a star with a mass concentrated in a very small volume (called "white dwarf") was found in the companion of Syrius, which has a mass of about 0.96, in unit of solar masses, and an intrinsic luminosity which is nearly 10^{-2} that of the sun.

The presence of a companion of Syrius was initially discovered only indirectly, by observing the periodical motion of Syrius itself. Only in a subsequent time it was discovered optically, because its luminosity is very faint. The strange and unusual fact was that, from a spectral study, it was deduced that its surface temperature was very high: how to reconcile high temperature and low intrinsic luminosity? The solution is found only in the case of a very small radius, so that also the emitting surface is small.

It was found in this way that its density is $\sim 10^6$ g/cm^3 (very high compared to that of a standard star like our sun, whose density is ~ 1 g/cm^3).

In this type of stars, the self gravitational forces are no longer balanced by the thermonuclear reactions, which have been exhausted, and their stability can be explained considering the quantum behaviour of the electronic gas and using the Pauli exclusion principle and the Fermi-Dirac statistics: it is the pressure of this degenerate gas which balances the gravitational attraction.

Without doing explicitly quantum computations and going into the details of the equilibrium and stability problems of a star (the intersted reader is referred to an astrophysics text-book) we wish to give here only a brief outline of this subject.

A star, when its nuclear fuel is exhausted, begins to cool and to contract, until the electrons are frozen into the lowest available energy levels, according to Pauli's exclusion principle.

As there are $4\pi p^2 (2\pi\hbar)^{-3}$ dp levels per unit volume with momentum between p and p + dp, for a degenerate Fermi gas the maximum momentum p_F is related to the electron density n by

$$n = \frac{8\pi}{(2\pi\hbar)^3}\int_0^{p_F} p^2 dp = \frac{p_F^3}{3\pi^2\hbar^3} \tag{9.1}$$

(the factor two is due to the spin of the electron, which can be in two different states with helicity $\pm 1/2$). For $N = n V$ electrons in a volume V, we have $p_F = \hbar\,(3\pi^2)^{1/3}(N/V)^{1/3}$, and the limiting (nonrelativistic) kinetic energy is then

$$E_F = \frac{p_F^2}{2m} = \frac{\hbar^2}{2m}(3\pi^2)^{2/3}\left(\frac{N}{V}\right)^{2/3} \tag{9.2}$$

(m is the electron mass). The total energy in a volume V for a degenerate electronic gas at a zero absolute temperature is then

$$E = \int_0^{p_F} \frac{p^2}{2m}\, V\frac{8\pi}{(2\pi\hbar)^3}\, p^2 dp = \frac{4\pi}{5(2\pi\hbar)^3}\,\frac{p_F^5}{m}\, V =$$

$$= \frac{3}{10}\left(\frac{3}{8\pi}\right)^{2/3}\frac{(2\pi\hbar)^2}{m}\left(\frac{N}{V}\right)^{2/3} N \tag{9.3}$$

and since the pressure P satisfies $PV = 2E/3$, we are led to the equation of state[1]

$$P = \frac{\hbar^2}{15m\pi^2}(3\pi^2)^{5/3}\left(\frac{N}{V}\right)^{5/3} \tag{9.4}$$

which holds for temperatures T very close to the absolute zero, and such that

$$kT << E_f \sim \frac{\hbar^2}{m}\left(\frac{N}{V}\right)^{2/3} \tag{9.5}$$

(k is the Boltzmann constant), and moreover in the nonrelativistic limit.

For an extremely relativistic gas, the kinetic energy is given by pc, and using the equation of state $PV = E/3$ we obtain

$$P = \frac{1}{12}(3\pi^2)^{4/3}\frac{\hbar c}{\pi^2}\left(\frac{N}{V}\right)^{4/3} \tag{9.6}$$

Therefore $P \propto V^{-5/3}$ for a nonrelativistic degenerate Fermi gas, $P \propto V^{-4/3}$ in the relativistic case.

This pressure of statistical origin is just at the origin of the mechanism that assures the stability of the star against the gravitaional collapse. However, if the mass of the star is larger than a certain critival value, the pressure is not sufficient to balance the gravitational forces, and the star collapse is not stopped.

Denoting with T(p) the kinetic energy of an electron, the total energy E is given by

$$E = \frac{V}{\pi^2 \hbar^3} \int_0^{p_F} T(p)\, p^2\, dp \qquad (9.7)$$

and the pressure is 2E/3V . In the relativistic case we have

$$T(p) = mc^2 \left[\left(1 + \frac{p^2}{m^2 c^2}\right)^{1/2} - 1 \right] \qquad (9.8)$$

and then, by putting

$$\frac{p}{mc} = \sinh \vartheta \quad , \quad \frac{p_F}{mc} = \sinh \vartheta_F \qquad (9.9)$$

the integral (9.7) gives for the pressure P

$$P = \frac{2E}{3V} = \frac{m^4 c^5}{3\pi^2 \hbar^3} \int_0^{\vartheta_F} \sinh^4 \vartheta \, d\vartheta \qquad (9.10)$$

from which

$$P = \frac{m^4 c^5}{3\pi^2 \hbar^3} \left[\frac{1}{4} \sinh^3 \vartheta \cosh \vartheta - \frac{3}{16} \sinh 2\vartheta + \frac{3}{8} \vartheta \right]_{\vartheta = \vartheta_F} \qquad (9.11)$$

Defining $x = p_F/mc$, eq.(9.11) becomes

$$P = \frac{m^4 c^5}{24 \pi^2 \hbar^3} \left[x (2x^2 - 3)(x^2 + 1)^{1/2} + 3 \sinh^{-1} x \right] \qquad (9.12)$$

and the density ρ of the star

$$\rho = n \mu m_p \qquad (9.13)$$

where μ is the number of nucleons per electron ($\mu \simeq 2$ for stars like white dwarfs) and m_p the mass of a proton, can be written using (9.1), as follows

$$\rho = n \mu m_p = \frac{m^3 c^3 m_p}{3\pi^2 \hbar^3} x^3 \mu \qquad (9.14)$$

The Newtonian equation which expresses the condition of hydrostatic equilibrium for a spherically symmetric distribution of matter

$$\frac{dP}{dr} = - \rho \frac{G M(r)}{r^2} \qquad (9.15)$$

where

$$M(r) = \int_0^r 4\pi \rho r'^2 dr' \qquad (9.16)$$

can also be written in the form

$$\frac{1}{r^2} \frac{d}{dr} \left(\frac{r^2}{\rho} \frac{dP}{dr} \right) = -4\pi G \rho \qquad (9.17)$$

Putting

$$y^2 = x^2 + 1 \quad , \quad \Phi = \frac{y}{y_0} \quad , \quad \eta = \frac{r}{\alpha} \qquad (9.18)$$

where x_0 and y_0 are the values of x and y at the center of the star, and

$$\alpha = \left(\frac{2 m^4 c^5}{24 \pi^2 \hbar^3 \pi G} \right)^{1/2} \frac{3\pi^2 \hbar^3}{m^3 c^2 \mu m_p y_0} \qquad (9.19)$$

and combining (9.12,17) we obtain the following differential equation

$$\frac{1}{\eta^2} \frac{d}{d\eta} \left(\eta^2 \frac{d\Phi}{d\eta} \right) = - \left(\Phi^2 - \frac{1}{y_0^2} \right)^{1/2} \qquad (9.20)$$

which governs the structure of a spherical distribution of degenerate relativistic gas in hydrostatic equilibrium.

The boundary condition $\Phi = 1$ at the center and $(d\Phi/d\eta) = 0$ for $\eta = 0$, toghether with a particular value of y_0, determine completely Φ and then the mass of the configuration.

It is important to stress that each mass has its own peculiar density distribution, and one finds, solving (9.20) that the mass M of the star (corresponding to a given value of y_0) grows monotonically when the central density is increasing, reaching the maximum value ($M_\odot$ is the mass of our sun)

$$M_c \sim \frac{5.7}{\mu^2} M_\odot \qquad (9.21)$$

when the density at the center $\rho(0) \longrightarrow \infty$. This maximum value of M is known as the Chandrasekhar limit, and one is led to conclude that stable white dwarfs can exist for $M < M_c$, because for $M > M_c$ no solution exist to the equilibrium equation.

There are other effects, however, which must be taken into account. For example, when $p_F \sim 5\ m$, it becomes energetically favourable, for the electrons, to be captured by the nuclear matter, according to the reaction

$$p + e^- \rightarrow n + \nu_e \qquad (9.22)$$

and producing neutrinos which escape from the star. This effect produces an increase of the number μ of nucleons per electron, so that the maximum value (9.21) is lowered, for a given central density. We can expect then that M increases toward the Chandrasekhar limit until a critical density where the mass reaches a maximum value , and then it begins to decrease. Detailed calculations show that the maximum is 1.2 $M_\odot$, near to the Chandrasekhar limit, which is 1.26 $M_\odot$ for $\mu = 56/26$ (i.e. for an iron white dwarf). We may conclude that this maximum denotes the transition from stability to instability, and that stable iron white dwarfs can exist only for $M < 1.2\ M_\odot$.

2.- Neutron stars

As discussed in the previous Section, if the mass of a star is larger than a given limit, its gravitational self-attraction cannot be balanced by the pressure of a cold, degenerate electron gas, and its collapse is not stopped.

We have then two possibilities: the star goes on collapsing for ever (approaching the black-hole configuration, see the following Section), or it becomes so hot during the collapse that it explodes becoming a supernova, and in this case it is believed that the highly compressed remnants survive in the phase of a superdense neutron star, with a radius ~ 10 Km and a density $\rho \sim 10^{13} \div 10^{14}$ g/cm^3 .

A neutron star consists almost entirely of cold, degenerate neutrons, with enough electrons and protons left to prevent, according to the Pauli exclusion principle, the neutron β decay. This condition sets a lower limit for the mass of a stable neutron star. On the other hand, there is also an upper limit, which can be deduced in the sama way as in the white dwarf case, replacing the electron pressure with that of the degenerate neutrons (m becomes m_p and $\mu = 1$).

There are other important differences, however, to be considered. First of all, the total energy density ρ for a white dwarf is determined by the rest mass density of its nonrelativistic nucleons, while in a neutron star the nucleons have a relativistic kinetic energy comparable to the energy of their rest masses. Another difference is that the surface gravitational potential, in a neutron star, is of the order of unit, $GM/R \sim c^2$, and then one should use the equations of general relativity instead of the Newtonian ones.

The first study of a neutron star in a general relativistic context is due to Oppenheimer and Volkoff[2] (Landau performed similar computations, without using general relativity, in the framework of the Newtonian theory, having however the merit of being the first to have investigated the stability of a neutron star).

Consider a static and spherically symmetric line-element

$$ds^2 = e^\nu dt^2 - e^\lambda dr^2 - r^2 d\vartheta^2 - r^2 \sin^2\theta \, d\varphi^2 \qquad (9.23)$$

where $\lambda = \lambda(r)$ and $\nu = \nu(r)$, and assume, like in Chapter VI, Sect.1 ,

$$T_1{}^1 = T_2{}^2 = T_3{}^3 = -p \quad , \; T_4{}^4 = \rho \qquad (9.24)$$

(ρ and p denotes respectively the proper energy density and pressure). If we write explicitly the Einstein field equations (3.37) (with zero cosmological constant), and th conservation equation $T^{\mu\nu}{}_{;\nu} = 0$, we are left with the following three independent equations (see Chapter VI, Sect.1

$$\chi p = e^{-\lambda}\left(\frac{\nu'}{r} + \frac{1}{r^2}\right) - \frac{1}{r^2} \qquad (9.25)$$

$$\chi \rho = e^{-\lambda}\left(\frac{\lambda'}{r} - \frac{1}{r^2}\right) + \frac{1}{r^2} \qquad (9.26)$$

$$p' = -\frac{1}{2}\nu'(p+\rho) \qquad (9.27)$$

(a prime denotes derivatives with respect to r).

These equations, toghether with the equation of state $\rho = \rho(p)$, determine the gravitational field and the matter distribution. The boundary of the matter distribution, i.e. the star surface, corresponds to the value $r = r_b$ of the radial coordinate, such that $p = 0$ for $r > r_b$ and $p > 0$ for $r < r_b$.

The exterior solution of the field equations (for $r > r_b$) is the well known Schwarzschild solution (see 4.29)

$$e^\nu = e^{-\lambda} = 1 - \frac{2m}{r} \qquad (9.28)$$

In the interior case, $r < r_b$, the integration of (9.27) gives

$$\nu(r) = \nu(r_b) - \int_0^{p(r)} \frac{2\,dp}{p + \rho(p)} \tag{9.29}$$

and then

$$e^{\nu(r)} = e^{\nu(r_b)} \exp\left\{ - \int_0^{p(r)} \frac{2\,dp}{p + \rho(p)} \right\} \tag{9.30}$$

The constant $\nu(r_b)$ is determined imposing the continuity of ν at $r = r_b$, i.e.

$$1 - \frac{2m}{r_b} = e^{\nu(r_b)} \exp\left\{ - \int_0^{p(r_b)} \frac{2\,dp}{p + \rho} \right\} = e^{\nu(r_b)} \tag{9.31}$$

(as $p(r_b) = 0$).

Introducing the new variable u such that

$$e^{-\lambda} = 1 - 2\frac{u}{r} \tag{9.32}$$

we have

$$u' = \frac{1}{2} r^2 \left[e^{-\lambda} \left(\frac{\lambda'}{r} - \frac{1}{r^2} \right) + \frac{1}{r^2} \right] \tag{9.33}$$

and the field equation (9.26) gives

$$u' = \frac{1}{2} r^2 \chi \rho(p) \tag{9.34}$$

Finally, combining (9.25,27) we obtain

$$\chi p = \left(1 - 2\frac{u}{r}\right) \left(- \frac{2}{p + \rho} p' \frac{1}{r} + \frac{1}{r^2} \right) - \frac{1}{r^2} \tag{9.35}$$

that is

$$p' = - \frac{p + \rho(p)}{r^2 - 2ur} \left(\frac{1}{2} \chi p r^3 + u \right) \tag{9.36}$$

The two differential equations (9.34,36) for u and p, toghether with the initial conditions $u = u_0$ and $p = p_0$ at $r = 0$, and the equation of state $\rho = \rho(p)$ determine completely the matter distribution. Using eqs.(9.32, 28), the continuity condition for u(r) at $r = r_b$ gives

$$1 - \frac{2u(r_b)}{r_b} = e^{-\lambda(r_b)} = 1 - \frac{2m}{r_b} \tag{9.37}$$

that is

$$u(r_b) \equiv u_b = \frac{r_b}{2}\left[1 - e^{-\lambda(r_b)}\right] = m \qquad (9.38)$$

and then u_b is related to the total mass of the spherical distribution of matter, as measured by an external observer.

The equation of state used by Oppenheimer and Volkoff was eq.(9.11), and the upper limit for the mass of a neutron star, obtained with the same procedure as in the white dwarf case, was $\sim 0.7\ M_\odot$.

The lower limit (under which neutrons can disintegrate into protons, electrons and neutrinos) is about $0.1\ M_\odot$, and then the mass of a neutron star must be $0.1 \leq M \leq 0.7$ in unit of solar mass.

However, it should be noted that the upper limit is depending on the particular equation of state used. Many theoretical models, corresponding to different equations of state, have been proposed (including the effect of nuclear forces, magnetic fields, quarks, gluons.....) , and today the currently accepted upper limits are running from two to three solar masses.

If the mass of a star is larger, the equilibrium is not possible. We may imagine then that the collapse of the star continues without limit, and the Schwarzschild radius is reached.

3.- Black-holes

A black-hole is an extremely collapsed body, whose spatial extension is smaller than its gravitational radius Since, as we shall see, no information may escape, classically, out of that radius, the presence of a black-hole could be revealed only indirectly by its interaction with the surrounding matter. Up to now no decisive evidence of their existence has been found, even if, according to some astrophysicists, a black-hole could be present in the Cygnus X-1 binary system: in fact that system is emitting X-rays, whose spectrum may be explained as due to the fall of matter, from one of the stars toward the black-hole, in the presence of a magnetic field.

In order to give some details on the kinematics of a static and spherical black-hole, we consider the Schwarzschild line element (4.29)

$$ds^2 = \left(1 - 2\frac{m}{r}\right)dt^2 - \frac{dr^2}{1 - 2\frac{m}{r}} - r^2\left(d\vartheta^2 + \sin^2\vartheta\, d\varphi^2\right) \qquad (9.39)$$

where m is related to the total mass M of the source of the gravitational field by $m = GM/c^2$.

At the Schwarzschild radius (also called gravitational radius)

$$r_g = 2m = 2\frac{GM}{c^2} \qquad (9.40)$$

we have $g_{44} \to 0$ and $g_{11} \to \infty$, and then we are in the presence of a singularity. In general one can say that this is not a true space-time singularity, but only a mathematical one which can be removed by a suitable transformation of coordinates. It is interesting to note, however, that for $r < r_g$, g_{44} becomes negative and g_{11} becomes positive, and then, according to the signature of the metric, the coordinate t acquires a space-like character, while the coordinate r a time-like character. The three dimensional surface $r = r_g$ is then a surface physically singular, because, when crossing this surface, the roles of the radial coordinate r and of the time coordinate t are exchanged. In this sense the Schwarzschild singularity cannot be considered as a fictitious singularity due to a bad choice of the coordinate system.

Considering the gravitational collapse of a spherically symmetric star, the first point to be stressed is that, for an external observer, the collapsing body remains always outside its gravitational surface. In fact, for an observer collapsing with the star surface, the proper time $d\tau$ is given by

$$d\tau = \frac{1}{c}\sqrt{g_{44}}\, dx^4 = \left(1 - 2\frac{m}{r}\right)^{1/2} dt \qquad (9.41)$$

and $d\tau \leq dt$ (the equality holds only at infinity where the gravitational field is vanishing). When $r \to r_g = 2m$, then $g_{44} \to 0$, and this means that the time intervals dt for a distant observer become infinite. In other words, the surface of the star is seen to approach the Schwarzshild surface in an infinite time.

An external observer therefore will never see the collapsing body inside its gravitational radius, as the collapsing surface, for him, will remain always outside (even if infinitesimally) the Schwarzschild surface.

Considering light propagation along the radial direction, we have (from $ds^2 = 0$), $dr/dt = \pm (1 - 2m/r)$, and this shows that the light cones become more and more narrow as $r \to 2m$, and degenerate to a straight line parallel to the t axis when $r = 2m = r_g$ (see Fig.9.1).

The fall inside the black-hole, for an observer at rest with the surface of the collapsing star, takes place however in a finite proper time interval.

Considering in fact the metric (9.39) we obtain

$$1 = \left(1 - \frac{2m}{r}\right)\dot{t}^2 - \frac{\dot{r}^2}{1 - \frac{2m}{r}} - r^2\left(\dot{\theta}^2 + \sin^2\theta\,\dot{\varphi}^2\right) \tag{9.42}$$

where a dot denotes derivatives with respect to the proper time $ds = c d\tau$. Inside the radius $r = 2m$, the only positive term in the right-hand side of (9.42) is the coefficient of $\dot{r}^2$, and this equation is satisfied if

$$-\dot{r}^2\left(1 - \frac{2m}{r}\right)^{-1} > 1 \tag{9.43}$$

or

$$\left(\frac{dr}{d\tau}\right)^2 > \frac{2m}{r} - 1 \tag{9.44}$$

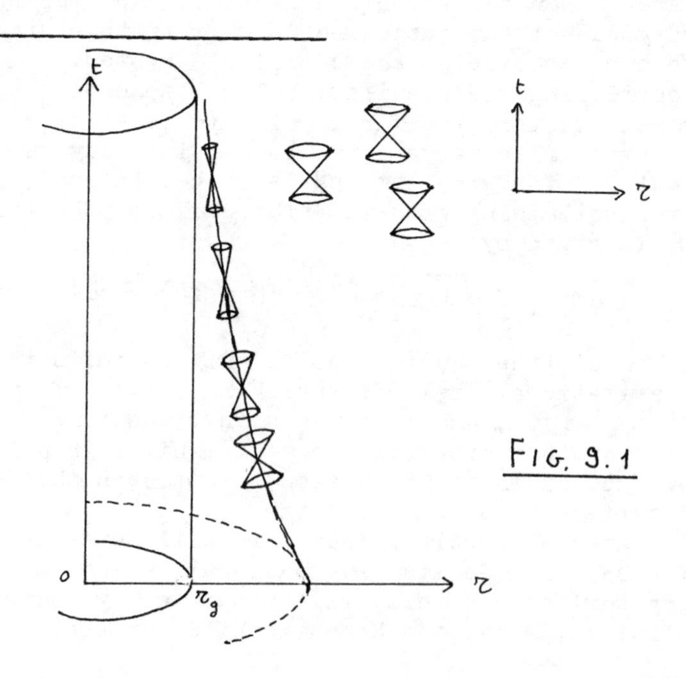

FIG. 9.1

As the radial velocity is negative (since the body is falling toward the center), we have

$$d\tau < -\frac{dr}{\left(\frac{2m}{r}-1\right)^{1/2}} \tag{9.45}$$

and the maximum interval of proper time, τ_M , required to reach the center starting, for example, from the edge of the black-hole r = 2m , is then

$$\tau_M = -\int_{2m}^{0} \frac{dr}{(2m/r-1)^{1/2}} = \pi m \tag{9.46}$$

showing that the time required for the collapse is finite, if measured in unit of proper time of the collapsed observer.

To investigate the physical behaviour of particles and light rays inside the Schwarzschild radius, it is convenient to introduce, as suggested by Finkelstein, a new coordinate t' related to the time coordinate t by

$$t' = t + 2m \ln\left(1-\frac{r}{2m}\right) \tag{9.47}$$

Differentiating this equation we have

$$dt = dt' + \frac{dr}{1-r/2m} \tag{9.48}$$

and putting

$$f = 2\frac{m}{r} \quad , \quad d\Omega^2 = r^2\left(d\theta^2 + \sin^2\theta \, d\varphi^2\right) \tag{9.49}$$

the line element (9.39) becomes

$$ds^2 = (1-f)\left(dt'^2 + \frac{dr}{1-\frac{1}{f}}\right)^2 - \frac{dr^2}{1-f} - d\Omega^2 =$$

$$= (1-f)\,dt'^2 + dr^2\left[\frac{1-f}{(1-1/f)^2} - \frac{1}{1-f}\right] + 2\,\frac{1-f}{1-1/f}\,dr\,dt' -$$

$$- d\Omega^2 =$$

$$= (1-f)\,dt'^2 - dr^2(1+f) - 2f\,dr\,dt' - d\Omega^2 =$$

$$= \left(1-2\frac{m}{r}\right)dt'^2 - \left(1+2\frac{m}{r}\right)dr^2 - 4\frac{m}{r}\,dr\,dt' -$$

$$- r^2\left(d\theta^2 + \sin^2\theta\,d\varphi^2\right) \tag{9.50}$$

It is easy to see now that the particles inside the Schwarzschild sphere cannot be at rest, and that the radial component of their velocity must be always nonvanishi and always directed toward the origin, i.e. $dr/dt' < 0$.

In fact the light cone is defined by $ds^2 = 0$: in order that a particle remain inside the light cone (i.e. the spatial components of its velocity be smaller than the local speed of light propagation, in agreement with the causality requirement), the space-time separation of two points connected by the particle world line must be positive, that is $ds^2 > 0$.

From (9.50), denoting with $v_r = dr/dt'$ the radial coo dinate velocity, we obtain then

$$g_{44} - \left(1 - \frac{2m}{r}\right) v_r^2 - \left(\frac{d\Omega}{dt'}\right)^2 - 4\frac{m}{r} v_r > 0 \qquad (9.51)$$

where $g_{44} = 1 - 2m/r$.

Inside the Schwarzschild radius, i.e. for $r < 2m$, we have $g_{44} < 0$. Since the first three terms of (9.51) are then negative, the only possibility to satisfy this disequality is provided by $v_r < 0$, i.e. by a nonvanishing radial velocity directed toward the origin.

This conclusion holds also in the case of photons, or other particles travelling at the speed of light. In fact, considering for simplicity the radial motion, and putting $ds^2 = 0$ we obtain, from (9.50),

$$(1-f)\,dt'^2 - (1+f)\,dr^2 - 2f\,dr\,dt' = 0 \qquad (9.52)$$

and the radial velocity $v_r = dr/dt'$ satisfies then the equation

$$1 - f - (1+f)\,v_r^2 - 2f\,v_r = 0 \qquad (9.53)$$

The two solutions are

$$v_r = \frac{2f \pm 2}{-2(1+f)} = \begin{cases} = -1 \\ = \dfrac{1-f}{1+f} \end{cases} \qquad (9.54)$$

and since $f = 2m/r > 1$ inside the Schwarzschild sphere, then v_r is always negative. The paths of the light rays converge then to the centre, and cannot escape from the interior of the Schwarzschild surface (it should be stressed that this result has been obtained in the framework of a classical theory of gravity: there is however the possibility that the energy can be slowly radiated away from a black-hole, through a mechanism of quantum tunneling[3]).

To conclude this Section, we note that the Schwarzschild line element considered here has led to the case of a static and spherically symmetric black-hole, but of course one may consider also more complicated models of black-holes, which for example are electrically charged and rotating, starting with more general solutions of the Einstein field equations.

REFERENCES

1) See for example L.Landau and E.Lifshitz: "Statistical physics" (Pergamon Press, London, 1959), Chapter V

2) J.R.Oppenheimer and G.M.Volkoff: Phys.Rev.55,374(1939)

3) S.W.Hawking: Scient.Am.236,34(1977)

CHAPTER X

THE EINSTEIN-CARTAN THEORY

1.- Physical motivations supporting a generalization of the Einstein theory

General relativity is the simplest theory of gravity which is in agreement with all the present day experimental data. Why, then one should try to formulate alternative or generalized version of this theory, without the need of explaining some experimental result which contradict the general relativistic predictions?

The motivations are of theoretical character, and they arise comparing general relativity with the theories of the other physical interactions.

In fact, strong, weak and electromagnetic forces are described by quantum relativistic fields interacting in a flat Minkowski space. The fields representing the interactions are defined over space-time, but are well distinguished from space-time, which is not affect by them. The gravitational interactions, on the contrary, modify the geometrical structure of space-time, and they are not represented by a new field, but by the deformation of the geometry itself.

Therefore, while 3/4 of modern physics (the physics of strong, weak and electromagnetic interactions acting at a microscopical level) are successful described at present in the framework of a flat and rigid space-time structure, the remaining 1/4 (the macroscopic physics of gravity) needs the introduction of a curved, dynamic geometrical background.

To overcome this unsatisfactory situation, it seems appropriate trying to extend the geometrical principles of general relativity also to microphysics, in order to establish a direct comparison, and possibly a connection, between gravity and the other interactions.

To this aim we note that in general relativity matter is represented by the energy-momentum tensor, which essentially provides a description of the mass density distribution in space and time. The mass-energy concept is therefore sufficient, in general relativity, to define the properties of the classical, macroscopic bodies.

But if we go down to a microscopic level, we find that matter is formed by elementary particles, which follows the laws of special relativity and quantum mechanics. Each particle is characterized not only by a mass, but also by a spin (intrinsic angular momentum) measured in units of $\hbar$. Mass and spin, at a microscopic level, are two elementary and independent objects, and just like mass distributions in space-time are descibed by the energy-momentum tensor, a spin distribution is described, in a field theory, by the spin density tensor.

Inside the macroscopical bodies, the elementary spins of the component particles are, in general, randomly oriented, so that the total resulting spin averages to zero. The spin density tensor of a macroscopical body is therefore vanishing, and this explains why the energy-momentum tensor only is enough for a dynamical characterization of the macroscopic matter, and the riemannian geometry is sufficient to describe the gravitational interactions.

It should be stressed, at this point, that the spin density tensor represents the intrinsic angular momentum of the particles, and not the classical orbital angular momentum, due to a macroscopic rotation. The fundamental difference is that the latter can be eliminated by a suitable transformation of coordinates, while a spin density can be eliminated at most at a given point only. The spin density tensor of a macroscopic body is nonvanishing only if, inside the body, the elementary spins are oriented, partially at least, along some preferred direction (like in the case of a ferromagnetic body), and it is not directly affected by the macroscopic rotation of the body.

At a microscopic level, therefore, the energy-momentum tensor alone is no longer sufficient to characterize dynamically the matter sources, but also the spin density tensor is needed (unless we are considering a system of scalar fields, corresponding to spinless particles).

If we wish to extend general relativity to include microphysics, we must take into account, therefore, that matter is described by mass and spin density; and since mass is related to curvature, by means of the energy-momentum tensor, in the framework of a generalized theory spin should be related, through the spin density tensor, to some other geometrical property of the space-time, according to the spirit of a geometric theory of gravity.

This requirement is satisfied by the Einstein-Cartan theory , also called E.C.S.K. theory (by Einstein, Cartan, Sciama, Kibble who mainly contributed to the foundations of this theory), or also briefly denoted as U_4 theory, where U_4 is a four dimensional Riemann-Cartan space-time (the U_4 geometry will be discussed in the following Section).

The new geometrical property of the space-time, related to the spin in the U_4 theory, is the torsion, $Q_{\alpha\beta}{}^{\mu}$, that is the antisymmetric part of the affine connection

$$Q_{\alpha\beta}{}^{\mu} = \Gamma_{[\alpha\beta]}{}^{\mu} = \frac{1}{2}\left(\Gamma_{\alpha\beta}{}^{\mu} - \Gamma_{\beta\alpha}{}^{\mu}\right) \qquad (10.1)$$

The torsion is a third-rank tensor, antisymmetric in the first two indices, with 24 independent components. When torsion is nonvanishing, the affine connection is no longer coincident with the Christoffel connection (as shown in Chapter 2, Sect.3), and the geometry is no loger riemannian, but we have a Riemann-Cartan space time U_4, with a nonsymmetric (but metric-compatible) connection.

The introduction of torsion represents a simple and very natural modification of the general relativity theory and of the riemannian geometry, and the existence of a relation between torsion and spin provides a physical motivation for introducing torsion, because in this way the theory can give a complete geometrical description of matter even at a microscopical level.

It should be noted that the possibility of relating the torsion of space to an intrinsic angular momentum of matter was first suggested by Cartan in 1923 . However, it was only in 1925 that the modern concept of spin was introduced by Uhlenbeck and Goutsmith. This perhaps may explain why the Cartan conjecture was ignored for many years, and why torsion, when introduced into the field equations of general relativity, was unsuccessful related not to the spin, but to the electromagnetic field, in the attempt of unifying electromagnetic and gravitational interactions.

2.- The geometry of a Riemann-Cartan manifold U_4

Consider a four dimensional differentiable manifold, with a metric tensor $g_{\mu\nu}(x)$ which provides the infinitesimal distance ds between two nearby points with coordinates

x^μ and $x^\mu + dx^\mu$, i.e.

$$ds^2 = g_{\mu\nu}\, dx^\mu dx^\nu \tag{10.2}$$

and also equipped with an affine connection , $\Gamma_{\alpha\beta}{}^\mu$, in general nonsymmetric in the first two indices, which defines the infinitesimal variation of a vector, B^μ, under parallel displacement from x^μ to $x^\mu + dx^\mu$,

$$\delta B^\mu = -\Gamma_{\alpha\beta}{}^\mu B^\beta dx^\alpha \tag{10.3}$$

The affine connection $\Gamma_{\alpha\beta}{}^\mu$ is a set of 64 coefficients transforming non-homogeneously under a general coordinate transformation, but its antisymmetric part, the torsion, is a tensor. In fact, starting from eq.(10.3), and following the same procedure as in the case of a Riemann space (see Chapter II, Sect.3) we obtain

$$dB^\mu - \delta B^\mu = \partial_\alpha B^\mu dx^\alpha + \Gamma_{\alpha\beta}{}^\mu B^\beta dx^\alpha \tag{10.4}$$

and we can define then a covariant derivative in terms of the generalized connection $\Gamma_{\alpha\beta}{}^\mu$ as follows

$$\nabla_\alpha B^\mu = \partial_\alpha B^\mu + \Gamma_{\alpha\beta}{}^\mu B^\beta \tag{10.5}$$

In a similar way, for a covariant vector, B_μ , we get

$$\nabla_\alpha B_\mu = \partial_\alpha B_\mu - \Gamma_{\alpha\mu}{}^\beta B_\beta \tag{10.6}$$

and so on for higher rank tensors. Note that $\nabla_\alpha B^\mu$ reduces to the usual riemannian covariant derivative $B^\mu{}_{;\alpha}$ if the connection $\Gamma_{\alpha\beta}{}^\mu$ is replaced by the Christoffel connection $\left\{{\mu \atop \alpha\beta}\right\}$.

Since $\nabla_\alpha B^\mu$ and B^μ behave like tensors under the general coordinate transformation $x^\mu \to x'^\mu$, we have (see also Chapter II, Sect.2)

$$\nabla_\alpha B^\mu = \frac{\partial x'^\beta}{\partial x^\alpha} \frac{\partial x^\mu}{\partial x'^\nu} \left(\nabla_\beta B^\nu \right)' \tag{10.7}$$

$$B^\mu = \frac{\partial x^\mu}{\partial x'^\nu} B'^\nu \tag{10.8}$$

and by taking the partial derivative of this last equation

$$\partial_\alpha B^\mu = \frac{\partial^2 x^\mu}{\partial x^\alpha \partial x'^\nu} B'^\nu + \frac{\partial x^\mu}{\partial x'^\nu}\frac{\partial x'^\beta}{\partial x^\alpha}\frac{\partial B'^\nu}{\partial x'^\beta} \qquad (10.9)$$

Using (10.7-10.9) in eq.(10.5) we obtain the transformatio law for the affine connection

$$\Gamma_{\alpha\beta}{}^\mu = \Gamma'_{\gamma\delta}{}^\nu \frac{\partial x^\mu}{\partial x'^\nu}\frac{\partial x'^\gamma}{\partial x^\alpha}\frac{\partial x'^\delta}{\partial x^\beta} + \frac{\partial^2 x'^\nu}{\partial x^\alpha \partial x^\beta}\frac{\partial x^\mu}{\partial x'^\nu} \qquad (10.10)$$

(the same transformation rule for the connection was requi red in Chapter II, in order to obtain a covariant deriva- tive tranforming like a tensor; conversely, in this Chapte this rule has been derived from the assumption that the co variant derivative is a tensor).

Considering the antisymmetric part of eq.(10.10) we ha

$$\Gamma_{[\alpha\beta]}{}^\mu = Q_{\alpha\beta}{}^\mu = Q'_{\gamma\delta}{}^\nu \frac{\partial x^\mu}{\partial x'^\nu}\frac{\partial x'^\gamma}{\partial x^\alpha}\frac{\partial x'^\delta}{\partial x^\beta} \qquad (10.11)$$

which shows that the torsion $Q_{\alpha\beta}{}^\mu$ transforms like a third rank tensor. Therefore it cannot be locally eliminated, be cause if a tensor is vanishing at a given point, it is va- nishing everywhere.

In order to esplain the geometric meaning of the anti- symmetric part of the connection, it must be noted that th torsion is related to the translation of a vector, like curvature is related to the rotation, when a vector is dis placed along an infinitesimal path in a Riemann-Cartan ma- nifold (see Fig.10.1)

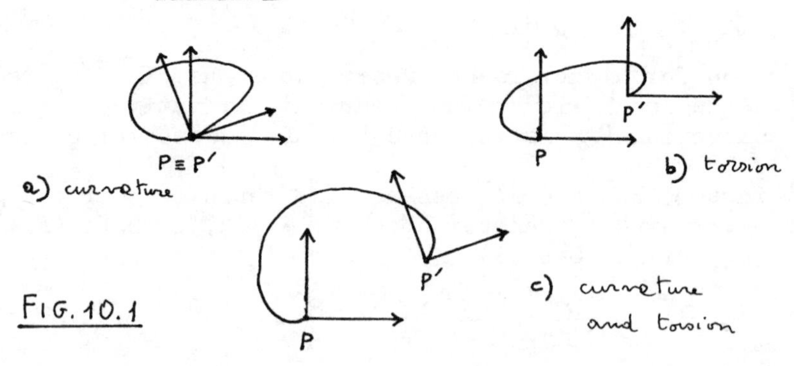

Fig.10.1. When a vector describes an infinitesimal closed path, and when this path is developed in the flat space ta gent to the manifold, we have a rotation (if there is only curvature) or a translation (if there is only torsion) or both (if there is curvature and torsion)

In other words, a closed contour (P, A_1, A_2, A_3) in an U_4 manifold becomes in general a non-closed contour (P, B_1, B_2, B_3) in the flat space E tangent to U_4 in P (see Fig.10.2)

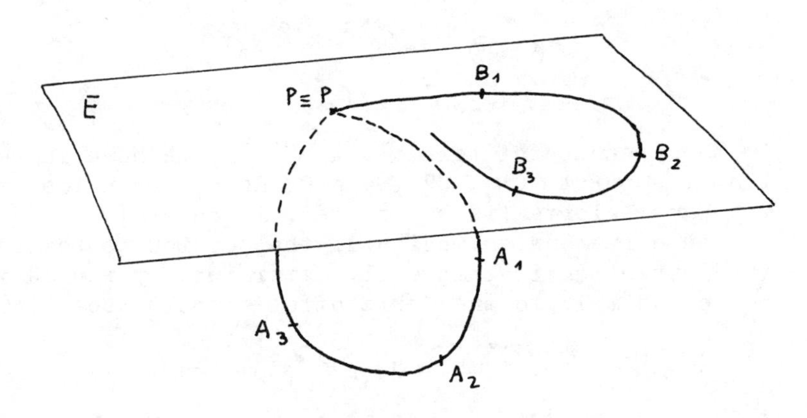

FIG. 10.2

Consider in fact the infinitesimal parallelogram PACB in the tangent space E, and put $\overline{PA} = dx'$, $\overline{PB} = dx''$ (see Fig.10.3)

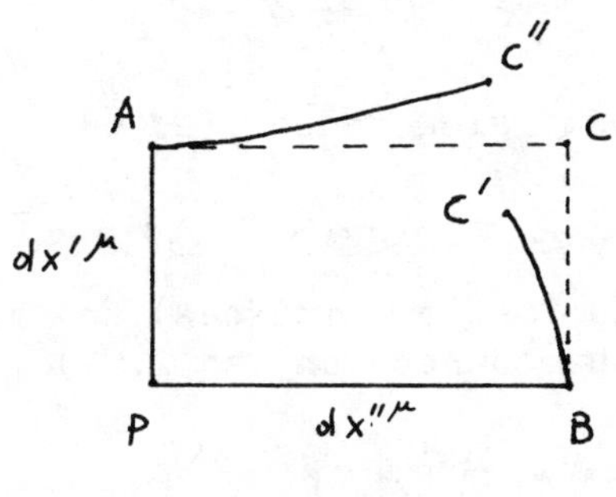

FIG. 10.3

Displacing dx' along dx" in the corresponding U_4 manifold we have

$$\delta\left(dx'^{\mu}\right) = -\ \Gamma_{\alpha\beta}{}^{\mu}\, dx'^{\beta}\, dx''^{\alpha} \qquad (10,12)$$

while for the displacement of dx" along dx'

$$\delta\left(dx''^{\mu}\right) = -\ \Gamma_{\alpha\beta}{}^{\mu}\, dx''^{\beta}\, dx'^{\alpha} \qquad (10,13)$$

Therefore

$$\delta(dx'^{\mu}) - \delta(dx''^{\mu}) = -\Gamma_{\alpha\beta}{}^{\mu}\left(dx'^{\beta}dx''^{\alpha} - dx'^{\alpha}dx''^{\beta}\right) =$$

$$= -\left(\Gamma_{\alpha\beta}{}^{\mu} - \Gamma_{\beta\alpha}{}^{\mu}\right) dx'^{\beta} dx''^{\alpha} =$$

$$= -2\, Q_{\alpha\beta}{}^{\mu}\, dx'^{\beta} dx''^{\alpha} \qquad (10.14)$$

in the presence of torsion, $Q_{\alpha\beta}{}^{\mu} \neq 0$, we have in general that the points C' , C" , and C do not coincide, so that the parallelogram is not closed, in general.

In a riemannian manifold, the torsion tensor is zero, the connection is completely determined by the metric and reduces simply to the Christoffel symbols (see 2.60)

$$\left\{ {\mu \atop \alpha\beta} \right\} = \frac{1}{2} g^{\mu\nu} \left(\partial_{\alpha} g_{\nu\beta} + \partial_{\beta} g_{\nu\alpha} - \partial_{\nu} g_{\alpha\beta} \right) \qquad (10.15)$$

In a general affine manifold A_4, however, the coefficient of the connection $\Gamma_{\alpha\beta}{}^{\mu}$ can be expressed in terms of the metric, of the torsion, and of the covariant derivative of the metric (as shown in Chapter II, Sect.3) , and we have, in general (see 2.65)

$$\Gamma_{\alpha\beta}{}^{\mu} = \left\{ {\mu \atop \alpha\beta} \right\} - K_{\alpha\beta}{}^{\mu} - V_{\alpha\beta}{}^{\mu} \qquad (10.16)$$

where $K_{\alpha\beta}{}^{\mu}$ is the contorsion tensor (2.66)

$$K_{\alpha\beta\mu} = -Q_{\alpha\beta\mu} - Q_{\mu\alpha\beta} + Q_{\beta\mu\alpha} = K_{\alpha[\beta\mu]} \qquad (10.$$

(antisymmetric in the last two indices) and $V_{\alpha\beta}{}^{\mu}$ is the non-metric part of the connection (eq.2.67)

$$V_{\alpha\beta\mu} = \frac{1}{2}\left(N_{\alpha\beta\mu} - N_{\mu\alpha\beta} - N_{\beta\mu\alpha} \right) = V_{(\alpha\beta)\mu} \qquad (1$$

given in terms of the non-metricity tensor

$$N_{\alpha\beta\mu} = \nabla_{\mu} g_{\alpha\beta} = N_{(\alpha\beta)\mu} =$$

$$= \partial_{\mu} g_{\alpha\beta} - \Gamma_{\mu\alpha}{}^{\nu} g_{\nu\beta} - \Gamma_{\mu\beta}{}^{\nu} g_{\alpha\nu} \qquad (10.19)$$

(remember that the general expression 10.16 can be obtaine starting from 10.19 and cyclically permuting indices , see Chapter II).

A Riemann-Cartan manifold U_4 is a particular case of

general affine manifold A_4 in which the metric tensor is covariantly constant, i.e.

$$N_{\alpha\beta\mu} = \nabla_\mu g_{\alpha\beta} = 0 \qquad (10.20)$$

This condition, which preserves scalar products (and then the invariance of lengths and angles) under parallel displacement, is usually called "metricity postulate", and secures a locally Minkowskian structure of the space-time, in agreement with special relativity. The connection, when the condition (10.20) is satisfied, is said to be "metric-compatible".

The connection of a Riemann-Cartan manifold U_4 is given then by

$$\Gamma_{\alpha\beta}{}^{\mu} = \left\{ {}^{\mu}_{\alpha\beta} \right\} - K_{\alpha\beta}{}^{\mu} =$$

$$= \left\{ {}^{\mu}_{\alpha\beta} \right\} + Q_{\alpha\beta}{}^{\mu} + Q^{\mu}{}_{\alpha\beta} - Q_{\beta}{}^{\mu}{}_{\alpha} \qquad (10,21)$$

It differs from a Riemann manifold V_4 only because of the presence of torsion, and reduces to the flat Minkowski space M_4 when torsion and curvature are both vanishing (see Fig.10.4).

The covariant derivatives in U_4 follow the same rules as in V_4, the only difference being the replacement of the Christoffel symbols with the generalized connection (10.21). It is important to note that, in U_4, the geodesic equation

$$\frac{d^2x^\mu}{ds^2} + \left\{ {}^{\mu}_{\alpha\beta} \right\} \frac{dx^\alpha}{ds} \frac{dx^\beta}{ds} \qquad (10.22)$$

can be obtained, as in V_4, by the variational principle

$$\delta \int (g_{\mu\nu}\, dx^\mu dx^\nu)^{1/2} = 0 \qquad (10.23)$$

(see Chapter II, Sect.4) . It defines curves of extremal length with respect to the metric, and it is different, in general, from the autoparallel equation

$$\frac{d^2x^\mu}{ds} + \Gamma_{\alpha\beta}{}^{\mu} \frac{dx^\alpha}{ds} \frac{dx^\beta}{ds} \qquad (10.24)$$

which defines curves along which a vector is displaced parallelely to itself, according to the affine connection of the manifold. In fact, taking the symmetric part of the

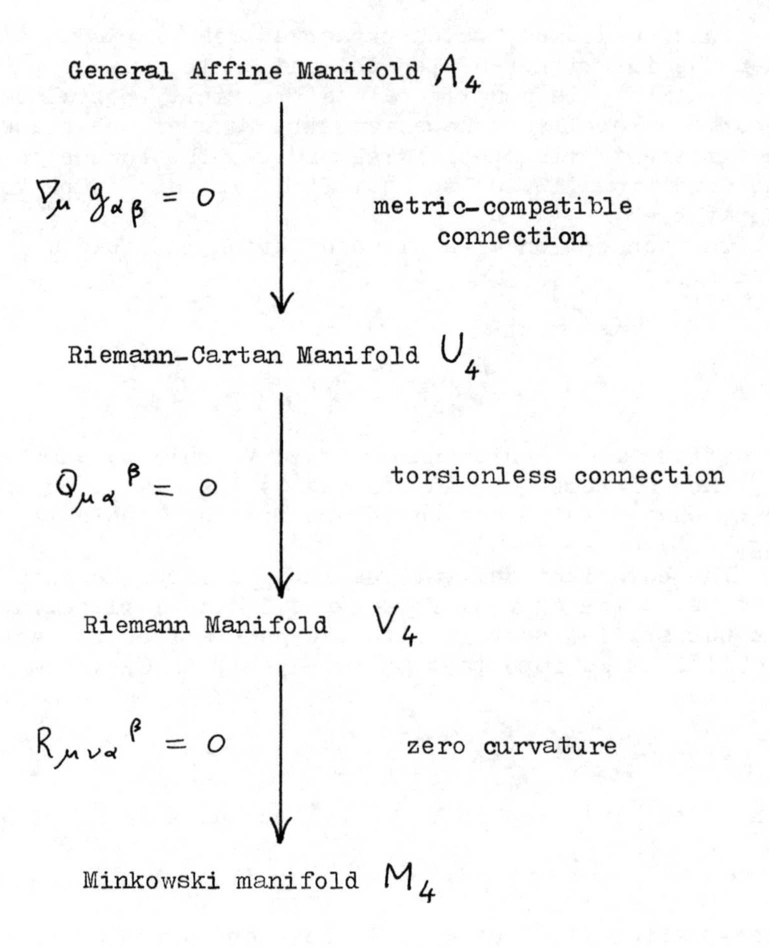

General Affine Manifold A_4
$\nabla_\mu g_{\alpha\beta} = 0$
metric-compatible connection
Riemann-Cartan Manifold U_4
$Q_{\mu\alpha}{}^{\beta} = 0$
torsionless connection
Riemann Manifold V_4
$R_{\mu\nu\alpha}{}^{\beta} = 0$
zero curvature
Minkowski manifold M_4

FIG. 10.4

connection (10.21) we have

$$\frac{d^2x^\mu}{ds^2} + \Gamma_{(\alpha\beta)}{}^\mu \frac{dx^\alpha}{ds}\frac{dx^\beta}{ds} =$$

$$= \frac{d^2x^\mu}{ds^2} + \left[-\left\{{\mu \atop \alpha\beta}\right\} + 2\,Q^\mu{}_{(\alpha\beta)}\right]\frac{dx^\alpha}{ds}\frac{dx^\beta}{ds} \qquad (10.25)$$

(because of the symmetry of the product $dx^\alpha dx^\beta = dx^\beta dx^\alpha$). This equation coincides with a geodesic only in a V_4 (where $Q_{\mu\nu\alpha} = 0$) , or in a U_4 manifold in which the torsion is a totally antisymmetric tensor,

$$Q_{\alpha\beta\mu} = Q_{[\alpha\beta\mu]} \qquad (10.26)$$

as in that case we have $Q_{\alpha(\beta\mu)} = 0$.

The curvature tensor is defined as in a Riemannian space, using however the U_4 connection instead of the Christoffel symbols. Therefore (see 3.7)

$$R_{\alpha\beta\mu}{}^\nu = \partial_\alpha \Gamma_{\beta\mu}{}^\nu - \partial_\beta \Gamma_{\alpha\mu}{}^\nu + \Gamma_{\alpha\lambda}{}^\nu \Gamma_{\beta\mu}{}^\lambda - \Gamma_{\beta\lambda}{}^\nu \Gamma_{\alpha\mu}{}^\lambda \qquad (10.27)$$

and using this explicit expression one may verify that the following identities are staisfied:

$$R_{(\alpha\beta)\mu\nu} = 0 = R_{\alpha\beta(\mu\nu)} \qquad (10.28)$$

$$R_{[\alpha\beta\mu]}{}^\nu = 2\,\nabla_{[\alpha} Q_{\beta\mu]}{}^\nu - 4\,Q_{[\alpha\beta}{}^\lambda Q_{\mu]\lambda}{}^\nu \qquad (10.29)$$

The Bianchi identity (3.20) is generalized as follows:

$$\nabla_{[\alpha} R_{\beta\mu]\nu}{}^\lambda = 2\,Q_{[\alpha\beta}{}^\rho R_{\mu]\rho\nu}{}^\lambda \qquad (10.30)$$

The Einstein tensor, as usual, is defined by

$$G_{\mu\nu} = R_{\mu\nu} - \tfrac{1}{2}\, g_{\mu\nu} R \qquad (10.31)$$

where $R_{\mu\nu} = R_{\alpha\mu\nu}{}^\alpha$ is the Ricci tensor, and $R = R_\mu{}^\mu$. It is important to stress, however, that in a Riemann-Cartan space $G_{\mu\nu}$ and $R_{\mu\nu}$ are no longer symmetric tensors. In fact, contracting (10.29) we have

$$\tfrac{3}{2} R_{[\alpha\mu\nu]}{}^\alpha = \tfrac{1}{2}\left(R_{\alpha\mu\nu}{}^\alpha + R_{\mu\nu\alpha}{}^\alpha + R_{\nu\alpha\mu}{}^\alpha\right) = R_{\alpha[\mu\nu]}{}^\alpha =$$

$$= R_{[\mu\nu]} = \nabla_\alpha T_{\mu\nu}{}^\alpha + 2\,Q_\alpha T_{\mu\nu}{}^\alpha \qquad (10.32)$$

(because $R_{\mu\nu\alpha}{}^{\alpha} = 0$) where Q_α is the trace of the torsion

$$Q_\alpha = Q_{\alpha\nu}{}^{\nu} \qquad (10.33)$$

and

$$T_{\mu\nu}{}^{\alpha} = Q_{\mu\nu}{}^{\alpha} + \delta^{\alpha}_{\mu} Q_\nu - \delta^{\alpha}_{\nu} Q_\mu \qquad (10.34)$$

is the so-called modified torsion tensor. By taking then the antisymmetric part of (10.31) , and using (10.32), we have

$$G_{[\mu\nu]} = R_{[\mu\nu]} = \nabla_\alpha T_{\mu\nu}{}^{\alpha} + 2 Q_\alpha T_{\mu\nu}{}^{\alpha} \qquad (10.35)$$

Therefore the Einstein tensor is in general nonsymmetric if the torsion is nonvanishing.

This geometrical property is physically related to the conservation of the total angular momentum in the presence of a nonzero intrinsic spin density, because, as we shall see in the following Section, this nonsymmetric Einstein tensor is proportional to the canonical energy-momentum tensor (remember the Appendix B of Chapter III, and see also Chapter XIV, eq.14.105).

Contracting the Bianchi identity (10.30), and using the definitions (10.31,34), we obtain

$$\nabla_\alpha G_\mu{}^{\alpha} = - 2 Q_\alpha G_\mu{}^{\alpha} - 2 Q_{\alpha\mu}{}^{\nu} G_\nu{}^{\alpha} + T_{\alpha\beta}{}^{\nu} R_{\mu\nu}{}^{\alpha\beta} \qquad (10.36)$$

In the absence of torsion ($Q_{\mu\nu}{}^{\alpha} = 0$) this condition is reduced to $G_\mu{}^{\nu}{}_{;\nu} = 0$. Therefore eq.(10.36) leads to a generalized conservation law for the matter energy-momentum tensor, as we shall see in Sect.4 of this Chapter.

Finally, we conclude this Section giving the expression for the Einstein tensor $G_{\mu\nu}$ and the scalar curvature R of an U_4 manifold, in which the usual riemannian parts $G_{\mu\nu}(\{\})$ and $R(\{\})$ are explicitly separated in order to put into evidence the torsion contributions.

Starting with the definition (10.27), inserting the explicit form (10.21) of the connection we obtain

$$G^{\mu\nu}(\Gamma) = G^{\mu\nu}(\{\}) + \overset{*}{\nabla}_\alpha (T^{\mu\nu\alpha} - T^{\nu\alpha\mu} + T^{\alpha\mu\nu}) +$$

$$+ 4 T^{\mu\alpha}{}_{[\beta} T^{\nu\beta}{}_{\alpha]} + 2 T^{\mu\alpha\beta} T^{\nu}{}_{\alpha\beta} - T^{\alpha\beta\mu} T_{\alpha\beta}{}^{\nu} -$$

$$- \frac{1}{2} g^{\mu\nu} \left(4 T_{\lambda}{}^{\alpha}{}_{[\beta} T^{\lambda\beta}{}_{\alpha]} + T^{\alpha\beta\lambda} T_{\alpha\beta\lambda} \right) \qquad (10.37)$$

and

$$\sqrt{-g}\, R(\Gamma) = \sqrt{-g}\, R(\{\}) + \sqrt{-g}\, T^{\alpha\mu\nu} K_{\nu\mu\alpha} +$$

$$+ \partial_\mu \left(2\sqrt{-g}\, K_\nu{}^{\mu\nu} \right) \qquad (10.38)$$

where we have defined, for simplicity,

$$\overset{*}{\nabla}_\alpha = \nabla_\alpha + 2 Q_\alpha \qquad (10.39)$$

and where $G^{\mu\nu}(\{\}) = G^{\nu\mu}(\{\})$ and $R(\{\})$ denote the Einstein tensor and the scalar curvature constructed with the Christoffel symbols $\left\{ {\mu \atop \alpha\beta} \right\}$. Obviously when $Q_{\mu\nu}{}^{\alpha} = 0$ then $T_{\mu\nu\alpha} = 0$ and $G_{\mu\nu} = G_{\mu\nu}(\{\})$, $R = R(\{\})$, i.e. in the absence of torsion all the U_4 geometrical objects are reduced to the usual riemannian ones.

3.- Field equations

The field equations of the Einstein-Cartan theory can be obtained from a variational principle, following the same procedure used in the case of general relativity (see Chapter III, Sect.3).

Suppose to have a Lagrangian density $\mathcal{L}_m$ for a classical field ψ , representing matter sources in the flat Minkowski space, and suppose that this Lagrangian involves only the first derivatives of the field, i.e. $\mathcal{L}_m = \mathcal{L}_m(\psi, \partial\psi, \eta)$.

When the gravitational interaction is introduced, the matter Lagrangian is to be generalized to become a scalar under general coordinate transformations. This can be arranged, according to the minimal coupling procedure, by replacing the Minkowski metric with the world metric tensor, $\eta_{\mu\nu} \rightarrow g_{\mu\nu}$, and the partial derivatives with the covariant ones, $\partial_\mu \rightarrow \nabla_\mu$ (this is not only an empirical procedure, as it finds also a deep theoretical motivation in the framework of a gauge theory of gravity, as we shall see in Chapter XII).

Moreover we must add, to the matter Lagrangian, a kinetic term for the gravitational field, $\mathcal{L}_g$: in analogy with general relativity, we choose $\mathcal{L}_g = R$, where R is the scalar curvature (10.38) for an U_4 manifold; therefo-

re $\mathcal{L}_g$ is different from the usual riemannian gravitationa
Lagrangian, even if formally similar.

This choice of $\mathcal{L}_g$, which at this point is completely arbitrary and may be justified only "a posteriori" comparing the theoretical predictions with the experimental dat is the simplest choice among all the other scalar objects that can be constructed using the curvature tensor. Other Lagrangians are of course possible (for example $R + R^2$), but in general they lead to gravitational theories with torsion which are different from the Einstein theory even in the limit of vanishing torsion.

The total action for the Einstein-Cartan theory is the

$$S = \int d^4x \sqrt{-g} \left\{ \mathcal{L}_m(\psi, \nabla\psi, g) - \frac{1}{2\chi} R(g, \partial g, Q) \right\} \tag{10.40}$$

Note that $\mathcal{L}_m$ contains the torsion because the covariant derivative is performed using the nonsymmetric U_4 connection (10.21).

It is also important to note that the coupling constan χ is the same both for the curvature and the torsion terms; it follows that the strength of the spin-torsion coupling is the same as that of the mass-curvature interaction (the possibility of different coupling constants, however, is not experimentally excluded at present, and this possibility has been considered by some authors).

The field equations are obtained by independent variation of the action with respect to ψ, $g_{\mu\nu}$ and $Q_{\mu\nu}{}^{\alpha}$. Instead of varying $Q_{\mu\nu\alpha}$, to obtain the torsion field equations, it is however convenient to perform the variation with respect to the contorsion $K_{\mu\nu\alpha}$, which appears explicitly in the gravitational Lagrangian (see 10.38) and also in the covariant derivatives of the matter fields: we obtain in this way equivalent field equations.

Performing the variations, we have the equations for the matter field

$$\frac{\delta(\sqrt{-g}\,\mathcal{L}_m)}{\delta\psi} = 0 \tag{10.41}$$

and for the gravitational field

$$\frac{\delta(\sqrt{-g}\,R)}{\delta g^{\mu\nu}} = 2\chi \frac{\delta(\sqrt{-g}\,\mathcal{L}_m)}{\delta g^{\mu\nu}} \tag{10.42}$$

$$\frac{\delta(\sqrt{-g}\,R)}{\delta K_{\alpha\beta\mu}} = 2\chi \frac{\delta(\sqrt{-g}\,\mathcal{L}_m)}{\delta K_{\alpha\beta\mu}} \tag{10.43}$$

where, in the Eulero-Lagrange formalism, the symbol $\delta A/\delta B$ denotes

$$\frac{\delta A}{\delta B} = \frac{\partial A}{\partial B} - \partial_\mu \frac{\partial A}{\partial(\partial_\mu B)} \tag{10.44}$$

as shown explicitly in the Appendix B of Chapter III.

The right-hand side of (10.42) is just proportional to the usual dynamic energy-momentum tensor (see 3.67)

$$T_{\mu\nu} = \frac{2}{\sqrt{-g}} \frac{\delta(\sqrt{-g}\,\mathcal{L}_m)}{\delta g^{\mu\nu}} \tag{10.45}$$

Therefore, if we define the dynamican spin density tensor $S_{\mu\beta\alpha}$ as follows

$$S^{\mu\beta\alpha} = \frac{1}{\sqrt{-g}} \frac{\delta(\sqrt{-g}\,\mathcal{L}_m)}{\delta K_{\alpha\beta\mu}} \tag{10.46}$$

(for the definition of the canonical spin density tensor see the Appendix B of Chapter III) the gravitational field equations (10.42,43) may be rewritten as

$$\frac{1}{\sqrt{-g}} \frac{\delta(\sqrt{-g}\,R)}{\delta g^{\mu\nu}} = \chi T_{\mu\nu} \tag{10.47}$$

$$\frac{1}{2\sqrt{-g}} \frac{\delta(\sqrt{-g}\,R)}{\delta K_{\alpha\beta\mu}} = \chi S^{\mu\beta\alpha} \tag{10.48}$$

in which the matter sources $T_{\mu\nu}$ and $S_{\mu\alpha\beta}$ appear explicitly. Note that the dynamical energy-momentum tensor is symmetric, $T_{\mu\nu} = T_{\nu\mu}$, and the spin density tensor is antisymmetric in the first two indices, $S_{\mu\beta\alpha} = S_{[\mu\beta]\alpha}$, as $K_{\alpha\beta\mu} = K_{\alpha[\beta\mu]}$.

Using the definition of the curvature tensor we can obtain directly

$$\frac{1}{\sqrt{-g}} \frac{\delta(\sqrt{-g}\,R)}{\delta g^{\mu\nu}} = G_{\mu\nu} - \overset{*}{\nabla}_\alpha \left(T_{\mu\nu}{}^{\alpha} - T_\nu{}^{\alpha}{}_\mu + T^{\alpha}{}_{\mu\nu}\right) \tag{10.49}$$

and from (10.38) we have

$$\frac{1}{2\sqrt{-g}} \frac{\delta(\sqrt{-g}\,R)}{\delta K_{\alpha\beta\mu}} = T^{\mu\beta\alpha} \tag{10.50}$$

(remember that a total divergence does not contribute to the variation of the action, as stressed in Chapter III). The field equations of the Einstein-Cartan theory are then

$$G^{\mu\nu} - \overset{*}{\nabla}_\alpha \left(T^{\mu\nu\alpha} - T^{\nu\alpha\mu} + T^{\alpha\mu\nu}\right) = \chi T^{\mu\nu} \tag{10.51}$$

$$T^{\mu\nu\alpha} = \chi S^{\mu\nu\alpha} \tag{10.52}$$

or, including the spin contributions directly into the energy-momentum tensor, introducing the modified source

$$\theta^{\mu\nu} = T^{\mu\nu} + \overset{*}{\nabla}_\alpha \left(S^{\mu\nu\alpha} - S^{\nu\alpha\mu} + S^{\alpha\mu\nu}\right) \tag{10.53}$$

they may be written

$$G^{\mu\nu} = \chi\, \theta^{\mu\nu} \tag{10.54}$$

$$T^{\mu\nu\alpha} = \chi S^{\mu\nu\alpha} \tag{10.55}$$

The tensor $\theta_{\mu\nu}$ can be shown to coincide with the canonical energy-momentum tensor (see Chapter 3, Appendix B, in particular eq.B32), which in curvilinear coordinates is defined as[1]

$$\sqrt{-g}\, \theta_\mu{}^\nu = \delta_\mu{}^\nu \mathcal{L}_m - \nabla_\mu \psi \frac{\partial \mathcal{L}_m}{\partial(\partial_\nu \psi)} \tag{10.56}$$

as which is related to $T_{\mu\nu}$ by a symmetrization procedure. The canonical tensor, unlike the dynamical tensor $T_{\mu\nu}$, is generally nonsymmetric: from the definition (10.53) we have in fact

$$\theta^{[\mu\nu]} = \overset{*}{\nabla}_\alpha S^{\mu\nu\alpha} = \nabla_\alpha S^{\mu\nu\alpha} + 2 Q_\alpha S^{\mu\nu\alpha} \tag{10.57}$$

which generalize eq.B.31 of Chapter 3 (note that this relation can also directly obtained from the U_4 identity (10.35), using the field equations (10.54,55)).

From (10.57,54,55) we can see clearly that it is spin that generates a nonsymmetric part in the canonical energy-momentum tensor and then, according to the Einstein-Cartan theory, produces a deviation from the Riemann geometry. The second field equation (10.55) can also be rewritten as

$$Q_{\alpha\beta}{}^{\mu} = \chi\left(S_{\alpha\beta}{}^{\mu} + \frac{1}{2}\delta_{\alpha}^{\mu} S_{\beta\nu}{}^{\nu} - \frac{1}{2}\delta_{\beta}^{\mu} S_{\alpha\nu}{}^{\nu}\right) \qquad (10,58)$$

and then, in the absence of spin, $S_{\mu\nu\alpha} = 0$, we have $Q_{\mu\nu}{}^{\alpha}=0$, $G_{\mu\nu}= G_{\mu\nu}(\{\})$ and $\theta_{\mu\nu} = T_{\mu\nu}$. Since (10.58) is an equation of algebraic type, and not a differential equation, no propagation of torsion is allowed, so that torsion can be nonzero only inside matter, in the presence of particles with aligned spins. In vacuum the Einstein-Cartan field equations are reduced then exactly to the equations of general relativity.

Eliminating the torsion in eq.(10.54), i.e. replacing everywhere $Q_{\mu\nu\alpha}$ with the spin tensor $S_{\mu\nu\alpha}$, according to (10.55), using eqs.(10.37,53), we obtain the following generalization of the Einstein equations in the case of spinning sources:

$$G^{\mu\nu}(\{\}) = \chi T^{\mu\nu} + \chi^2 \tau^{\mu\nu} \qquad (10,59)$$

where

$$\tau^{\mu\nu} = -4 S^{\mu\alpha}{}_{[\beta} S^{\nu\beta}{}_{\alpha]} - 2 S^{\mu\alpha\beta} S^{\nu}{}_{\alpha\beta} + S^{\alpha\beta\mu} S_{\alpha\beta}{}^{\nu} +$$

$$+ \frac{1}{2} g^{\mu\nu}\left(4 S_{\alpha}{}^{\beta}{}_{[\gamma} S^{\alpha\gamma}{}_{\beta]} + S^{\alpha\beta\gamma} S_{\alpha\beta\gamma}\right) \qquad (10,60)$$

represents a correction to the dynamical energy-momentum, which takes into account the spin contributions to the geometry of the manifold, corresponding to a spin-spin contact interaction (which is however extremely weak, being proportional to $\chi^2 = 64\pi^2 G^2/c^8$).

It should be noted, therefore, that even if the torsion in the Einstein-Cartan theory cannot propagate outside matter, the spin density acts as a source of gravitation, producing a modified energy-momentum tensor (a simple example is considered in Chapter XI, Sect.4). A modification of the spin distribution implies, through the field equations (10.60), a modification of the metric tensor (at least in pronciple) which propagates to infinity. The spin content of a body can influence then the geometry also outside the body, even if indirectly(through the metric tensor) and very weakly.

4.- Bianchi identities and equations of motion

In a Riemann-Cartan manifold U_4 , the equations of motion cannot coincide in general with the geodesic equation (10.22), otherwise the trajectories of the test particles would not be influenced by the torsion, and then dynamics would be identical to that we have in a Riemann manifold.

Nor can be represented by the equation of the autoparallel curves (10.24), because according to those equation all the particles are affected in the same way by torsion, regardless of their spin content, so that even the spinless particles would follow a trajectory which deviates from a geodesic.

The correct equation of motion for the U_4 theory shou contain a term due to the spin-torsion interaction, so tha only spinning particles can "feel" torsion, and then modify their geodesic path.

In order to obtain such equation, we can start from th Bianchi identity and the energy-momentum conservation law, following the procedure developed by Papapetrou in general relativity[2] .

In general relativity, the contracted Bianchi identity (3.34)

$$G^{\mu\nu}(\{\,\})_{;\nu} = 0 \qquad (10.61)$$

implies, through the field equations (3.40)

$$G^{\mu\nu}(\{\,\}) = \chi T^{\mu\nu} \qquad (10.62)$$

the covariant conservation law

$$T^{\mu\nu}{}_{;\nu} = 0 \qquad (10.63)$$

which can be written explicitly as

$$\partial_\nu T^{\mu\nu} + \left\{{}^{\mu}_{\nu\alpha}\right\} T^{\alpha\nu} + \left\{{}^{\nu}_{\nu\alpha}\right\} T^{\mu\alpha} =$$

$$= \partial_\nu T^{\mu\nu} + \left\{{}^{\mu}_{\nu\alpha}\right\} T^{\alpha\nu} + T^{\mu\alpha}\partial_\alpha (\ln\sqrt{-g}) = 0 \qquad (10.6$$

(remember eq.3.29 for the Christoffel symbols).

Multiplying by $\sqrt{-g}$ we have then

$$\partial_\nu (\sqrt{-g}\, T^{\mu\nu}) + \sqrt{-g}\left\{{}^{\mu}_{\nu\alpha}\right\} T^{\alpha\nu} = 0 \qquad (10,65)$$

Now we integrate this equation over a three-dimensional space-like hypersurface (t = const); considering the energy-momentum tensor of a test particle, different from zero only inside a world tube of very small spatial section, we can apply the Gauss theorem to obtain

$$\int d^3x' \, \partial_\nu (\sqrt{-g}\, T^{\mu\nu}) + \int d^3x' \sqrt{-g} \left\{ {\mu \atop \nu\alpha} \right\} T^{\alpha\nu} =$$

$$= \frac{d}{dt} \int d^3x' \sqrt{-g}\, T^{\mu 4} + \int d^3x' \sqrt{-g} \left\{ {\mu \atop \nu\alpha} \right\} T^{\alpha\nu} = 0 \tag{10.66}$$

Denoting with x^μ the coordinates of the center of mass of the test particle, and expanding the external gravitational field in power series inside the particle, we have

$$\left\{ {\mu \atop \nu\alpha} \right\} = \left\{ {\mu \atop \nu\alpha} \right\}_o + \partial_\beta \left\{ {\mu \atop \nu\alpha} \right\}_o (x'^\beta - x^\beta) + \cdots \tag{10.67}$$

where the subscript "o" denotes the value of the Christoffel symbols at the position of the particle, $x'^\mu = x^\mu$.

In first approximation (the so-called "pole-particle" approximation) we suppose that

$$\int d^3x' \sqrt{-g}\, T^{\mu\nu} \neq 0 \tag{10.68}$$

but

$$\int d^3x' \sqrt{-g}\, (x'^\beta - x^\beta)\, T^{\mu\nu} = 0 \tag{10.69}$$

From (10.66) we obtain then (omitting the subscript "o")

$$\frac{d}{dt} \int d^3x' \sqrt{-g}\, T^{\mu 4} + \left\{ {\mu \atop \nu\alpha} \right\} \int d^3x' \sqrt{-g}\, T^{\alpha\nu} = 0 \tag{10.70}$$

Defining

$$p^\mu u^\nu = \frac{dt}{ds} \int d^3x' \sqrt{-g}\, T^{\mu\nu} \tag{10.71}$$

where s is the proper time along the particle world line, p^μ the four momentum vector of the particle, and

$$u^\mu = \frac{dx^\mu}{ds} \tag{10.72}$$

is the four velocity, and multiplying (10.70) by $u^4 = dt/ds$, we have

$$\frac{d}{ds} \left(\frac{ds}{dt} p^\mu u^4 \right) + \left\{ {\mu \atop \nu\alpha} \right\} p^\alpha u^\nu =$$

$$= \frac{dp^\mu}{ds} + \left\{ {\mu \atop \nu\alpha} \right\} p^\alpha u^\nu = 0 \qquad (10.73)$$

which is the equation of motion obtained, in first approximation, from the covariant conservation law of the energy-momentum tensor.

In general relativity $p^\mu = m\,u^\mu$, and then (10.73) reduces to the geodesic equation

$$\frac{du^\mu}{ds} + \left\{ {\mu \atop \nu\alpha} \right\} u^\nu u^\alpha = 0 \qquad (10.74)$$

In second approximation, for a "dipole-particle" (wher the integrals 10.69 are no longer vanishing), one obtains with the same procedure a force due to the coupling betwe the curvature tensor and the intrinsic angular momentum of the test body[2] .

In the Einstein-Cartan theory, on the contrary, the interaction between spin and curvature is directly obtained already in first approximation.

From the contracted Bianchi identity (10.36)

$$\nabla_\nu G^{\mu\nu} + 2\,Q_\nu G^{\mu\nu} + 2\,Q_\alpha{}^{\mu\nu} G_\nu{}^\alpha - T_{\alpha\beta\nu} R^{\mu\nu\alpha\beta} = 0$$

and the field equations (10.54,55)

$$G^{\mu\nu} = \chi\, \Theta^{\mu\nu}$$

$$T^{\mu\nu\alpha} = \chi\, S^{\mu\nu\alpha}$$

we obtain in fact the following generalized conservation law

$$\nabla_\nu \Theta^{\mu\nu} + 2\,Q_\nu \Theta^{\mu\nu} + 2\,Q_\alpha{}^{\mu\nu} \Theta_\nu{}^\alpha - S_{\alpha\beta\nu} R^{\mu\nu\alpha\beta} = 0 \qquad (1$$

The U_4 covariant derivative is given explicitly by

$$\nabla_\nu \Theta^{\mu\nu} = \partial_\nu \Theta^{\mu\nu} + \Gamma_{\nu\alpha}{}^\mu \Theta^{\alpha\nu} + \Gamma_{\nu\alpha}{}^\nu \Theta^{\mu\alpha} =$$

$$= \partial_\nu \Theta^{\mu\nu} + \left\{ {\mu \atop \nu\alpha} \right\} \Theta^{\alpha\nu} + \left\{ {\nu \atop \nu\alpha} \right\} \Theta^{\mu\alpha} - K_{\nu\alpha}{}^\mu \Theta^{\alpha\nu}$$

$$- K_{\nu\alpha}{}^\nu \Theta^{\mu\alpha} \qquad (10.76)$$

Multiplying (10.75) by $\sqrt{-g}$ we have then

$$\partial_\nu (\sqrt{-g}\,\theta^{\mu\nu}) + \left\{ {\mu \atop \nu\alpha} \right\} \theta^{\alpha\nu}\sqrt{-g} + \sqrt{-g}\Big[- K_{\nu\alpha}{}^{\mu}\theta^{\alpha\nu} - K_{\nu\alpha}{}^{\nu}\theta^{\mu\alpha} + 2Q_\nu\theta^{\mu\nu} + 2Q^{\nu\mu\alpha}\theta_{\alpha\nu} - S_{\alpha\beta\nu}R^{\mu\nu\alpha\beta}\Big] = 0 \qquad (10.77)$$

which generalizes the previous eq.(10.65).

This equation can be simplified. In fact, from the definition (10.17) of the contorsion tensor we have

$$K_{\nu\alpha}{}^{\nu} = -Q_{\nu\alpha}{}^{\nu} - Q^{\nu}{}_{\nu\alpha} + Q_{\alpha}{}^{\nu}{}_{\nu} = 2Q_{\alpha\nu}{}^{\nu} = 2Q_\alpha \qquad (10.78)$$

and then the sum of the second and third term in the square brackets of (10.77) is vanishing. Moreover we note that

$$-K^{\nu\alpha\mu}\theta_{\alpha\nu} + 2Q^{\nu\mu\alpha}\theta_{\alpha\nu} = \left\{-K^{(\nu\alpha)\mu} - 2Q^{\mu(\nu\alpha)}\right\}\theta_{(\alpha\nu)} + \left\{-K^{[\nu\alpha]\mu} - 2Q^{\mu[\nu\alpha]}\right\}\theta_{[\alpha\nu]} \qquad (10.79)$$

and that

$$K^{(\nu\alpha)\mu} = -Q^{(\nu\alpha)\mu} - Q^{\mu(\nu\alpha)} - Q^{\mu(\alpha\nu)} = -2Q^{\mu(\nu\alpha)} \qquad (10.80)$$

$$K^{[\nu\alpha]\mu} = -Q^{[\nu\alpha]\mu} - Q^{\mu[\nu\alpha]} - Q^{\mu[\alpha\nu]} = -Q^{[\nu\alpha]\mu} = -Q^{\nu\alpha\mu} \qquad (10.81)$$

so that

$$-K^{[\nu\alpha]\mu} - 2Q^{\mu[\nu\alpha]} = Q^{\nu\alpha\mu} - Q^{\mu\nu\alpha} + Q^{\mu\alpha\nu} = -Q^{\mu\nu\alpha} - Q^{\alpha\mu\nu} + Q^{\nu\alpha\mu} = K^{\mu\nu\alpha} \qquad (10.82)$$

Therefore (10.79) gives

$$-K^{\nu\alpha\mu}\Theta_{\alpha\nu}+2Q^{\nu\mu\alpha}\Theta_{\alpha\nu}=K^{\mu\nu\alpha}\Theta_{[\alpha\nu]} \qquad (10.83)$$

and the conservation equation (10.77) reduces simply to

$$\partial_\nu(\sqrt{-g}\,\Theta^{\mu\nu})+\sqrt{-g}\left\{{\mu \atop \nu\alpha}\right\}\Theta^{(\alpha\nu)}+\sqrt{-g}\,K^{\mu}{}_{\nu\alpha}\Theta^{[\alpha\nu]}-$$

$$-\sqrt{-g}\,S_{\alpha\beta\nu}R^{\mu\nu\alpha\beta}=0 \qquad (10.84)$$

The antisymmetric part of the canonical energy-momentum can be expressed in terms of the spin density, according to the U_4 relation (19.57)

$$\Theta^{[\mu\nu]}=\overset{*}{\nabla}_\alpha S^{\mu\nu\alpha} \qquad (10.85)$$

The conservation law can be written then in the final form

$$\partial_\nu(\sqrt{-g}\,\Theta^{(\mu\nu)})+\sqrt{-g}\left\{{\mu \atop \nu\alpha}\right\}\Theta^{(\alpha\nu)}+\sqrt{-g}\,\Phi^{\mu}=0 \qquad (10.86)$$

where

$$\Phi^{\mu}=\frac{1}{\sqrt{-g}}\partial_\mu(\sqrt{-g}\,\Theta^{[\mu\nu]})+K^{\mu}{}_{\nu\alpha}\Theta^{[\alpha\nu]}-$$

$$-S_{\alpha\beta\nu}.R^{\mu\nu\alpha\beta}=$$

$$=\frac{1}{\sqrt{-g}}\partial_\nu\left(\sqrt{-g}\,\overset{*}{\nabla}_\alpha S^{\mu\nu\alpha}\right)+K^{\mu}{}_{\nu\alpha}\overset{*}{\nabla}_\beta S^{\alpha\nu\beta}-S_{\alpha\beta\nu}R^{\mu\nu\alpha\beta}$$

$$(10.87)$$

represents the spin contributions to the equation of motion.

Following the same procedure as before, i.e. integrating over a space-like section of the world tube of the test particle and expanding the gravitational field in power series, we obtain from (10.86), in first approximation,

$$\int d^3x'\sqrt{-g}\,\partial_\nu(\sqrt{-g}\,\Theta^{(\mu\nu)})+\int d^3x'\sqrt{-g}\left\{{\mu \atop \nu\alpha}\right\}\Theta^{(\alpha\nu)}+\int d^3x'\sqrt{-g}\,\Phi^{\mu}=$$

$$=\frac{d}{dt}\int d^3x'\sqrt{-g}\,\Theta^{(\mu4)}+\left\{{\mu \atop \nu\alpha}\right\}\int d^3x'\sqrt{-g}\,\Theta^{(\alpha\nu)}+\int d^3x'\sqrt{-g}\,\Phi^{\mu}$$

$$(10.88)$$

Defining

$$p^{\mu}u^{\nu} = \frac{dt}{ds}\int d^3x'\sqrt{-g}\;\theta^{(\mu\nu)} \tag{10.89}$$

and multiplying (10.88) by dt/ds, we obtain finally the equation of motion for the Einstein-Cartan theory in the form

$$\frac{dp^{\mu}}{ds} + \left\{ {\mu \atop \nu\alpha} \right\} p^{\alpha}u^{\nu} + F^{\mu} = 0 \tag{10.90}$$

where

$$F^{\mu} = \frac{dt}{ds}\int d^3x'\sqrt{-g}\;\Phi^{\mu} \tag{10.91}$$

represents a force term arising from the interaction of the spin with the geometry of the U_4 manifold. If the test particle is spinless (or if the macroscopic test body is unpolarized) then $S_{\mu\nu\alpha} = 0$, $\Phi^{\mu} = 0$, $\theta^{\mu\nu} = T^{\mu\nu}$, and the equation of motion reduces to the usual geodesic equation (10.74).

5.- The Dirac equation in the U_4 theory

The Lagrangian density for a Dirac spinor with mass m, in the flat Mikowski space, is given by (we put $\hbar = c = 1$)

$$\mathcal{L} = \frac{i}{2}\left(\bar{\psi}\gamma^{k}\partial_{k}\psi - \partial_{k}\bar{\psi}\gamma^{k}\psi\right) - m\bar{\psi}\psi \tag{10.92}$$

where γ^k are the four constant Dirac matrices, and $\bar{\psi} = \psi^{+}\gamma^{4}$ (in this Section, Latin indices run from 1 to 4 and denote Lorentz indices, i.e. indices transforming covariantly under a Lorentz transformation in flat space).

By varying the action with respect to $\bar{\psi}$ (see Chapter III, Appendix B)

$$\frac{\delta\mathcal{L}}{\delta\bar{\psi}} = \frac{\partial\mathcal{L}}{\partial\bar{\psi}} - \partial_{k}\frac{\partial\mathcal{L}}{\partial(\partial_{k}\bar{\psi})} \tag{10.93}$$

we obtain the Dirac equation

$$i\gamma^{k}\partial_{k}\psi - m\psi = 0 \tag{10.94}$$

and varying with respect to ψ we have

$$i\,\partial_{k}\bar{\psi}\gamma^{k} + m\bar{\psi} = 0 \tag{10.95}$$

This equation may be obtained also directly by taking the hermitian conjugate of (10.94) and multipliying from the right by γ^4. Remember that γ^4 anticommutes with $\gamma^1, \gamma^2, \gamma^3$ and that, with our conventions, $(\gamma^4)^+ = \gamma^4$, $(\gamma^1)^+ = -\gamma^1$, $(\gamma^2)^+ = -\gamma_2$, $(\gamma^3)^+ = -\gamma_3$.

Multiplying from the left (10.94) by $(i\gamma^i \partial_i + m)$ we get

$$(\gamma^i \gamma^K \partial_i \partial_K + m^2)\psi = 0 \tag{10.96}$$

and the Klein-Gordon equation is obtained

$$(\Box + m^2)\psi = 0 \tag{10.97}$$

(which expresses the relativistic constraint $-E^2 + p^2 + m^2 = 0$ in terms of differential operators) provided that the following anticommutation rules (which define the Dirac matrices) are satisfied

$$\gamma^{(i}\gamma^{K)} = \frac{1}{2}(\gamma^i\gamma^K + \gamma^K\gamma^i) = \eta^{iK} \tag{10.98}$$

This condition holds in the flat Minkowski space; in the presence of a gravitational foeld, however, (10.98) is to be generalized by introducing new matrices $\gamma^\mu(x)$ such that

$$\gamma^{(\mu}\gamma^{\nu)} = g^{\mu\nu} \tag{10.99}$$

(Greek letters denote world indices transforming covariantly as tensor indices under a general coordinate transformation).

To this aim, the usual procedure is to introduce, at each point of the space-time manifold, a set of four vector fields, $\{e_\mu^i(x)\}$, called "tetrads" or "vierbeins", spanning the local Minkowski space tangent to the U_4 manifold at the given point. The anholonomic Lorentz indice "i" of e_μ^i denotes the vector, while the holonomic world indice "μ" denotes the covariant components of the vector e^i in the curved U_4 manifold. The following orthonormality relations are satisfied by the vierbein field e_μ^i

$$g_{\mu\nu} = e_\mu{}^i e_\nu{}^K \eta_{iK}$$
$$\eta^{iK} = e_\mu{}^i e_\nu{}^K g^{\mu\nu} \tag{10.100}$$

and its inverse $e^{\mu}{}_{i}$

$$g^{\mu\nu} = e^{\mu}{}_{i}\, e^{\nu}{}_{k}\, \eta^{ik}$$
$$\eta_{ik} = e^{\mu}{}_{i}\, e^{\nu}{}_{k}\, g_{\mu\nu} \tag{10.101}$$

where

$$e^{\mu}{}_{i} = g^{\mu\nu}\, \eta_{ik}\, e_{\nu}{}^{k} \tag{10.102}$$

Note that Latin indices are raised and lowered by the Minkowski metric η^{ik}, while Greek indices by the metric $g_{\mu\nu}$ of the world manifold U_4.

In general, given a world tensor $B_{\mu\nu}$, its corresponding components B_{ik} in the flat tangent manifold can be obtained directly contracting the indices with the vierbein fields as follows

$$B_{ik} = e^{\mu}{}_{i}\, e^{\nu}{}_{k}\, B_{\mu\nu}$$
$$B^{ik} = e_{\mu}{}^{i}\, e_{\nu}{}^{k}\, B^{\mu\nu} \tag{10.103}$$

and vice-versa

$$B_{\mu\nu} = e_{\mu}{}^{i}\, e_{\nu}{}^{k}\, B_{ik}$$
$$B^{\mu\nu} = e^{\mu}{}_{i}\, e^{\nu}{}_{k}\, B^{ik} \tag{10.104}$$

It is important to stress that, if $B_\mu B_\nu$ is a world tensor, i.e. a tensor under general coordinate transformations, then $B_i B_k$ is a world scalar, but it transforms like a tensor with respect to the local Lorentz transformations which rotate the vierbein frame in the local tangent space. Obviously

$$B_\mu B^\mu = g^{\mu\nu} B_\mu B_\nu = g^{\mu\nu} e_{\mu}{}^{i} e_{\nu}{}^{k} B_i B_k = \eta^{ik} B_i B_k = B_i B^i \tag{10.105}$$

is both a world scalar and a Lorentz scalar.

In the absence of gravity, the world metric tensor reduces to the Minkowski metric, $g_{\mu\nu} = \eta_{\mu\nu}$, and the vierbein field is given by $e_{\mu}{}^{i} = \delta_{\mu}{}^{i}$, $e^{\mu}{}_{i} = \delta^{\mu}{}_{i}$ (see 10.100,101).

Using the vierbein field, the Dirac matrices $\gamma^\mu(x)$ for the U_4 manifold can be defined as

$$\gamma^{\mu} = e^{\mu}{}_{k}\, \gamma^{k} \tag{10.106}$$

where γ^{k} are the (constant) flat-space Dirac matrices. We have in fact, from (10.101),

$$\gamma^{(\mu}\gamma^{\nu)} = e^{(\mu}{}_{i}\, e^{\nu)}{}_{k}\, \gamma^{i}\gamma^{k} = e^{\mu}{}_{i}\, e^{\nu}{}_{k}\, \gamma^{(i}\gamma^{k)} =$$

$$= e^{\mu}{}_{i}\, e^{\nu}{}_{k}\, \eta^{ik} = g^{\mu\nu} \tag{10.107}$$

In order to obtain the Dirac equation in an U_4 manifold starting with the flat-space Lagrangian (10.92), we must replace the Dirac matrices γ^{k} with γ^{μ}, and, according to the minimal couplig procedure, the partial derivatives ∂_{k} with the covariant ones ∇_{μ} . Finally, the action is to be completed by adding the Lagrangian for the gravitational field, $\sqrt{-g}\, R / 2\chi$. We are led then to the following total Lagrangian density

$$\sqrt{-g}\, \mathcal{L} = \sqrt{-g} \left[\frac{i}{2} \left(\bar{\psi} \gamma^{\mu} \nabla_{\mu} \psi - \nabla_{\mu} \bar{\psi} \gamma^{\mu} \psi \right) - m \bar{\psi} \psi - \right.$$
$$\left. - \frac{1}{2\chi} R \right] \tag{10.108}$$

where R is given by (10.38), and the factor

$$\sqrt{-g} = \left[- \det g_{\mu\nu} \right]^{1/2} = \left[- \det \left(e_{\mu}{}^{i}\, e_{\nu}{}^{k}\, \eta_{ik} \right) \right]^{1/2} =$$

$$= \left[- (\det e_{\mu}{}^{i})^{2} \det \eta_{ik} \right]^{1/2} = \det e_{\mu}{}^{i} \equiv e \tag{10.109}$$

has to be introduced to make the Lagrangian a scalar density (see Chapter III, Sect.3).

At this point, an explicit expression for the covariant derivative of a spinor field, $\nabla_{\mu}\psi$, is needed.

To this aim, we note that the transformation properties of a spinor field can be locally defined on the flat Minkowski space tangent to the world manifold (it should be mentioned, however, that recently the possibility of obtaining Dirac-like equations directly in curved space, in terms of infinite components "world spinors", has been considered by Ne'eman and Sijački[3]). Consider a vector Lorentz transformation in the flat tangent space:

$$x'^{i} = \Lambda^{i}{}_{k}\, x^{k} \tag{10.110}$$

where $\Lambda^{i}{}_{k}$ is a Lorentz matrix which satisfies

$$\eta_{ij}\Lambda^{i}{}_{k}\Lambda^{j}{}_{l} = \eta_{kl} \tag{10.111}$$

The corresponding transformation law for a spinor field can be written as follows

$$\psi'(x') = U(\Lambda)\,\psi(x) \tag{10.112}$$

where $U(\Lambda)$ is a 4x4 constant matrix representing the Lorentz transformation (the spinor indices are not written explicitly). The Dirac equation (10.94) is transformed as

$$i\gamma^{i}\frac{\partial}{\partial x'^{i}}\psi'(x') - m\psi'(x') =$$

$$= i\gamma^{i}\Lambda_{i}{}^{k}\partial_{k}U\psi - mU\psi = 0 \tag{10.113}$$

Multiplying from the left by U^{-1}, and imposing on the Dirac equation to be invariant in form under a Lorentz transformation, we obtain the following condition for the matrix U :

$$U^{-1}\gamma^{i}U = \gamma^{k}\Lambda^{i}{}_{k} \tag{10.114}$$

(we have used the property 10.111).

Consider now an infinitesimal transformation

$$\Lambda_{ik} \simeq \eta_{ik} + \omega_{ik} \tag{10.115}$$

where $\omega_{ik} = \omega_{[ik]}$, so that (10.111) is satisfied, and parametrize the matrix U as follows

$$U = 1 + \frac{1}{2}\omega_{ik}S^{ik} \tag{10.116}$$

The ω_{ik} are six constant infinitesimal parameters, and S^{ik} are the generators of the infinitesimal Lorentz transformation. The condition (10.114), obtained performing a Lorentz transformation on a spinor field, is then satisfied by putting

$$S^{ik} = \frac{1}{2}\gamma^{[i}\gamma^{k]} = \frac{1}{4}\left(\gamma^{i}\gamma^{k} - \gamma^{k}\gamma^{i}\right) \tag{10.117}$$

The infinitesimal variation of a spinor under the Lorentz transformation is then

$$\delta\psi = \psi' - \psi = \frac{1}{2}\omega_{ik} S^{ik} = \frac{1}{4}\omega_{ik}\gamma^{[i}\gamma^{k]}\psi \tag{10.118}$$

while the infinitesimal variation of the vierbein $e_\mu{}^i$, which tranforms like a Lorentz vector with respect to the Latin index, is, from (10.115),

$$\delta e_\mu{}^i = e'_\mu{}^i - e_\mu{}^i = \omega^i{}_k e_\mu{}^k \tag{10.119}$$

Consider now two nearby points, x_1 and x_2, and denote with $\psi(x_1)$ and $\psi(x_2)$ the spinor field referred respectively to local vierbein fields $e_\mu{}^i(x_1)$ and $e_\mu{}^i(x_2)$. If the vierbein field $e_\mu{}^i(x_1)$ is parallelely displaced from x_1 to x_2, we obtain, at x_2, a new vierbein field which we denote with $e'_\mu{}^i(x_2)$, and $\psi'(x_2)$ will be the spinor field at the point x_2, referred to the vierbein $e'_\mu{}^i(x_2)$.

We can define then the covariant differential of a spinor field as

$$D\psi = dx^k \nabla_k \psi = \psi'(x_2) - \psi(x_1) \tag{10.120}$$

which can be written also as

$$D\psi = \psi(x_2) - \psi(x_1) - [\psi(x_2) - \psi'(x_2)] \tag{10.121}$$

In this way we have separated the term due to a translation $\psi(x_2) - \psi(x_1)$, from the part relative to a local rotation of the tetrads, $\psi'(x_2) - \psi(x_2)$.

We have then explicitly

$$\psi(x_2) - \psi(x_1) = dx^k \partial_k \psi \tag{10.122}$$

and, from (10.118),

$$\psi(x_2) - \psi'(x_2) = -\frac{1}{4}\omega_{ik}\gamma^{[i}\gamma^{k]}\psi \tag{10.123}$$

so that

$$D\psi = dx^k \partial_k \psi + \frac{1}{4}\omega_{ik}\gamma^{[i}\gamma^{k]}\psi \tag{10.124}$$

Now we must give an explicit expression for the infinitesimal Lorentz matrix ω_{ik}, which relates $e'_\mu{}^i(x_2)$ and $e_\mu{}^i(x_2)$ according to (10.119).

To this aim we observe that the derivative of a geometrical object carrying anholonomic Lorentz indices can be made covariant under local Lorentz rotations, provided that a new, tangent space connection $\omega_\mu{}^{ik}$ is introduced (also called the "spin connection" or anholonomic connection). Consider for example a Lorentz controvariant vector A^i, which transforms as $A'^i = \Lambda^i{}_k A^k$. If we have a "local" Lorentz rotation, represented by matrices which are position dependent, $\Lambda^i{}_k = \Lambda^i{}_k(x)$, then the partial derivative of a vector does not transforms like a vector, in fact

$$(\partial_\mu A^i)' = \partial_\mu (\Lambda^i{}_k A^k) = \Lambda^i{}_k \partial_\mu A^k + A^k \partial_\mu \Lambda^i{}_k \tag{10.125}$$

(remember that the world index "μ" is a scalar under Local Lorentz rotations). We can define however a Lorentz covariant derivative

$$D_\mu A^i = \partial_\mu A^i + \omega_\mu{}^i{}_k A^k \tag{10.126}$$

which transforms correctly as

$$(D_\mu A^i)' = \Lambda^i{}_k(x) D_\mu A^k \tag{10.127}$$

provided that the connection $\omega_\mu{}^{ik}$ transforms inhomogeneously as follows

$$\omega'_\mu{}^{ik} = \Lambda^i{}_j \omega_\mu{}^j{}_l (\Lambda^{-1})^{lk} - (\partial_\mu \Lambda)^i{}_j (\Lambda^{-1})^{jk} \tag{10.128}$$

(for more details see Chapter XII, Sect.3). Of course, requiring that

$$D_\mu (B_i A^i) = \partial_\mu (B_i A^i) \tag{10.129}$$

(because $B_i A^i$ is a Lorentz scalar) we obtain that the covariant derivative of a Lorentz covariant vector is given by

$$D_\mu B_i = \partial_\mu B_i - \omega_\mu{}^k{}_i B_k \tag{10.130}$$

The total covariant derivative of a geometrical object carrying both flat and curvilinear indices is to be performed using both the anholonomic and holonomic connection, $\omega_\mu{}^{ik}$ and $\Gamma_{\mu\nu}{}^\alpha$. The resulting derivative is then covariant under local Lorentz and general coordinate transformations.

The total covariant derivative of the vierbein field,

fór example, is

$$\nabla_\mu e_\nu{}^i = \partial_\mu e_\nu{}^i + \omega_\mu{}^i{}_k e_\nu{}^k - \Gamma_{\mu\nu}{}^\alpha e_\alpha{}^i =$$

$$= D_\mu e_\nu{}^i - \Gamma_{\mu\nu}{}^\alpha e_\alpha{}^i \qquad (10.131)$$

Note that ω acts only on the flat indices, while Γ on the curved ones. This expression transforms loke a covariant tensor, with indices μ, ν, under a general coordinate transformation $x^\mu \to x'^\mu$

$$\nabla_\mu e_\nu{}^i \longrightarrow \frac{\partial x^\alpha}{\partial x'^\mu}\frac{\partial x^\beta}{\partial x'^\nu}\nabla_\alpha e_\beta{}^i \qquad (10.132)$$

and like a controvariant Lorentz vector, with index i, under a local Lorentz transformation $x^i \to x'^i = \Lambda^i{}_k x^k$

$$\nabla_\mu e_\nu{}^i \longrightarrow \Lambda^i{}_k \nabla_\mu e_\nu{}^k \qquad (10.133)$$

In the Einstein-Cartan theory the vierbein field is assumed to satisfy the condition

$$\nabla_\mu e_\nu{}^i = 0 \qquad (10.134)$$

which provides a relation between the two connections ω and Γ. Moreover, from the metricity condition $\nabla_\alpha g_{\mu\nu} = 0$, the spin connection is constrained to be antisymmetric in the last two indices: we have in fact

$$\nabla_\alpha g_{\mu\nu} = \nabla_\alpha (e_\mu{}^i e_\nu{}^k \eta_{ik}) = e_\mu{}^i e_\nu{}^k \nabla_\alpha \eta_{ik} =$$

$$= e_\mu{}^i e_\nu{}^k (\partial_\alpha \eta_{ik} - \omega_\alpha{}^j{}_i \eta_{jk} - \omega_\alpha{}^j{}_k \eta_{ij}) =$$

$$= - e_\mu{}^i e_\nu{}^k 2\omega_{\alpha(ik)} = 0 \qquad (10.135)$$

Using this property, it is easy to obtain an explicit expression for $\omega_\mu{}^{ik}$.

Antisymmetrizing (10.134) we have in fact

$$\partial_{[\mu} e_{\nu]}{}^i + \omega_{[\mu}{}^i{}_{\nu]} - \Gamma_{[\mu\nu]}{}^i = 0 \qquad (10.136)$$

and contracting with $e^\mu{}_a e^\nu{}_b$ we get

$$C_{ab}{}^i + \frac{1}{2}(\omega_a{}^i{}_b - \omega_b{}^i{}_a) - Q_{ab}{}^i = 0 \qquad (10.137)$$

where

$$C_{ab}{}^{i} = e^{\mu}{}_{a} e^{\nu}{}_{b} \partial_{[\mu} e_{\nu]}{}^{i} = C_{[ab]}{}^{i} \qquad (10.138)$$

are the so-called Ricci rotation coefficients, and $Q_{ab}{}^{i} = Q_{[ab]}{}^{i}$ the torsion tensor. Lowering all the indices, and cyclically permuting them we obtain

$$C_{abi} + \frac{1}{2}(\omega_{aib} - \omega_{bia}) - Q_{abi} = 0$$

$$-C_{bia} - \frac{1}{2}(\omega_{iba} - \omega_{iab}) + Q_{bia} = 0 \qquad (10.139)$$

$$-C_{iab} - \frac{1}{2}(\omega_{bai} - \omega_{abi}) + Q_{iab} = 0$$

By adding these three equations and using the property $\omega_{abc} = \omega_{a[bc]}$ we have finally

$$\omega_{iba} = C_{abi} - C_{bia} - C_{iab} - Q_{abi} + Q_{bia} + Q_{iab} \qquad (10.140)$$

or

$$\omega_{\mu}{}^{ba} = \gamma_{\mu}{}^{ba} - K_{\mu}{}^{ab} \qquad (10.141)$$

where $\gamma_{\mu}{}^{ba}$ is the riemannian part of the connection, given in terms of the Ricci rotation coefficients

$$\gamma_{\mu}{}^{ba} = e_{\mu i}(C^{abi} - C^{bia} - C^{iab}) = \gamma_{\mu}{}^{[ba]} \qquad (10.142)$$

and $K_{\mu}{}^{ab}$ is the contorsion tensor (see 10.17)

$$K_{\mu}{}^{ab} = e_{\mu i} K^{iab} =$$

$$= e_{\mu i}(-Q^{iab} - Q^{bia} + Q^{abi}) = K_{\mu}{}^{[ab]} \qquad (10.143)$$

Note that the non-riemannian part of the tangent space connection is obtained simply by contracting the indices of the contorsion tensor with the vierbein field,

$$K_{\mu}{}^{ab} = e^{\alpha a} e^{\beta b} K_{\mu\alpha\beta} = e^{\alpha a} e^{\beta b}(-Q_{\mu\alpha\beta} - Q_{\beta\mu\alpha} + Q_{\alpha\beta\mu}) \qquad (10.144)$$

but the total connection ω, which is an affinity and not a tensor, cannot be obtained from the holonomic U_4 connection Γ directly in this way: the general relation between ω and Γ is provided by (10.134), that is (remember 10.2]

$$\omega_\mu{}^{ik} = e_\alpha{}^i e^{\nu k} \Gamma_{\mu\nu}{}^\alpha - e^{\nu k} \partial_\mu e_\nu{}^i =$$

$$= e_\alpha{}^i e^{\nu k} \left\{ {\alpha \atop \mu\nu} \right\} - K_\mu{}^{ki} - e^{\nu k} \partial_\mu e_\nu{}^i \qquad (10.145$$

Combining this equation with (10.141), we note, incidentally, that the Christoffel coefficients can be expressed in terms of the derivatives of the vierbein field as

$$\left\{ {\alpha \atop \mu\nu} \right\} = e^\alpha{}_i e_{\nu k} \gamma_\mu{}^{ik} + e^\alpha{}_i \partial_\mu e_\nu{}^i \qquad (10.146)$$

where $\gamma_\mu{}^{ik}$ is given in eq.(10.142).

Using the anholonomic connection ω, the local Lorentz rotation of the vierbein field (eq.10.119) can now be expressed in the tangent space as

$$\delta e_\nu{}^i = dx^\mu \omega_\mu{}^i{}_k e_\nu{}^k \qquad (10.147)$$

Therefore $\omega^{ik} = dx^\mu \omega_\mu{}^{ik}$, and the covariant differential of the spinor field (10.124) can be expressed in terms of the anholonomic connection as

$$D\psi = dx^\mu \left(\partial_\mu \psi + \frac{1}{4} \omega_\mu{}^{ab} \gamma_{[a} \gamma_{b]} \right) \psi \qquad (10.148)$$

The covariant derivative is then

$$\nabla_\mu \psi = \left(\partial_\mu + \frac{1}{4} \omega_\mu{}^{ab} \gamma_{[a} \gamma_{b]} \right) \psi =$$

$$= \left(\partial_\mu + \frac{1}{2} \omega_\mu{}^{ab} S_{ab} \right) \psi \qquad (10.149)$$

(the spinor generators are defined by (10.117)). Performing the hermitian conjugation, and multiplying by γ^4 from the right we obtain

$$(\nabla_\mu \psi)^+ \gamma^4 \equiv \nabla_\mu \bar\psi = \partial_\mu \bar\psi - \frac{1}{4} \omega_\mu{}^{ab} \bar\psi \gamma_{[a} \gamma_{b]} =$$

$$= \partial_\mu \bar\psi - \frac{1}{2} \omega_\mu{}^{ab} \bar\psi S_{ab} \qquad (10.150)$$

The spinor field is coupled then to the metric and to the torsion, both contained implicitly in the connection

(10.141). The torsionic part can be easily separated from the rest of the gravitational interaction, as from (10.141) we can put

$$\nabla_a \psi = e^{\mu}{}_a \nabla_\mu \psi = \psi_{;a} - \frac{1}{4} K_{acb} \gamma^{[b} \gamma^{c]} \psi \tag{10.151}$$

where

$$\psi_{;a} = e^{\mu}{}_a \psi_{;\mu} = \partial_a \psi + \frac{1}{4} \gamma_{abc} \gamma^{[b} \gamma^{c]} \psi \tag{10.152}$$

is the covariant derivative of a Dirac field in a Riemann manifold V_4.

Noting that $\gamma^\mu \nabla_\mu = \gamma^\kappa e^{\mu}{}_\kappa \nabla_\mu = \gamma^\kappa \nabla_\kappa$, we can rewrite explicitly the matter part of the Lagrangian (10.108) as

$$\begin{aligned} \sqrt{-g}\,\mathcal{L}_m = \sqrt{-g}\left[\frac{i}{2}\left(\bar\psi \gamma^\kappa \psi_{;\kappa} - \bar\psi_{;\kappa}\gamma^\kappa\psi\right) - m\bar\psi\psi\right] \\ - \frac{i}{8}\sqrt{-g}\,K_{acb}\left(\bar\psi\gamma^a\gamma^{[b}\gamma^{c]}\psi + \bar\psi\gamma^{[b}\gamma^{c]}\gamma^a\psi\right) \end{aligned} \tag{10.153}$$

and the gravitational part

$$\mathcal{L}_g = -\frac{\sqrt{-g}}{2\chi}\left[R(\{\}) + T^{abc} K_{cba}\right] \tag{10.154}$$

(we have used (10.138), neglecting a total divergence which does not contribute to the field equations).

Moreover, by considering the totally antisymmetrized product of three Dirac matrices, and using the anticommutative property $\gamma^a\gamma^b = -\gamma^b\gamma^a + 2\eta^{ab}$, we obtain

$$\begin{aligned} \gamma^{[a}\gamma^b\gamma^{c]} &= \frac{1}{6}\left(\gamma^a\gamma^b\gamma^c + \gamma^b\gamma^c\gamma^a + \gamma^c\gamma^a\gamma^b - \right. \\ &\quad \left. - \gamma^b\gamma^a\gamma^c - \gamma^a\gamma^c\gamma^b - \gamma^c\gamma^b\gamma^a\right) = \\ &= \gamma^a\gamma^b\gamma^c - \eta^{ab}\gamma^c - \eta^{bc}\gamma^a + \eta^{ac}\gamma^b \end{aligned} \tag{10.155}$$

Therefore

$$\begin{aligned} \gamma^a\gamma^{[b}\gamma^{c]} &= \gamma^{[a}\gamma^b\gamma^{c]} + 2\eta^{a[b}\gamma^{c]} \\ \gamma^{[b}\gamma^{c]}\gamma^a &= \gamma^{[b}\gamma^c\gamma^{a]} + 2\eta^{a[c}\gamma^{b]} \end{aligned} \tag{10.156}$$

and

$$\gamma^a\gamma^{[b}\gamma^{c]} + \gamma^{[b}\gamma^{c]}\gamma^a = 2\gamma^{[a}\gamma^b\gamma^{c]} \tag{10.157}$$

The total Lagrangian density can be rewritten then as follows

$$\sqrt{-g}\mathcal{L}_m + \mathcal{L}_g = \sqrt{-g}\Big[\frac{i}{2}(\bar{\psi}\gamma^k\psi_{;k} - \bar{\psi}_{;k}\gamma^k\psi) - m\bar{\psi}\psi + \frac{i}{4}K_{abc}\bar{\psi}\gamma^{[a}\gamma^b\gamma^{c]}\psi - \frac{1}{2\chi}R(\{\}) - \frac{1}{2\chi}T^{abc}K_{cba}\Big] \tag{10.158}$$

The torsionic contributions are represented explicitly by the terms containing the contorsion tensor K_{abc}, and in the limit $K_{abc} \to 0$ we obtain the general relativistic Lagrangian for a Dirac particle in a Riemann space V_4.

It is important to note that the Dirac spinor couples only to the totally antisymmetric part of the torsion tensor,

$$K_{[abc]} = -Q_{[abc]} - Q_{[cab]} + Q_{[bca]} = -Q_{[abc]} \tag{10.159}$$

because of the totally antisymmetric product of Dirac matrices in (10.158). Therefore, in the absence of other sources of torsion, an U_4 manifold with a Dirac field will be characterized by a totally anitymmetric non-riemannian part of the connection.

In order to obtain the generalized Dirac equation by a variational procedure, it is convenien to rewrite the matter part of the Lagrangian putting into evidence explicitly the terms containing the partial derivatives of ψ. We have then, using (10.149,150,157)

$$\sqrt{-g}\mathcal{L}_m = \sqrt{-g}\Big[\frac{i}{2}(\bar{\psi}\gamma^\mu\nabla_\mu\psi - \nabla_\mu\bar{\psi}\gamma^\mu\psi) - m\bar{\psi}\psi\Big] = \sqrt{-g}\Big[\frac{i}{2}(\bar{\psi}\gamma^a e^\mu{}_a\partial_\mu\psi - \partial_\mu\bar{\psi}e^\mu{}_a\gamma^a\psi) + \frac{i}{4}\omega_{abc}\bar{\psi}\gamma^{[a}\gamma^b\gamma^{c]}\psi - m\bar{\psi}\psi\Big] \tag{10.160}$$

Performing the variation with respect to $\bar{\psi}$ we have

$$\frac{\partial(\sqrt{-g}\mathcal{L}_m)}{\partial\bar{\psi}} = \sqrt{-g}\Big[\frac{i}{2}\gamma^a\partial_a\psi + \frac{i}{4}\omega_{abc}\gamma^{[a}\gamma^b\gamma^{c]}\psi - m\psi\Big] \tag{10.161}$$

$$\partial_\mu\frac{\partial(\sqrt{-g}\mathcal{L}_m)}{\partial(\partial_\mu\bar{\psi})} = -\frac{i}{2}\gamma^a\partial_\mu(\sqrt{-g}\,e^\mu{}_a)\psi - \frac{i}{2}\sqrt{-g}\,\gamma^a\partial_a\psi \tag{10.162}$$

This last equation can be rewritten in another form, more

useful for subsequent computations. First of all, remembering (3.29),

$$\partial_a (\ln \sqrt{-g}) = e^\mu{}_a \left\{ {\alpha \atop \mu\alpha} \right\} \qquad (10.163)$$

and then, using the relation (10.146) between Christoffel's symbols and the vierbein field, it follows that

$$\partial_a (\ln \sqrt{-g}) = e^\mu{}_a \left(e^\nu{}_i \, e_{\nu k} \, \gamma_\mu{}^{ik} + e^\nu{}_i \, \partial_\mu e_\nu{}^i \right) =$$

$$= e^\mu{}_a \left(e^\nu{}_i \, \partial_\mu e_\nu{}^i \right) \qquad (10.164)$$

From the definition (10.138) of the Ricci rotation coefficients we have then

$$C_{ab}{}^b = e^\mu{}_a \, e^\nu{}_b \, \frac{1}{2} \left(\partial_\mu e_\nu{}^b - \partial_\nu e_\mu{}^b \right) =$$

$$= \frac{1}{2} \partial_a (\ln \sqrt{-g}) - \frac{1}{2} e^\mu{}_a \, e^\nu{}_b \, \partial_\nu e_\mu{}^b \qquad (10.165)$$

Using the identity

$$\partial_\nu (e^\nu{}_b \, e_\mu{}^b) = e^\nu{}_b \, \partial_\nu e_\mu{}^b + e_\mu{}^b \, \partial_\nu e^\nu{}_b \equiv \partial_\nu (\delta^\nu_\mu) = 0 \qquad (10.166)$$

eq.(10.165) can be rewritten as

$$C_{ab}{}^b = \frac{1}{2} \frac{1}{\sqrt{-g}} \partial_a (\sqrt{-g}) + \frac{1}{2} e^\mu{}_a \, e_\mu{}^b \, \partial_\nu e^\nu{}_b$$

from which

$$2\sqrt{-g} \, C_{ab}{}^b = \partial_a (\sqrt{-g}) + \sqrt{-g} \, \partial_\nu e^\nu{}_a = \partial_\nu (\sqrt{-g} \, e^\nu{}_a) \qquad (10.167)$$

The definition (10.142) of the riemannian part of the anholonomic connection gives

$$\gamma_b{}^b{}_a = C_{ab}{}^b - C_{ba}{}^b = 2 C_{ab}{}^b \qquad (10.168)$$

and then (10.162) can be rewritten as

$$\partial_\mu \frac{\partial(\sqrt{-g} \mathcal{L}_m)}{\partial(\partial_\mu \bar{\psi})} = -\frac{i}{2} \sqrt{-g} \, \gamma^a \partial_a \psi - \frac{i}{2} \sqrt{-g} \, \gamma_b{}^b{}_a \gamma^a \psi \qquad (10.169)$$

The Eulero-Lagrange equation for the Dirac field are obtained equating (10.169) and (10.161)

$$i\left[\gamma^a \partial_a \psi + \frac{1}{4}\omega_{abc}\,\gamma^{[a}\gamma^b\gamma^{c]}\psi + \frac{1}{2}\gamma_b{}^b{}_a\,\gamma^a\psi\right] - m\psi = 0$$

that is (10.170)

$$i\left[\gamma^a \nabla_a \psi + \frac{1}{4}\omega_{abc}\left(\gamma^{[a}\gamma^b\gamma^{c]} - \gamma^a\gamma^{[b}\gamma^{c]}\right)\psi + \frac{1}{2}\gamma_b{}^b{}_a\,\gamma^a\psi\right] = m\psi \qquad (10.171)$$

Using (10.156) and the explicit definition (10.141) of the Lorentz connection we obtain

$$i\left[\gamma^a \nabla_a \psi + \frac{1}{4}\left(\gamma_{abc} - K_{acb}\right)\left(-2\eta^{a[b}\gamma^{c]}\right)\psi + \frac{1}{2}\gamma_b{}^b{}_a\,\gamma^a\psi\right] - m\psi =$$

$$= i\left[\gamma^a \nabla_a \psi + \frac{1}{2}K_{ac}{}^a\gamma^c\psi\right] - m\psi = 0 \qquad (10.172)$$

Finally, from (10.151,156) we have

$$i\left[\gamma^a \psi_{;a} - \frac{1}{4}K_{acb}\,\gamma^a\gamma^{[b}\gamma^{c]}\psi + \frac{1}{2}K_{ac}{}^a\gamma^c\psi\right] - m\psi =$$

$$= i\gamma^a\psi_{;a} + \frac{i}{4}K_{abc}\,\gamma^{[a}\gamma^b\gamma^{c]}\psi - m\psi = 0 \qquad (10.173)$$

This is the generalized Dirac equation in a Riemann-Cartan manifold, following from the Lagrangian (10.158) minimally coupled to the metric and to the torsion.

By taking the hermitian conjugate and multiplying from the right by γ^4 we obtain the adjoint equation for $\bar{\psi}$

$$i\bar{\psi}_{;a}\gamma^a - \frac{i}{4}K_{abc}\,\bar{\psi}\gamma^{[a}\gamma^b\gamma^{c]} + m\bar{\psi} = 0 \qquad (10.174)$$

It should be stressed that, in the absence of torsion, the Dirac equation in a curved space could be obtained also directly from the special relativistic equation (10.94) by the minimal coupling procedure, i.e. replacing the partial derivative with the covariant one,

$$i\gamma^k\psi_{;k} - m\psi = 0 \qquad (10.175)$$

In the presence of torsion, however, (i.e. in a Riemann-Cartan geometrical background) this procedure, applied directly to the field equations, does not lead to the correct result, because the spin of a Dirac particle is a totally antisymmetric tensor, as we shall see, and then a Dirac particle does not couple to the full nonriemannian part of the connection, but only to its totally antisymmetric part.

In fact, using the definition (10.46), we obtain, from

the matter part of the Lagrangian (10.158), that the spin density tensor of a Dirac field is given by

$$S^{abc} = \frac{1}{\sqrt{-g}} \frac{\delta(\sqrt{-g}\mathcal{L}_m)}{\delta K_{cba}} = -\frac{i}{4}\bar{\psi}\gamma^{[a}\gamma^{b}\gamma^{c]}\psi = S^{[abc]} \tag{10.176}$$

and then it is a totally antisymmetric tensor, as expected (the same expression was obtained in Chapter III, Appendix B, using the definition of canonical spin tensor).

The second field equation (10.55) of the Einstein-Cartan theory, using the definition (10.34) of $T_{\mu\nu\alpha}$, gives then

$$T^{abc} = \chi S^{abc} = T^{[abc]} = Q^{[abc]} = -K^{[abc]} \tag{10.177}$$

The Dirac equation (10.173) can be rewritten in the so-called second-order formalism, in which the contorsion tensor is given explicitly in terms of the spin sources. From (10. 176,177) we have

$$i\gamma^a\psi_{;a} - \frac{\chi}{16}\bar{\psi}\gamma_a\gamma_b\gamma_c\psi\gamma^{[a}\gamma^{b}\gamma^{c]}\psi - m\psi = 0 \tag{10.178}$$

Using the following property of the Dirac matrices

$$i\gamma^5\gamma^a = \frac{1}{3!}\varepsilon^{abcd}\gamma_b\gamma_c\gamma_d \tag{10.179}$$

where

$$\gamma^5 = \frac{i}{4!}\varepsilon^{abcd}\gamma_a\gamma_b\gamma_c\gamma_d \tag{10.180}$$

we obtain

$$i\gamma^5\gamma^a\varepsilon_{aijk} = -\gamma_{[i}\gamma_j\gamma_{k]} \tag{10.181}$$

and then

$$\gamma_i\gamma_j\gamma_k\gamma^{[i}\gamma^{j}\gamma^{k]} = 6(\gamma^5\gamma_a)(\gamma^5\gamma^a) \tag{10.182}$$

The Dirac equation (10.178) reduces then to the following nonlinear spinor equation[4]

$$i\gamma^a\psi_{;a} - \frac{3}{8}\chi\bar{\psi}\gamma^5\gamma_a\psi\gamma^5\gamma^a\psi - m\psi = 0 \tag{10.183}$$

(note that $\bar{\psi}\gamma^5\gamma_a\psi = J^5{}_a$ is the axial-vector spinor current). It is important to stress that, in the case of massless spinor fields satisfying the Weyl condition

$$\frac{1}{2}(1+\gamma^5)\psi = \psi \qquad (10.184)$$

(which defines two-components Weyl neutrinos) the nonlinea correction induced by torsion is identically vanishing[5], and (10.183) becomes simply $\gamma^K \psi_{;K} = 0$.

In conclusion of this Section, we note that, inserting eqs.(10.176,77) into the Lagrangian (10.158), we can re- write, in the second order formalism, the action for the Dirac field in a Riemann-Cartan manifold as follows:

$$S = \int d^4x \sqrt{-g} \left\{ \frac{i}{2}(\bar{\psi}\gamma^k\psi_{;k} - \bar{\psi}_{;k}\gamma^k\psi) - m\bar{\psi}\psi - \right.$$

$$\left. - \frac{1}{2\chi} R(\{\}) + \frac{1}{2}\chi\, S_{[abc]} S^{[abc]} \right\} \qquad (10.185)$$

This shows explicitly thet the only difference induced by torsion, with respect to the standard general relativistic theory for a Dirac particle in a gravitational field, is represented by the spin-spin contact interaction $S_{[abc]}S^{[abc]} \propto J^5{}_K J^{5K}$.

6.- Propagating torsion

In the previous Sections we have seen that, in the Ein- stein-Cartan theory, the torsion tensor is nonvanishing on- ly inside matter, and it is the source of a contact intera- tion: a spinning particle cannot influence directly, by menas of the torsion of the manifold, another spinning par- ticle (unless the two particles collide and interpenetrate because at a classical level torsion is a non-propagating field, and it disappears immediately outside the spinning bodies.

This is one of the main characteristics of the Einstei Cartan theory, and in this way torsion becomes physically interesting only at a microscopical level or , macroscopi- cally, when considering extremely collapsed matter.

In order to obtain torsion propagation in vacuum, ma- ny modifications of the U_4 theory have been proposed, in- troducing a higher order gravitational Lagrangian differen from the simple scalar curvature, in order to obtain, for the torsion tensor, a differential field equation instead of an algebraic one.

In this Section we wish to display that an alternative possibility of introducing a propagating torsion, without modifying the simple gravitational Lagrangian of the U_4 theory, is provided by the hypothesis that the torsion field can be expressed in terms of a potential[5]. The predictions of a theory constructed on this assumption are not in disagreement with the present day experimental results, and moreover in this case torsion may be coupled also to the electromagnetic field without breaking the electromagnetic gauge invariance (unlike in the standard Einstein-Cartan theory).

Consider the interaction between torsion and a spinor field, described by the Lagrangian (10.158). In this case, as shown previously, we have a totally antisymmetric torsion tensor, and defining the axial vector Q_a

$$Q_a = \frac{1}{3!}\varepsilon_{abcd}Q^{bcd} \qquad (10.186)$$

we have

$$\varepsilon^{abcd}Q_d = Q^{[abc]} = T^{[abc]} = -K^{[abc]} \qquad (10.187)$$

so that the total Lagrangian density (10.158) can be rewritten, using (10.187,179), as

$$\sqrt{-g}\mathcal{L} = \sqrt{-g}\mathcal{L}_m + \mathcal{L}_g = \sqrt{-g}\Big[-\frac{1}{2\chi}R(\{\}) + \frac{1}{\chi}3Q_\mu Q^\mu + \frac{i}{2}(\bar\psi\gamma^\mu\psi_{;\mu} - \bar\psi_{;\mu}\gamma^\mu\psi) - m\bar\psi\psi - \frac{3}{2}Q_\mu\bar\psi\gamma^5\gamma^\mu\psi\Big] \qquad (10.188)$$

Suppose that the axial part of the torsion tensor can be expressed in terms of a pseudo-scalar potential φ :

$$Q_\mu = \partial_\mu\varphi \qquad (10.189)$$

The action becomes then

$$S = \int d^4x\sqrt{-g}\Big[-\frac{1}{2\chi}R(\{\}) + \frac{3}{\chi}\partial_\mu\varphi\partial^\mu\varphi + \frac{i}{2}(\bar\psi\gamma^\mu\psi_{;\mu} - \bar\psi_{;\mu}\gamma^\mu\psi) - m\bar\psi\psi - \frac{3}{2}\partial_\mu\varphi\bar\psi\gamma^5\gamma^\mu\psi\Big] \qquad (10.190)$$

and the variation with respect to $\bar\psi$, ψ, and φ gives the following equations for the Dirac field and the torsion potential:

$$i\gamma^\mu\psi_{;\mu} - \frac{3}{2}\partial_\mu\varphi\gamma^5\gamma^\mu\psi - m\psi = 0 \qquad (10.191)$$

$$i\bar{\psi}_{;\mu}\gamma^{\mu} - \frac{3}{2}\partial_{\mu}\varphi\bar{\psi}\gamma^{\mu}\gamma^{5} + m\bar{\psi} = 0 \qquad (10.192)$$

$$\varphi^{,\mu}{}_{;\mu} = \frac{1}{2}\chi\,(\bar{\psi}\gamma^{5}\gamma^{\mu}\psi)_{;\mu} \qquad (10.193)$$

We have then a propagating torsion field, because its potential φ satisfies a wave-equation.

In order to investigate some physical consequences of this theory, suppose to neglect the usual riemannian terms depending on the metric and the curvature ($\varphi_{;\mu} \to \partial_{\mu}\varphi$), as we are interested only in the spin-torsion interaction, and consider the limit of static spin sources. In this limit $\bar{\psi}\gamma^{5}\gamma^{4}\psi \simeq 0$ and $\vec{J}^{5} \simeq \vec{\Sigma}$ is the spin density vector of the source; eq.(10.193) gives then

$$\nabla^{2}\varphi = \frac{1}{2}\chi\,\vec{\nabla}\cdot\vec{\Sigma} \qquad (10.194)$$

If we have an extended macroscopic body composed of static spinning particles, with an averaged spin density $\vec{\Sigma}(\vec{x})$ the solution of (10.194) is

$$\varphi(\vec{x}) = -\frac{G}{c^{3}}\int_{V} d^{3}x'\,\frac{\vec{\nabla}\cdot\vec{\Sigma}(\vec{x}')}{|\vec{x}-\vec{x}'|} \qquad (10.195)$$

being V the volume of the body. At a large distance from the source, the solution can be expanded in this way ($r = |\vec{x}|$)

$$\varphi(\vec{x}) = -\frac{G}{c^{3}}\Big\{\frac{1}{r}\int_{V} d^{3}x'\,\vec{\nabla}\cdot\vec{\Sigma}(\vec{x}') + \frac{\vec{x}}{r^{3}}\cdot\Big[\int_{V} d^{3}x'\,\vec{x}'\,(\vec{\nabla}\cdot\vec{\Sigma}(\vec{x}'))\Big] + \cdots\Big\} \qquad (10.196)$$

By using the Gauss theorem, the first integral can be shown to be vanishing (because $\Sigma = 0$ outside the body), and the second dipolar term gives

$$\varphi(\vec{x}) = \frac{G}{c^{3}}\,\frac{\vec{x}}{r^{3}}\cdot\Big[\int_{V} d^{3}x'\,\vec{\Sigma}(\vec{x}')\Big] \qquad (10.197)$$

If there is no preferred direction for the spin alignment inside the body, then the total spin

$$\vec{S} = \int_{V} d^{3}x'\,\vec{\Sigma}(\vec{x}') \qquad (10.198)$$

is vanishing, and there is no torsion field, outside the body, in the dipolar approximation.

If however $\vec{S} \neq 0$, because spins are (at least partially) oriented along some preferred direction, then outside the body the following dipolar torsion field is produced;

$$\vec{Q} = \vec{\nabla}\varphi = \frac{G}{c^3 r^5}\left[\vec{S} r^2 - 3\vec{x}\,(\vec{S}\cdot\vec{x})\right] \qquad (10,199)$$

The accelēration of a test body in the torsion field produced by another macroscopical source can be obtained starting with the Lagrangian

$$\mathcal{L}' = \mathcal{L}_P - \frac{3}{2}\sqrt{g}\, g^{\mu\nu}\partial_\mu \varphi J^5{}_\nu \qquad (10,200)$$

where $\mathcal{L}_P$ is the test body Lagrangian, and the second term describes the interaction with the external torsion field (remember 10.188).

Performing the variation of (10.200) with respect to $g^{\mu\nu}$, we obtain that the total energy-momentum, including the contributions of the spin-torsion interaction energy, is given by

$$T_{\mu\nu} - \frac{3}{2}\left(\partial_\mu \varphi J^5{}_\nu - \frac{1}{2} g_{\mu\nu} J^{5\alpha}\partial_\alpha \varphi\right) \qquad (10,201)$$

where $T_{\mu\nu} = \delta\mathcal{L}_P / \delta g^{\mu\nu}$. Imposing the conservation of the total energy, in the approximation in which the gravitational field is locally vanishing, i.e. $g^{\mu\nu} = \eta^{\mu\nu}$, we get

$$\partial^\nu T_{\mu\nu} = \frac{3}{4}\partial_\nu \partial_\mu \varphi J^{\nu 5} \qquad (10,202)$$

modulo terms depending on the divergenge of the spin density, which disappear, in the static case, after a volume integration. Defing the four momentum vector as (see Appendix B, Chapter VIII)

$$P_\mu = \int d^3x'\, T_\mu{}^4 \qquad (10,203)$$

and integrating (10.202) over a space-like 3-dimensional section of the world tube of the test particle, developing in power series the external torsion field (see Section 4 of this Chapter), we obtain, in first approximation and in the static limit,

$$\dot{\vec{P}} = -\frac{3}{4}\left(\int_{V'} d^3x'\, \vec{\Sigma}'\right)\cdot\vec{\nabla}(\vec{\nabla}\varphi) \qquad (10,204)$$

A dot denotes the derivative with respect to time, $\vec{\Sigma}'$ is the spin density of the test body, V' its volume, and φ is the external torsion potential.

If the total spin of the test body is vanishing, $\vec{S}' = \int_{V'} d^3x' \vec{\Sigma}' = 0$, then $\vec{P} = 0$: this implies that unpolarized test bodies, also with different internal composition, in a given gravitational field must fall with the same acceleration (in the dipolar approximation), even if an external torsion field is present, besides the gravitational one.

Therefore there is no disagreement between this hypothesis of torsion propagation and the well known Eötvös-Dicke-Braginsky experimental tests of the principle of equivalence, performed with unpolarized aluminum and platinum test bodies in the solar and terrestrial gravitational fields, even if these celestial bodies are the sources of an external torsion field.

According to the hypothesis considered in this Section however, test bodies with different polarizations should fall with different accelerations in the gravitational field of a polarized source.

In fact, putting in first approximation $\dot{\vec{P}} = m \ddot{\vec{X}}$, where m is the mass and $\vec{X}$ the coordinates of the center of mass of the test body, it follows from (10.204) that in the local frame in which the Christoffel symbols are vanishing, a test body with total spin $\vec{S}'$ experiences an acceleration

$$\ddot{\vec{X}} = -\frac{3}{4m} (\vec{S}' \cdot \vec{\nabla}) \vec{Q} \qquad (10.205)$$

where $\vec{Q} = \vec{\nabla} \varphi$ is the external torsion field. It is easy to note the formal analogy between this equation and the expression for the electromagnetic force acting on a dipole magnetic moment in a non-homogeneous magnetic field.

In conclusion, it is interesting to observe that in the hypothesis of torsion determined by potential, according to (10.189), we are allowed to couple torsion to the electromagnetic potential A_μ , without breaking the local gauge invariance. Consider in fact the following photon-torsion interaction Lagrangian:

$$\mathcal{L}_I = \alpha \, \varepsilon^{\mu\nu\alpha\beta} A_\mu F_{\nu\alpha} Q_\beta \qquad (10.206)$$

where $F_{\mu\nu} = \partial_\mu A_\nu - \partial_\nu A_\mu$ and α is an adimensional coupling constant (a coupling of this form can be obtained also through an explicit computation, taking into account the minimal microscopic interactions between torsion and the virtual e^+-e^- pairs associated to a propagating electromagnetic field on a given torsion background[7]).

Under the local gauge transformation

$$A_\mu \to A'_\mu = A_\mu + \partial_\mu \Lambda(x) \qquad (10.207)$$

the interaction Lagrangian (10.126) becomes

$$\begin{aligned} \mathcal{L}_I \to \mathcal{L}_I &+ \alpha\, \varepsilon^{\mu\nu\alpha\beta} \partial_\mu \Lambda F_{\nu\alpha} Q_\beta + \\ &+ \alpha\, \varepsilon^{\mu\nu\alpha\beta} A_\mu \partial_\nu \partial_\alpha \Lambda\, Q_\beta = \\ = \mathcal{L}_I &+ \partial_\mu (\alpha\, \varepsilon^{\mu\nu\alpha\beta} \Lambda F_{\nu\alpha} Q_\beta) - \\ &- \alpha\, \varepsilon^{\mu\nu\alpha\beta} (\partial_\mu F_{\nu\alpha} Q_\beta + F_{\nu\alpha} \partial_\mu Q_\beta) \end{aligned} \qquad (10.208)$$

But $\partial_{[\mu} F_{\nu\alpha]} = \partial_{[\mu}\partial_\nu A_{]} = 0$ and if $Q_\beta = \partial_\beta \varphi$ we have also $\partial_{[\mu} Q_{\beta]} = \partial_{[\mu}\partial_{\beta]} \varphi = 0$. In the case of a propagating axial torsion derived from the pseudoscalar potential φ , the photon-torsion coupling (10.206) is then invariant under a local gauge transformation of the electromagnetic potential (modulo a total divergence which does not contribute to the field equations derived from the action by a variational principle, see Chapter III, Appendix B).

REFERENCES

1) See for example F.W.Hehl, P.von der Heyde, G.D.Kerlick and J.M.Nester: Rev.Mod.Phys.48,393(1976)

2) A.Papapetrou: Proc.R.Soc.A209,248(1951)

3) Y.Ne'eman and Dj.Sijacki: Proc.Nat.Acad.Sci.USA 76, 561(1979) ; Ann.Phys.120,262(1979)

4) F.W.Hehl and B.K.Datta: J.Math.Phys.12,1334(1971)

5) B.Kuchowicz: Phys.Lett.50A,267(1974)

6) V.De Sabbata and M.Gasperini, in "Unified field theo-

ries of more than four dimensions", ed. by V.De Sabbata and E.Schmutzer (World Scientific, Singapore,1983) p.152 and references therein.

7) V.De Sabbata and M.Gasperini: Phys.Lett.77A,300(1980)

CHAPTER XI

THE STRONG GRAVITY THEORY

1.- The tensor mesons dominance hypothesis

From the point of view of quantum field theory, the gravitational interaction between two particles should be produced by the exchange of a massless, spin-two boson, the graviton (see Chapter VIII, Sect.2), according to the Feyman diagram of Fig.11.1

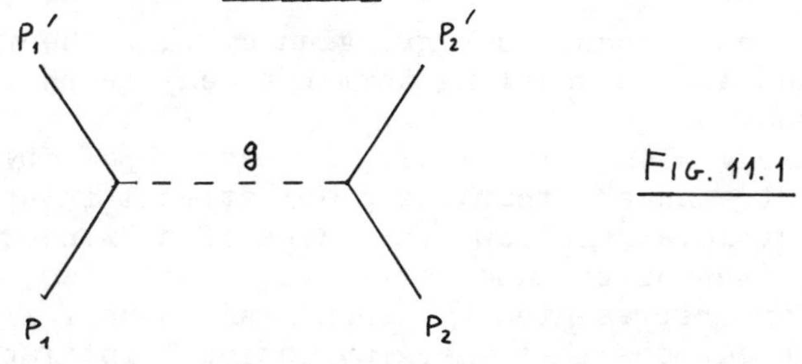

FIG. 11.1

The graviton, represented by a dashed line, is supposed to mediate the gravitational interactions, just like electromagnetic forces are mediated by the exchange of photons in quantum electrodynamics.

Of course the notion of graviton find a complete justification only in the framework of a quantum theory of gravity, and in this Section it will be used only as a guiding idea to introduce the strong gravity theory, which will be formulated however as a classical theory.

Shortly before the development of gauge theories and of unified theories of strong and electroweak interactions, it was suggested that charged leptons (i.e. particles without strong interactions, like electrons, muons,..) interact electromagnetically with the exchange of photons (γ), while hadrons (i.e. strongly interacting particles like protons, neutrons...) have electromagnetic interactions which take place with the exchange of a particle which has the same quantum numbers as the photon, but it is massive (the ρ meson, with spin = 1, parity = -1).

This hypothesis is called "vector meson dominance" (because particles with spin 1, i.e. vector particles, play a fundamental role in this scheme) and implies that electromagnetic interactions between hadrons (h) and leptons (ℓ) are described by diagrams like the one of Fig.11.2.

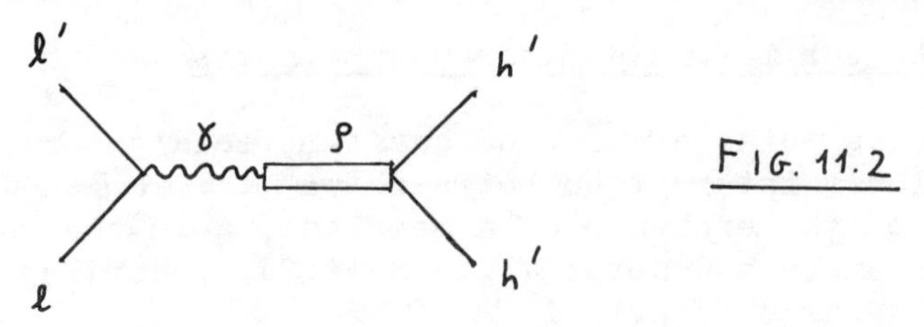

FIG. 11.2

The total electromagnetic Lagrangian of this theory must include then also a coupling term between the photon and the ρ meson.

By analogy with this theory, it was proposedby Isham, Salam and Strathdee[1] that the gravitational interactions, at a nuclear level, follow the rules of a "tensor meson dominance" (tensor because particles of spin two, like the graviton, are represented by second rank tensor fields). That is one suppose that the gravitational interaction between leptons takes place with the exchange of gravitons while in the case of hadrons the exchanged particle is a sort of heavy graviton (denoted by f) which has the same spin and quantum numbers as the graviton, but it is massive (see Fig.11.3).

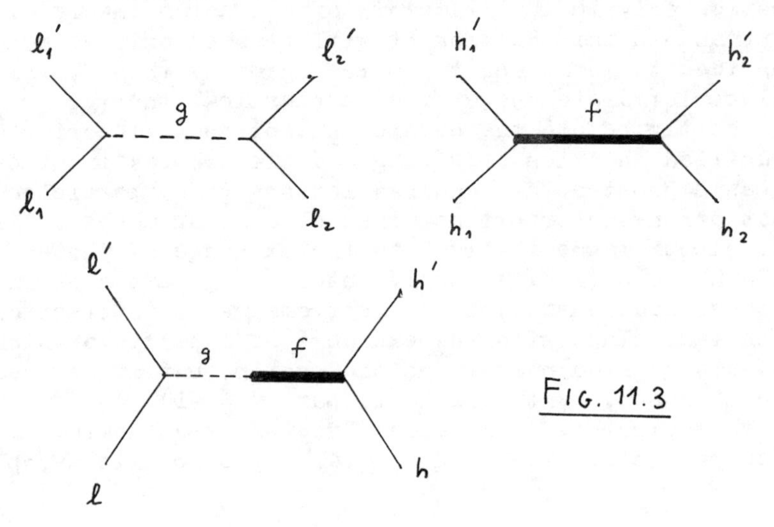

FIG. 11.3

The lightest tensor particle satisfying these requirements is the f-meson, which has a mass $M \sim 1200$ Mev: therefore this theory is also called "f-gravity" theory , or "strong gravity" theory, because in this case the strength of the gravitational interactions among hadrons is of the same order as that of strong interactions.

From a classical and geometrical point of view, the tensor meson dominance hypothesis means that the gravitational field is described by two metric tensors: besides $g_{\mu\nu}$, corresponding to the ordinary gravitational interaction, with an infinite range because it corresponds to massless exchanged particles, and coupled to matter by the Newton coupling constant G, one must introduce a second metric tensor, $f_{\mu\nu}$, representing strong gravity acting at the hadronic level, with a range of the order of the nuclear radius (remember that the range r is related to the mass of the exchanged particle, m_f , by $r \sim \hbar/m_f c$) , and with a coupling constant $G_f \sim 10^{39} \div 10^{40}$ G .

2.- Field equations

Consider a special relativistic matter Lagrangian L_m which can be written as the sum of two terms, one containing only leptons, ℓ , and the other only hadrons, h :

$$L_m = L_\ell(\ell, \partial\ell, \eta) + L_h(h, \partial h, \eta) \qquad (11.1)$$

Introducing gravity according to the minimal coupling procedure, we use however the metric tensor $g_{\mu\nu}$ for the leptonic part and another tensor , $f_{\mu\nu}$, for the hadronic part of the Lagrangian. Therefore

$$L_\ell \longrightarrow \mathcal{L}_\ell = \sqrt{-g}\, L_\ell(\ell, \ell_{;\mu}, g_{\mu\nu}) \qquad (11.2)$$

$$L_h \longrightarrow \mathcal{L}_h = \sqrt{-f}\, L_h(h, h_{|\mu}, f_{\mu\nu}) \qquad (11.3)$$

where $g = \det g_{\mu\nu}$, $f = \det f_{\mu\nu}$. Note that we have denoted with a semicolon the usual Riemann covariant derivative for the connection $\Gamma_{\alpha\beta}{}^{\mu}(g)$ corresponding to $g_{\mu\nu}$, and with a bar the derivative for the connection $\Gamma_{\alpha\beta}{}^{\mu}(f)$ corresponding to $f_{\mu\nu}$. Assuming that the connections are symmetric and metric-compatible (see Chapter II, Sect.3) we have then

$$\Gamma_{\alpha\beta}{}^{\mu}(g) = \left\{ {\mu \atop \alpha\beta} \right\}(g) = \frac{1}{2} g^{\mu\nu} (\partial_\alpha g_{\nu\beta} + \partial_\beta g_{\nu\alpha} - \partial_\nu g_{\alpha\beta})$$

$$\Gamma_{\alpha\beta}{}^{\mu}(f) = \left\{ {\mu \atop \alpha\beta} \right\}(f) = \frac{1}{2} f^{\mu\nu} (\partial_\alpha f_{\nu\beta} + \partial_\beta f_{\nu\alpha} - \partial_\nu f_{\alpha\beta}) \tag{11.4}$$

As regards the gravitational part of the Lagrangian, $\mathcal{L}_f + \mathcal{L}_g$, we choose, in analogy to the general relativity theory, the two scalar curvatures R(f) and R(g), with different coupling constants however:

$$\mathcal{L}_g = -\frac{\sqrt{-g}}{2\chi_g} R(g) \quad , \quad \mathcal{L}_f = -\frac{\sqrt{-f}}{2\chi_f} R(f) \tag{11.5}$$

where $\chi_g = 16\pi G/c^4$, $\chi_f = 16\pi G_f/c^4$.

Finally, according to the tensor meson dominance hypothesis, an interaction Lagrangian $\mathcal{L}_{fg}$, which couples the two metric tensors, is to be added (a possible explicit form of this coupling will be presented in a moment: now we assume only that no derivatives of the metric tensors are contained in the mixing term $\mathcal{L}_{fg}$). The total action for the theory is then

$$S = \int d^4x \, (\mathcal{L}_\ell + \mathcal{L}_h + \mathcal{L}_g + \mathcal{L}_f + \mathcal{L}_{fg}) \tag{11.6}$$

(note that in this Chapter the square roots of the determinants of the metric tensors are included into the definitions of the Lagrangians $\mathcal{L}$, which are then scalar densities). According to this action leptons feel g-gravity, hadrons f-gravity, but an indirect hadron-lepton gravitational interaction at a microscopical level is allowed through the generally nonlinear mixing term $\mathcal{L}_{fg}$.

The variation of the action with respect to $g^{\mu\nu}$ and $f^{\mu\nu}$ provides the field equations for the weak and strong gravity, respectively:

$$G_{\mu\nu}(g) = \chi_g T_{\mu\nu}(\ell) + \frac{2\chi_g}{\sqrt{-g}} \frac{\partial \mathcal{L}_{fg}}{\partial g^{\mu\nu}} \tag{11.7}$$

$$G_{\mu\nu}(f) = \chi_f T_{\mu\nu}(h) + \frac{2\chi_f}{\sqrt{-f}} \frac{\partial \mathcal{L}_{fg}}{\partial f^{\mu\nu}} \tag{11.8}$$

where $G_{\mu\nu}$ is the usual Einstein tensor (3.35)

$$G_{\mu\nu}(g) = \frac{1}{\sqrt{-g}} \frac{\delta(\sqrt{-g}\,R(g))}{\delta g^{\mu\nu}} \qquad (11.9)$$

$$G_{\mu\nu}(f) = \frac{1}{\sqrt{-f}} \frac{\delta(\sqrt{-g}\,R(f))}{\delta f^{\mu\nu}} \qquad (11.10)$$

and $T_{\mu\nu}$ is the dynamical energy-mometum tensor for the leptonic and hadronic matter, respectively $T_{\mu\nu}(\ell)$ and $T_{\mu\nu}(h)$:

$$T_{\mu\nu}(\ell) = \frac{2}{\sqrt{-g}} \frac{\delta \mathcal{L}_\ell}{\delta g^{\mu\nu}} \qquad (11.11)$$

$$T_{\mu\nu}(h) = \frac{2}{\sqrt{-f}} \frac{\delta \mathcal{L}_h}{\delta f^{\mu\nu}} \qquad (11.12)$$

(see Chapter III, Sect.3).

In the limit $\mathcal{L}_{fg} \rightarrow 0$, the field equations of the theory reduce to the decoupled Einstein equations, i.e. the two leptonic and hadronic worlds are non-communicating (from the gravitational point of view).

The interaction Lagrangian $\mathcal{L}_{fg}$ is chosen so as to satisfy the following requirements: to be a scalar density; to contain no derivatives; to give a mass, M , to one of the two tensor fields, corresponding to the short range nature of strong gravity. The most commonly used mixing term is the following (different choices are however possible)

$$\mathcal{L}_{fg} = \frac{\sqrt{-f}}{K_f^2} \frac{M^2}{4} (f^{\alpha\beta} - g^{\alpha\beta})(f^{\mu\nu} - g^{\mu\nu})(g_{\alpha\mu} g_{\beta\nu} - g_{\alpha\beta} g_{\mu\nu}) \qquad (11.13)$$

where $k_f^2 = 2\chi_f$, $k_g^2 = 2\chi_g$, and $k_f \sim 10^{20} k_g$.

In this case we have

$$\frac{\partial \mathcal{L}_{fg}}{\partial g^{\mu\nu}} = -\frac{M^2}{2K_f} B^\alpha{}_\mu \, g_{\alpha\nu} \qquad (11.14)$$

$$\frac{\partial \mathcal{L}_{fg}}{\partial f^{\mu\nu}} = \left(-\frac{1}{2}\delta^\alpha{}_\mu \mathcal{L}_{fg} + \frac{M^2}{2K_f} B^\alpha{}_\mu\right) f_{\alpha\nu} \qquad (11.15)$$

where

$$B^\alpha{}_\mu = \frac{\sqrt{-f}}{K_f} \left[g_{\mu\rho} f^{\rho\sigma} g_{\sigma\nu} f^{\nu\alpha} - g_{\mu\rho} f^{\rho\alpha} (g_{\nu\sigma} f^{\sigma\nu} - 3) \right] \qquad (11.16)$$

Moreover in this case it follows that

$$\left(\frac{\partial \mathcal{L}_{fg}}{\partial g^{\mu\nu}}\right)^{;\nu} + \left(\frac{\partial \mathcal{L}_{fg}}{\partial f^{\mu\nu}}\right)^{|\nu} = 0 \qquad (11.17)$$

and since the contracted Bianchi identities give $G_\mu{}^\nu(g)_{;\nu} = 0 = G_\mu{}^\nu(f)_{|\nu}$, we have, from the field equations (11.7,8), that the hadronic and leptonic energy-momentum tensors are not, in general, separately conserved, i.e.

$$T_{\mu\nu}(\ell)^{;\nu} \neq 0 \quad , \quad T_{\mu\nu}(h)^{|\nu} \neq 0 \qquad (11.18)$$

If, however, from the matter Lagrangian one has $T_{\mu\nu}(\ell)^{;\nu} = 0 = T_{\mu\nu}(h)^{|\nu}$, then the field equations imply

$$\left(\frac{\partial \mathcal{L}_{fg}}{\partial g^{\mu\nu}}\right)^{;\nu} = 0 = \left(\frac{\partial \mathcal{L}_{fg}}{\partial f^{\mu\nu}}\right)^{|\nu} \qquad (11.19)$$

and these four conditions are to be imposed on the tensors $g_{\mu\nu}$ and $f_{\mu\nu}$ as a constraint.

3.- Linear approximation

It is easy to show , from the field equations (11.7,8) in vacuum, that if $R_{\mu\nu}(g) = 0$, then $g_{\mu\nu} = f_{\mu\nu}$. Therefore, in a coordinate system in which $g_{\mu\nu} = \eta_{\mu\nu}$, we have also $f_{\mu\nu} = \eta_{\mu\nu}$. This result is important as it implies that, if a linear approximation for $g_{\mu\nu}$ is used, the same approximation holds also for $f_{\mu\nu}$. We are allowed then to expand both the two metrics, in the weak field limit, as follows

$$\begin{aligned} g_{\mu\nu} &= \eta_{\mu\nu} + K_g h_{\mu\nu} \\ f_{\mu\nu} &= \eta_{\mu\nu} + K_f F_{\mu\nu} \end{aligned} \qquad (11.20)$$

Using these expressions in the total action (11.6), we can also approximate the matter lagrangian as

$$\mathcal{L}_\ell = K_g h_{\mu\nu} T^{\mu\nu}(\ell) \quad , \quad \mathcal{L}_h = K_f F_{\mu\nu} T^{\mu\nu}(h) \qquad (11.21)$$

and we retain, in the total Lagrangian, at most quadratic terms in the field variables $h_{\mu\nu}$ and $F_{\mu\nu}$. The mixing term (11.13) becomes then

$$\frac{M^2}{4K_f^2}\left(K_f F^{\mu\nu} - K_g h^{\mu\nu}\right)\left(K_f F^{\alpha\beta} - K_g h^{\alpha\beta}\right)\left(\eta_{\mu\alpha}\eta_{\nu\beta} - \eta_{\mu\nu}\eta_{\alpha\beta}\right) \qquad (11.22)$$

Defining

$$F'^{\mu\nu} = \cos\vartheta\, F^{\mu\nu} - \sin\vartheta\, h^{\mu\nu}$$
$$h'^{\mu\nu} = \sin\vartheta\, F^{\mu\nu} + \cos\vartheta\, h^{\mu\nu} \qquad (11.23)$$

where

$$\cos\vartheta = \frac{K_f}{(K_f^2 + K_g^2)^{1/2}} \;, \qquad \sin\vartheta = \frac{K_g}{(K_f^2 + K_g^2)^{1/2}} \qquad (11.24)$$

and imposing the gauge conditions (see Chapter VIII)

$$\partial_\nu (h'^{\mu\nu} - \tfrac{1}{2}\eta^{\mu\nu} h') = 0 \quad , \quad \partial_\nu F'^{\mu\nu} = 0 \qquad (11.25)$$

we obtain finally the following linearized equations

$$\Box(h'^{\mu\nu} - \tfrac{1}{2}\eta^{\mu\nu}h') = \frac{1}{2}\frac{K_f K_g}{(K_f^2+K_g^2)^{1/2}}\left[T^{\mu\nu}(\ell) + T^{\mu\nu}(h)\right] \qquad (11.26)$$

$$\Box F'^{\mu\nu} - \partial^\mu\partial^\nu F' + M'^2 (F'^{\mu\nu} - \eta^{\mu\nu}F') =$$
$$= \frac{1}{2}\frac{K_f^2}{(K_f^2+K_g^2)^{1/2}}\left[T^{\mu\nu}(h) - \frac{K_g^2}{K_f^2}T^{\mu\nu}(\ell)\right] \qquad (11.27)$$

where $M'^2 = M^2(1 + K_g^2/K_f^2)$, and $F' = \eta^{\mu\nu}F'_{\mu\nu}$, $h' = \eta^{\mu\nu}h'_{\mu\nu}$. As $k_f \sim 10^{20}\, k_g$, then $M' \simeq M$, and in practice the linear equations reduce to

$$\Box(h'_{\mu\nu} - \tfrac{1}{2}\eta_{\mu\nu}h') = \tfrac{1}{2}K_g\left[T_{\mu\nu}(\ell) + T_{\mu\nu}(h)\right] \qquad (11.28)$$

$$\Box F'_{\mu\nu} - \partial_\mu\partial_\nu F' + M^2(F'_{\mu\nu} - \eta_{\mu\nu}F') = \tfrac{1}{2}K_f T_{\mu\nu}(h) \qquad (11.29)$$

Eq.(11.28) describes a massless spin-two field, coupled to hadrons and leptons with the same strength (the Newton coupling constant). It is natural then to interpret $h'_{\mu\nu}$ as the field representing ordinary, long range, gravitational forces.

The other equation corresponds to a spin-two particle of mass M, coupled only to hadronic matter with a strong interaction constant. The solutions of eq.(11.29), in the static limit, show the typical exponential falloff with the distance outside the source, so that, outside a radius $R \sim \hbar/Mc \sim 10^{-14}$ cm, the strong field $F'_{\mu\nu}$ is pratically vanishing.

The path of a macroscopical test-body in the linear approximation is influenced then only by the weak field $h'_{\mu\nu}$, quite independently from its content of hadrons or leptons. In fact, in the case of distances greater than the nuclear radius, $F'_{\mu\nu} = 0$, and then, from (11.20,23,24), $F_{\mu\nu} \simeq (k_g/k_f)h$, and

$$g_{\mu\nu} = \eta_{\mu\nu} + K_g h_{\mu\nu} \simeq \eta_{\mu\nu} + K_g h'_{\mu\nu}$$

$$f_{\mu\nu} = \eta_{\mu\nu} + K_f F_{\mu\nu} \simeq g_{\mu\nu} \tag{11.30}$$

The two metric coincides, so that hadrons and leptons move along the same geodesic. The strong gravity theory is then in agreement with the classical, macroscopical tests of general relativity, performed in the case of weak gravitational fields.

4.- Strong gravity with torsion

In order to take into account spin effects according to the Einstein-Cartan formalism, it is interesting to generalize the strong gravity field equations, formulating the theory using nonsymmetric affine connections[3].

Supposing that the matter sources are spinor fields, the introduction of a vierbein field is needed, as shown in Chapter X, Sect.5. In this case we must introduce however two sets of vectors, $e_\mu{}^k$ and $a_\mu{}^k$, as we have two metric tensors, so that

$$g_{\mu\nu} = e_\mu{}^i e_\nu{}^k \eta_{ik}$$

$$f_{\mu\nu} = a_\mu{}^i a_\nu{}^k \eta_{ik} \tag{11.31}$$

(therefore $e_\mu{}^i$ is the "weak" vierbein, $a_\mu{}^i$ the "strong" one) The inverse fields are then defined raising and lowering indices as follows

$$e_{\mu i} = \eta_{ik} e_\mu{}^k \quad , \qquad e^{\mu i} = g^{\mu\nu} e_\nu{}^i \tag{11.32}$$

$$a_{\mu i} = \eta_{ik} a_\mu{}^k \quad , \qquad a^{\mu i} = f^{\mu\nu} a_\nu{}^i \tag{11.33}$$

Assuming, for the weak and strong gravity, a Riemann-Cartan geometrical structure, we generalize the Christoffel connections (11.4) as follows (see Chapter X):

$$\Gamma_{\alpha\beta}{}^{\mu}(g) = \left\{{}^{\mu}_{\alpha\beta}\right\}(g) - K_{\alpha\beta}{}^{\mu}(g) \tag{11.34}$$

$$\Gamma_{\alpha\beta}{}^{\mu}(f) = \left\{{}^{\mu}_{\alpha\beta}\right\}(f) - K_{\alpha\beta}{}^{\mu}(f) \tag{11.35}$$

where $K_{\mu\nu}{}^{\alpha}(f)$ is the contorsion tensor relative to the strong metric and the hadronic matter, while $K_{\mu\nu}{}^{\alpha}(g)$ is related to the leptonic part of the gravitational interaction.

The Lagrangian density for the gravitational field is then

$$\mathcal{L}_G = \mathcal{L}_f + \mathcal{L}_g + \mathcal{L}_{fg} \tag{11.36}$$

where, separating the torsion contributions according to eq.(10.38)

$$\mathcal{L}_g = -\frac{e}{2\chi_g}\left[R(\{\})(g) + T_{\alpha\mu\nu}(g)\,K^{\nu\mu\alpha}(g)\right] \tag{11.37}$$

$$\mathcal{L}_f = -\frac{a}{2\chi_f}\left[R(\{\})(f) + T_{\alpha\mu\nu}(f)\,K^{\nu\mu\alpha}(f)\right] \tag{11.38}$$

$T_{\mu\nu\alpha}$ is the modified torsion tensor defined by eq.(10.34), $e = \det e_{\mu}{}^{\kappa} = \sqrt{-g}$, and $a = \det a_{\mu}{}^{\kappa} = \sqrt{-f}$. Finally $\mathcal{L}_{fg}$ is a coupling term between the two metric tensors, like that of eq.(11.13), which can be expressed in terms of the two vierbein fields, using (11.31) (note that also $R(\{\})$ can be expressed in terms of the vierbeins, see for example Chapter XII, Section 3).

It should be stressed that $\mathcal{L}_{fg}$ represents the interaction of the two metric tensors: more generally, however, we could consider also a coupling of the two torsion tensors; a particular possible form of such mixing term will be presented and briefly discussed at the end of this Section.

The matter Lagrangians (11.2,3) become

$$\mathcal{L}_{\ell} = e\,L_e\left[\ell, \nabla_{\mu}(g)\ell, e_{\mu}{}^{\kappa}\right] \tag{11.39}$$

$$\mathcal{L}_h = a\,L_h\left[h, \nabla_{\mu}(f)h, a_{\mu}{}^{\kappa}\right] \tag{11.40}$$

where $\nabla_\mu(g)$ and $\nabla_\mu(f)$ denote the covariant derivatives relative to the Riemann-Cartan connections (11.34) and (11.35) respectively.

Performing the variation of the total action corresponding to the Lagrangian $\mathcal{L}_g + \mathcal{L}_f + \mathcal{L}_{fg} + \mathcal{L}_l + \mathcal{L}_h$, with respect to the vierbein fields $e_\mu{}^i$, $a_\mu{}^i$ and the two contorsion K(g), K(f), we get the following field equations:

$$-\frac{\delta \mathcal{L}_g}{\delta e_\mu{}^k} e_\nu{}^k = \frac{\delta \mathcal{L}_l}{\delta e_\mu{}^k} e_\nu{}^k + \frac{\delta \mathcal{L}_{fg}}{\delta e_\mu{}^k} e_\nu{}^k$$

$$-\frac{\delta \mathcal{L}_g}{\delta K^{\alpha\beta\mu}(g)} = \frac{\delta \mathcal{L}_l}{\delta K^{\alpha\beta\mu}(g)} \qquad (11.41)$$

for the leptonic part, and

$$-\frac{\delta \mathcal{L}_f}{\delta a_\mu{}^k} a_\nu{}^k = \frac{\delta \mathcal{L}_h}{\delta a_\mu{}^k} a_\nu{}^k + \frac{\delta \mathcal{L}_{fg}}{\delta a_\mu{}^k} a_\nu{}^k$$

$$-\frac{\delta \mathcal{L}_f}{\delta K^{\alpha\beta\mu}(f)} = \frac{\delta \mathcal{L}_h}{\delta K^{\alpha\beta\mu}(f)} \qquad (11.42)$$

for the hadronic one.

We must remember now that by varying the gravitational Lagrangian with respect to K we obtain the modified torsion tensor, see eq.(10.50), while the variation of the matter Lagrangian gives the spin density tensor, eq.(10.46). Moreover the canonical energy-momentum tensor can be defined in terms of the tetrad variation of the matter Lagrangian as[2])

$$\Theta^\mu{}_\nu(l) = \frac{2}{e} \frac{\delta \mathcal{L}_l}{\delta e_\mu{}^k} e_\nu{}^k$$

$$\Theta^\mu{}_\nu(h) = \frac{2}{a} \frac{\delta \mathcal{L}_h}{\delta a_\mu{}^k} a_\nu{}^k \qquad (11.43)$$

while the tetrad variation of the field Lagrangian gives the Einstein tensor of the U_4 theory

$$G^\mu{}_\nu(g) = -\frac{2\chi_g}{e} \frac{\delta \mathcal{L}_g}{\delta e_\mu{}^k} e_\nu{}^k$$

$$G^\mu{}_\nu(f) = -\frac{2\chi_f}{a} \frac{\delta \mathcal{L}_f}{\delta a_\mu{}^k} a_\nu{}^k \qquad (11.44)$$

The field equations (11.41,42) become then

$$G_{\mu\nu}(g) = \chi_g \, \Theta_{\mu\nu}(\ell) + X_{\mu\nu}$$

$$T_{\alpha\beta\mu}(g) = \chi_g \, S_{\alpha\beta\mu}(\ell) \qquad (11.45)$$

$$G_{\mu\nu}(f) = \chi_f \, \Theta_{\mu\nu}(h) + Y_{\mu\nu}$$

$$T_{\alpha\beta\mu}(f) = \chi_f \, S_{\alpha\beta\mu}(h) \qquad (11.46)$$

where

$$X_{\mu\nu} = \frac{2\chi_g}{e} \frac{\delta \mathcal{L}_{fg}}{\delta e^{\mu}{}_{k}} e_{\nu}{}^{k}$$

$$Y_{\mu\nu} = \frac{2\chi_f}{a} \frac{\delta \mathcal{L}_{fg}}{\delta a^{\mu}{}_{k}} a_{\nu}{}^{k} \qquad (11.47)$$

This system of Einstein-Cartan equations, coupled by the interaction terms X and Y, generalizes the field equations (11.7,8) of the strong gravity theory, so as to include the dynamical spin contributions according to the geometry of the U_4 manifold.

In order to stress that spin, in this theory, is expected to play a fundamental role in the determination of the hadronic geometry (unlike in the case of ordinary gravity, where the torsion corrections are neglegible), we can develop a simple semiclassical application of the field equations, and evaluate directly the spin contributions to the strong gravity potential.

Considering the gravitational field only in the immediate vicinity of the hadronic source, we can neglect in first approximation the finite range of strong gravity, equating to zero the mass of the f-meson.

In this limit $\mathcal{L}_{fg} \to 0$, the interaction terms X and Y disappear from the field equations, so that the hadronic and leptonic part become decoupled. The strong gravity field can be determined then by solving the Einstein-Cartan equations for $f_{\mu\nu}$, i.e.

$$G_{\mu\nu}(f) = \chi_f \, \Theta_{\mu\nu}(h)$$

$$T_{\alpha\beta\mu}(f) = \chi_f \, S_{\alpha\beta\mu}(h) \qquad (11.48)$$

which can also be written in the form (10.59)

$$G_{\mu\nu}(\{\})(f) = \chi_f T_{\mu\nu}(h) + \chi_f^2 \Upsilon_{\mu\nu}(h) \qquad (11.49)$$

In vacuum the Einstein-Cartan field equations become those of general relativity (because the torsion is vanishing); nevertheless the metric, as stressed in Chapter X, Sect.3, is indirectly influenced by the spin content of the source.

Consider, for simplicity, a static and spherically symmetric matter configuration, and introduce the following modified stress-energy tensor

$$\tilde{T}_{\mu\nu} = T_{\mu\nu} + \chi_f \Upsilon_{\mu\nu} \qquad (11.50)$$

Comparing the mass parameter M of the Schwarzschild solution with the internal solution of (11.49) we get

$$M = \frac{1}{c^2}\int_V d^3x\, \tilde{T}_4{}^4 = \frac{4\pi}{c^2}\int_0^R dr\, r^2\, \tilde{T}_4{}^4 \qquad (11.51)$$

where V is the volume of the source, and R its radius. Describing the source, in a semiclassical approximation, as a Weyssenhoff spinning fluid [5], from (11.50) and (10.60) one obtains, after performing a suitable averaging procedure (see for example ref.6)

$$M = 4\pi\int_0^R r^2 dr \left(\rho - \frac{2\pi}{c^4} G_f \sigma^2\right) \qquad (11.52)$$

where ρ is the proper mass density and σ the averaged spin density of the fluid, $\sigma^2 = \langle S_{\mu\nu} S^{\mu\nu}\rangle$. For a sphere of mass m, spin $\hbar/2$ and radius $\hbar/mc$

$$\rho = \frac{m}{V}\ , \qquad \sigma^2 = \frac{\hbar^2}{4V^2}\ , \qquad V = \frac{4}{3}\pi\left(\frac{\hbar}{mc}\right)^3 \qquad (11.53)$$

and one is led to the following strong gravitational potential

$$\phi = \frac{G_f M}{r c^2} = \frac{G_f}{r c^2}\left(m - \frac{3}{8} G_f \frac{m^3}{\hbar c}\right) \qquad (11.54)$$

In this approximation, the spin effect produce only a correction of the usual linear relation existing between mass and static gravitational potential. Putting $\phi = \phi_0 + \phi_1$,

$$\phi_0 = G_f \frac{m}{r c^2}\ , \qquad \phi_1 = \frac{3}{8}\frac{G_f^2}{r}\frac{m^3}{\hbar c^3} \qquad (11.55)$$

and taking for m the mass of a nucleon, we have

$$\frac{\phi_1}{\phi_0} = \frac{3 G_f m^2}{8 \hbar c} \sim 1 \qquad (11.56)$$

Therefore, in the framework of the strong gravity theory, the spin contributions to the metric are of the same order as the mass contributions, at a distance of the order of the nuclear radius. In the usual Einstein-Cartan theory, on the contrary, replacing G_f with G in (11.56) one has

$$\frac{\phi_1(g)}{\phi_0(g)} = \frac{3 G m^2}{8 \hbar c} \sim 10^{-39} \qquad (11.57)$$

so that the spin contributions are neglegible.

Finally, to conclude this Section, we wish to discuss briefly the possible role of a propagating torsion in the context of a strong gravity theory[7].

Supposing that the sources are a leptonic Dirac spinor, ℓ, of mass m, and a hadronic Dirac spinor, h, of mass M, the matter Lagrangians (11.39,40), separating the torsion contributions, can be written as follows:

$$\mathcal{L}_\ell = \sqrt{-g}\,\Big[\frac{i}{2} g^{\mu\nu}(\bar{\ell}\gamma_\mu \ell_{;\nu} - \bar{\ell}_{;\mu}\gamma_\nu \ell) + + \frac{i}{4} K_{\alpha\beta\mu}(g)\,\bar{\ell}\gamma^{[\alpha}\gamma^\beta\gamma^{\mu]}\ell - m\bar{\ell}\ell\Big] \qquad (11.58)$$

$$\mathcal{L}_h = \sqrt{-f}\,\Big[\frac{i}{2} f^{\mu\nu}(\bar{h}\gamma_\mu h_{|\nu} - \bar{h}_{|\mu}\gamma_\nu h) + + \frac{i}{4} K_{\alpha\beta\mu}(f)\,\bar{h}\gamma^{[\alpha}\gamma^\beta\gamma^{\mu]}h - M\bar{h}h\Big] \qquad (11.59)$$

(remember 10.158).

Because of the axial character of the spin tensor of a Dirac field, only the totally antisymmetric part of the torsion interact with matter, as shown in Chapter X, Sect.5, and then, in the absence of other sources beside the two spinors, we can put (see 10.187)

$$\begin{aligned} K_{\alpha\beta\mu}(g) &= \varepsilon_{\nu\alpha\beta\mu} Q^\nu(g) = - Q_{[\alpha\beta\mu]}(g) \\ K_{\alpha\beta\mu}(f) &= \varepsilon_{\nu\alpha\beta\mu} Q^\nu(f) = - Q_{[\alpha\beta\mu]}(f) \end{aligned} \qquad (11.60)$$

where Q_ν is the axial-vector part of the torsion tensor.

Following the spirit of the original strong gravity

theory, where the strong gravitational field is a short range interaction falling down to zero very fast outside the particle radius ($\sim 10^{-13}$ cm), while the weak gravitational field is propagating also at a macroscopical level with an infinite range, it is natural to suppose that the weak torsion is determined by a pseudoscalar potential, like in Chapter X, Sect.6,

$$Q_\mu(g) = \partial_\mu \varphi \tag{11.61}$$

and then it is a propagating field.

The spin torsion interactions for the the hadronic and leptonic case become then

$$\frac{i}{4} K_{\alpha\beta\mu}(g)\, \bar{\ell}\, \gamma^{[\alpha}\gamma^{\beta}\gamma^{\mu]}\, \ell = -\frac{3}{2}\, g^{\mu\nu}\, \partial_\mu \varphi\, J^5_\nu(\ell)$$

$$\frac{i}{4} K_{\alpha\beta\mu}(f)\, \bar{h}\, \gamma^{[\alpha}\gamma^{\beta}\gamma^{\mu]}\, h = -\frac{3}{2}\, f^{\mu\nu}\, Q_\nu\, J^5_\nu(h) \tag{11.62}$$

where $J^5_\mu = \bar{\psi}\gamma^5\gamma_\mu\psi$, and henceforth we denote with Q_μ the axial part of the strong torsion, i.e. $Q_\mu = Q\ (f)$.

In the generalized case in which, besides the two metrics, we have also two torsions, it is natural to introduce an interaction Lagrangian describing the coupling of these two tensors; in particular, one can choose the following mixing term (the physical motivations for this choice will be explained in a moment):

$$\mathcal{L}(\varphi, Q) = -\frac{3}{2}\left[\sqrt{-f}\, f^{\mu\nu}\, Q_\mu\, J^5_\nu(\ell) + \sqrt{-g}\, g^{\mu\nu}\, \partial_\mu \varphi\, J^5_\nu(h)\right] \tag{11.63}$$

The total Lagrangian is obtained then by adding to the field Lagrangians (11.37,38) and to the matter Lagrangians (11.58,59) the interaction Lagrangians $\mathcal{L}_{fg}$ and $\mathcal{L}(\varphi, Q)$. The result, $\mathcal{L}_T$, can be written in this way

$$\mathcal{L}_T = \mathcal{L}(g,\ell) + \mathcal{L}(f,h) + \mathcal{L}(g,f,\ell,h,\varphi,Q) \tag{11.64}$$

where the weak, Riemannian part is

$$\mathcal{L}(g,\ell) = \sqrt{-g}\left[-\frac{R(\{\})(g)}{2\chi_g} + \frac{i}{2}\, g^{\mu\nu}(\bar{\ell}\gamma_\mu \ell_{;\nu} - \bar{\ell}_{;\mu}\gamma_\nu \ell) - m\bar{\ell}\ell\right] \tag{11.65}$$

the strong, Riemannian part is

$$\mathcal{L}(f,h) = \sqrt{-f}\left[-\frac{R(\{\})(f)}{2\chi_f} + \frac{i}{2} f^{\mu\nu}(\bar{h}\gamma_\mu h_{|\nu} - \bar{h}_{|\mu}\gamma_\nu h) - M\bar{h}h\right] \tag{11.66}$$

and the interaction Lagrangian, containing also the torsionic corrections, is

$$\mathcal{L}(g,f,h,\ell,\varphi,Q) = \mathcal{L}_{fg} + \sqrt{-g}\, g^{\mu\nu}\partial_\mu\varphi\left\{\frac{3}{\chi_g}\partial_\nu\varphi - \right.$$
$$\left. - \frac{3}{2}\left[J^5_\nu(\ell) + J^5_\nu(h)\right]\right\} + \sqrt{-f}\, f^{\mu\nu} Q_\mu \left\{\frac{3}{\chi_f} Q_\nu - \right.$$
$$\left. - \frac{3}{2}\left[J_\nu{}^5(\ell) + J_\nu{}^5(h)\right]\right\} \tag{11.67}$$

(note that in the limit in which the torsion disappears we recover the usual strong gravity theory).

By independent variation of φ and Q we get the equations for the weak and strong torsion field:

$$\varphi^{;\mu}{}_{;\mu} = \frac{1}{2}\chi_g\left[J^{\mu 5}(\ell)_{;\mu} + J^{\mu 5}(h)_{;\mu}\right] \tag{11.68}$$

$$Q^\mu = \frac{1}{2}\chi_f\left[J^{\mu 5}(\ell) + J^{\mu 5}(h)\right] \tag{11.69}$$

The spin-torsion mixing term of eq.(11.63) implies then that the hadronic and leptonic spin densities contribute in the same way to the torsion of space-time, as shown by (11.68,69) (this seems physically acceptable, because the spin of a lepton is the same as that of an hadron, $\hbar/2$, unlike their energy content).

At a microscopical level is produced a strong torsion field, which is algebraically related to the total spin density and hence it is different from zero only inside a matter distribution (eq.11.69). Outside matter it survives a weak torsion field, determined by a propagating potential and active also at a macroscopical level (eq. 11.68).

Therefore, according to this theory, the spin-torsion interaction, very weak in empty space, should become a strong interaction when two spinning particles are separated by a distance shorter than their characteristic radius (for instance in high energy scattering), and this should occur both for hadrons and leptons.

APPENDIX A

Strong gravity and the large numbers hypothesis

Using the fundamental parameters of microphysics (such as the Planck constant h , the mass of the nucleon m,...) and of cosmology (such as the Hubble constant H, the mass of the universe M,...) combined with the gravitational constant G and the light velocity c, we can construct large dimensionless numbers of the order of D^n , where $D \sim 10^{40}$, and n = 1,2,3,...

For example, the ratio between the present age of the universe ($\sim H^{-1}$, see Chapter V) and the time needed by ligh to cross an elementary particle ($\sim r/c$, where $r \sim \hbar/mc$), is given by

$$\frac{mc^2}{\hbar H} \sim D \tag{A.1}$$

where m is a typical hadronic mass.

The ratio between the Compton wave-length of the proton and its gravitational radius is

$$\frac{\hbar c}{2Gm^2} \sim D \tag{A.2}$$

The ratio between the mass M contained inside the Hubble radius (in the case of a universe with critical density, see Chapter V),

$$M \sim \frac{Rc^2}{G} \tag{A.3}$$

(where $R \sim cH^{-1}$ is the Hubble radius), and the mass of the proton is

$$\frac{M}{m} \sim \frac{c^3}{GHm} \sim D^2 \tag{A.4}$$

and so on.

The ratio in eq.(A.1) is not a constant, but is growing linearly in time (because the age of the universe is growing): if we believe that the equality of all these ratios is not a numerical coincidence, characteristic only of our present epoch, we are led then to conclude that when dimensionless number is of the order of D^n , it should be variable in time as t^n.

This is the so-called large numbers hypothesis of Dirac[8].

An immediate consequence of this assumption is that, in the system of units in which the light velocity, the Planck constant and the proton mass are constants, the Newton constant of gravity must decrease in time, from (A.2), like $G \propto t^{-1}$. Moreover, from (A.4), it follows that the number of nucleons contained within the hubble radius, N = = M/m, must grow in time like $N \propto t^2$.

To reconcile these effects with the standard Einstein theory, which requires a constant value of G, one may assume that general relativity is valid in a system of units different from that based on the atomic constants. In this case it follows, however, that the proper time interval ds_E of the Einstein theory is different from the experimentally measured interval ds_A. In fact, all the laboratory measurements of times and distances are based on atomic units, such as the wave-length of the spectral lines or the frequency of monochromatic radiation, so that what one measures is ds_A (the light velocity is assumed to be the same in the two systems). The other time interval, ds_E, is supposed to correspond to phenomena governed by the gravitational interaction, such as planetary motion.

In order to obtain a relation between the two intervals ds_E and ds_A, consider the motion of a planet around the sun, in the Newton approximation. The equation of motion for a circular orbit of radius r is $GM_\odot = rv^2$. In Einstein units then

$$G_E M_E = r_E v_E^2 \qquad (A.5)$$

in atomic units

$$G_A M_A = r_A v_A^2 \qquad (A.6)$$

The velocity v is the same in the two systems, because the ratio v/c is adimensional, and c is constant in both units. In the Einstein units G_E, M_E are constants, and then r_E = const. In the atomic units, on the contrary, $G_A \propto t^{-1}$, and if the mass M_A is constant, then $r_A \propto t^{-1}$.

With respect to a system of atomic units, the radius of the orbit decreases then in time, while it is constant in Einstein units. It follows that $ds_E = t\, ds_A$ so that, denoting with τ the Einstein time and with t the atomic time, one gets $\tau = t^2/2$.

Therefore one could try to test this hypothesis, for example by accurate observations of planetary motions. But the point we wish to stress here is that the strong gravity theory may provide a physical interpretation of this large numbers coincidences[a].

Suppose in fact that the universe and a hadron are two physical systems governed by similar internal laws, differing only for a scale factor which transforms the weak gravitational field into the strong gravity field, and assume that the strong gravity potential of a hadron, ϕ (h) $\sim G_f m/rc^2$, for $r \sim \hbar/mc$, is of the same order of magnitude as the "gravitational potential" of the universe, ϕ (u) $\sim GM/Rc^2$, where $R = c/H$. It follows that

$$\frac{G_f}{G} \sim \frac{M}{m}\frac{r}{R} \sim D \qquad (A.7)$$

in agreement with the value suggested for the strong gravity coupling constant in Sect.1 of this Chapter.

If the Newton coupling constant varies in time as $G \propto t^{-1}$, from (A.7) we have that, at the epoch $t_0 = (HD)^{-1} \sim \hbar/mc^2 \sim 10^{-24}$ sec , it was $G = G_f$, so that the strength of ordinary gravity and of strong gravity coincided. One may speculate then that the strong gravitational interactions have been frozen with their promordial strength, but confined inside the hadronic matter, while the other interaction, becoming weaker and weaker in time , have reached the present value $G \sim 10^{-40} G_f$. It is natural then to identify the atomic proper time interval of Dirac, ds_A , with that of strong gravity, which is effective at a nuclea level only, and the Einstein interval, ds_E , with that of ordinary macroscopic gravity. In this sense the strong gravity theory provides a physical basis for the large numbers hypothesis.

It is worth stressing that, starting from the analogy between the gravitational interaction at a cosmological and at a hadronic level, and given the ratio of the two coupling constants, one can deduce the value of the Hubble radius , or of the critical mass of the universe, directly from the values of the corresponding microphysical quantities. Using (A.4,7) one has, for example,

$$\frac{R}{r} \sim D \qquad (A.8)$$

If we introduce spin and torsion into the strong gravity field equations, according to the Einstein-Cartan theory, then the gravitational potential of a nucleon is modified by the spin contributions , as shown in Section 4 of this Chapter (see 11.55). In particular, at a distance $r \sim \hbar/mc$, we can write

$$\phi_0(h) \sim G_f \frac{m^2}{\hbar c} \tag{A.9}$$

$$\phi_1(h) \sim G_f^2 \frac{\sigma^2 r^2}{c^6} \tag{A.10}$$

where $\sigma \sim \hbar/r^3$ is the spin density . Since the assumption $\phi_0(h) \sim \phi_0(u)$ provides the large numbers correlations between microphysics and cosmology, for the case of mass, radius and coupling strenght (see A.7), let us suppose also $\phi_1(h) \sim \phi_1(u)$. Eq.(A.10) gives then

$$G_f \sigma r \sim G \Sigma R \tag{A.11}$$

where $\Sigma \sim S/R^3$ is the density of intrinsic angular momentum of the universe. Using (A.7,8) we obtain

$$\sigma \sim \Sigma \tag{A.12}$$

which means that the angular momentum density of the universe and of an elementary particles are the same, or , in other words,

$$S \sim \sigma R^3 \sim (R/r)^3 \hbar \sim \hbar D^3 \tag{A.13}$$

It should be noted that interpreting this angular momentum as due to a rigid rotation of the universe as a whole, that is putting $\Sigma \sim M\omega/R$, one obtains that the angular velocity is, from (A.12,4,8)

$$\omega \sim \frac{R}{M}\frac{\hbar}{r^3} \sim \frac{c}{R} \sim H \tag{A.14}$$

and is perhaps remarkable that an angular velocity of the same order of magnitude is predicted also by Gödel's rotating model of universe[10],

$$\omega_G = 2\sqrt{2\pi G\rho} \sim \left(\frac{GM}{R^3}\right)^{1/2} \sim \frac{c}{R} \sim H \tag{A.15}$$

Therefore, in the framework of strong gravity with torsion, one is led to construct the following large numbers

$$\frac{G_f}{G} \sim \frac{R}{r} \sim D \qquad \text{related to coupling strenghts}$$

$$\frac{M}{m} \sim \left(\frac{R}{r}\right)^2 \sim D^2 \qquad \text{related to mass}$$

$$\frac{S}{\hbar} \sim \left(\frac{R}{r}\right)^3 \sim D^3 \qquad \text{related to spin}$$

According to the Dirac hypothesis ($D^n \propto t^n$) from the first relation we have $G \propto t^{-1}$, from the second $M \propto t^2$ and from the third $S \propto t^3$. If one supposes that the angular momentum of the universe is due to a classical rotation, $S \sim M \omega R^2$, then the angular velocity should be decreasing in time like $\omega \propto t^{-1}$; if, however, S is the sum of elementary spins, $S \sim n\hbar$, it follows that their number within the Hubble radius should grow in time like $n \propto t^3$.

We conclude this Chapter giving the following Table, which shows how are modified energy, length, time, temperature and density related to the Planck mass $m_P = (\hbar c/G)^{1/2}$ (which are typical of the usual gravitational interaction), if the Newton constant is replaced by the strong gravity coupling constant, $G \rightarrow G_f$ (in the definition of the temperature T, we have denoted with k the Boltzmann constant).

G	G_f
$m_p = (\frac{\hbar c}{G})^{1/2} \sim 10^{-5}$ g	$\sim 10^{-24}$ g $\sim$ nucleon mass
$E = (\frac{\hbar c^5}{G})^{1/2} \sim 10^{19}$ GeV	~ 1 GeV $\sim$ nucleon rest energy
$\ell = (\frac{\hbar G}{c^3})^{1/2} \sim 10^{-33}$ cm	$\sim 10^{-14}$ cm $\sim$ nucleon Compton wave-length
$t = (\frac{\hbar G}{c^5})^{1/2} \sim 10^{-43}$ sec	$\sim 10^{-24}$ sec $\sim$ typical hadron life-time
$T = (\frac{\hbar c^5}{GK^2})^{1/2} \sim 10^{32}$ ^oK	$\sim 10^{13}$ ^oK $\sim$ limiting Hagedron temperature
$\rho = \frac{c^5}{\hbar G^2} \sim 10^{94}$ g/cm^3	$\sim 10^{16}$ g/cm^3 $\sim$ typical nucleon density

REFERENCES

1) C.J.Isham,A.Salam and J.Strathdee: Phys.Rev.D3,867(197

2) P.C.Aichelburg: Phys.Rev.D8,377(1973)

3) V.De Sabbata and M.Gasperini, in "Spin,torsion, rotation and supergravity", ed.by P.G.Bergamann and V.De Sabbata (Plenum, New York, 1980) p.139 and references therein

4) See for example F.W.Hehl, in "Spin, torsion, rotation and supergravity", ed. by P.G.Bergamann and V.De Sabbata (Plenum, New York 1980) p.5

5) W.Arkuszewski,W.Kopczynski and V.N.Ponomariev: Ann.Inst Henri Poincarè XXI,89(1974)

6) F.W.Hehl,P.von der Heyde and G.D.Kerlick: Phys.Rev.D10 1066(1974)

7) V.De Sabbata and M.Gasperini: Lett.Nuovo Cimento 32, 469(1981)

8) See for example P.M.Dirac: Proc.R.Soc.365A,19(1979)

9) C.Sivaram and K.Sinha: Phys.Lett.60B,181(1976)

10) K.Gödel: Rev.Mod.Phys.21,447(1949)

CHAPTER XII

GAUGE THEORY OF GRAVITY

1.- Gauge invariance and compensating fields

The Weinberg-Salam unified theory of weak and electromagnetic interactions is a gauge theory. Quantum chromodynamics (the theory of strong interactions) is a gauge theory. The unified theorie of strong and electroweak interactions (the so-called grand unified theories) are gauge theories. We are then motivated to consider the possibility of formulating a gauge theory of gravity, in analogy with the other theories, and in view of a possible unification of gravity with the other fundamental interactions.

In this Chapter we discuss the possible formulation of the Einstein-Cartan theory as a gauge theory with a local Lorentz invariance, after a short introduction to the general principles and to the formalism of the gauge theories.

Consider a matter Lagrangian density $L(\psi, \partial\psi)$ for a matter field ψ in the flat space. With the word "matter" we denote, in this context, the sources of the interactions, that is spinor fields (like quarks and leptons) which are believed to be the fundamental components of matter, or scalar fields, like the Higgs particles, which are used to introduce mass in a gauge theory (and which however could be composed of more fundamental fermion fields).

Photons, gluons and W,Z mesons, corresponding respectively to electromagnetic, strong and weak interactions, are not matter fields but gauge fields, and in this context may be regarded as derived from the fundamental fermion fields.

Gauge fields are bosons (vectors, tensors) and gauge theories provide a rather sharp conceptual separation between the sources of the interactions (matter fields, prevalently fermions) and the interaction themselves (gauge fields, exclusively bosons). This separation between bosons and fermions is lost however in the framework of the supersymmetric theories, as we shall see in Chapter XIII.

Considering then the matter Lagrangian $L(\psi, \partial\psi)$, suppose that the corresponding action

$$S = \int d^4x\, L(\psi, \partial\psi) \qquad (12.1)$$

is invariant under a group of non-abelian trasformations depending on n constant parameters

$$\psi' = U\psi \quad , \qquad U = e^{-i\varepsilon^A X_A} \qquad (12.2)$$

where ε^A are n real constant parameters (A,B,C = 1,2,..n), and $X_A = X_A^+$ are n constant matrices representing the generators of the transformation group, which satisfy the commutation rules of the Lie algebra

$$[X_A, X_B] = i f_{AB}{}^C X_C \qquad (12.3)$$

($f_{AB}{}^C = - f_{BA}{}^C$ are the structure constants of that group). An infinitesimal transformation can be written

$$\delta\psi = \psi' - \psi = -i\varepsilon^A X_A \psi \qquad (12.4)$$

Consider for example the following Lagrangian ($\hbar$=c=1)

$$L = (\partial_\mu \varphi)^+ \partial^\mu \varphi - m^2 \varphi^+ \varphi \qquad (12.5)$$

where φ is a multiplet of N complex scalar fields

$$\varphi = \begin{pmatrix} \varphi_1 \\ \varphi_2 \\ \vdots \\ \varphi_N \end{pmatrix} , \qquad \varphi^+ = (\varphi_1^*, \varphi_2^*, \cdots \varphi_N^*) \qquad (12.6)$$

In this case the generators are N×N matrices, $(X_A)^\alpha{}_\beta$, with α, β = 1,2,...,N and an infinitesimal transformation becomes

$$\delta\varphi^\alpha = -i\varepsilon^A (X_A)^\alpha{}_\beta \varphi^\beta$$
$$\delta\varphi^{+\alpha} = i\varepsilon^A \varphi^{+\beta} (X_A)^\alpha{}_\beta \qquad (12.7)$$

(henceforth, for simplicity, the representation indices α, β, will not be written explicitly). From this transformation rule we have, for the Lagrangian

$$\delta L = L' - L = (\partial_\mu \varphi)^+ \partial^\mu \delta\varphi + \partial_\mu (\delta\varphi)^+ \partial^\mu \varphi -$$
$$- m^2 \varphi^+ \delta\varphi - m^2 \delta\varphi^+ \varphi =$$
$$= -i\varepsilon^A X_A \left[(\partial_\mu \varphi)^+ \partial^\mu \varphi - (\partial_\mu \varphi)^+ \partial^\mu \varphi - m^2 \varphi^+ \varphi + \right.$$
$$\left. + m^2 \varphi^+ \varphi \right] \equiv 0 \qquad (12.8)$$

and the action is invariant under global gauge transformations (global because the parameters ε are constant).

If we consider however local gauge transformations, in which the parameters are functions of the time and of the position, $\varepsilon^A = \varepsilon^A(\vec{x},t) = \varepsilon^A(x)$,

$$U(x) = e^{-i\varepsilon^A(x) X_A} \qquad (12.9)$$

$$\delta\varphi = -i\varepsilon^A(x) X_A \varphi \qquad (12.10)$$

the Lagrangian is no loger invariant; we have in fact

$$\delta L = (\partial_\mu \varphi)^+ (-i X_A \varphi \partial^\mu \varepsilon^A) + (i \varphi^+ X_A \partial_\mu \varepsilon^A) \partial^\mu \varphi =$$
$$= -J^\mu{}_A \partial_\mu \varepsilon^A \neq 0 \qquad (12.11)$$

where

$$J^\mu{}_A = i \left[\varphi^+ X_A \partial^\mu \varphi - (\partial^\mu \varphi)^+ X_A \varphi \right] \qquad (12.12)$$

Note that the invariance is broken, $\delta L \neq 0$, because

$$\delta(\partial_\mu \varphi) = -i\varepsilon^A X_A \partial_\mu \varphi - i X_A \varphi \partial_\mu \varepsilon^A \qquad (12.13)$$

that is the derivative of the field transforms differently from the field itself (compare 12.10 and 12.13).

In order to preserve gauge invariance even in the case of local transformations, $\varepsilon = \varepsilon(x)$, we must replace then the partial derivatives, in the matter Lagrangian (12.5), with generalized derivatives which transform like the fields, i.e. $\partial_\mu \varphi \longrightarrow D_\mu \varphi$, where

$$(D_\mu \varphi)' = U(x) D_\mu \varphi \qquad (12.14)$$

so that

$$\delta\,(D_\mu\varphi) = -i\,\varepsilon^A X_A D_\mu \varphi \qquad (12.15)$$

To this aim we introduce, corresponding to each generator X_A of the group we are considering, a covariant vector field, $A_\mu{}^A$, which , with its transformation properties, may compensate the variation of the Lagrangian, so that $\delta L(\varphi, \partial\varphi, A) = 0$ even in the case $\varepsilon^A = \varepsilon^A(x)$. This vector field is called then compensating field, or gauge potential.

The generalized derivative, called gauge covariant derivative, is defined as follows:

$$D_\mu \varphi = (\partial_\mu - i g A_\mu{}^A X_A)\,\varphi \qquad (12.16)$$

where g is an arbitrary coupling constant. The transformation rules for the new field $A_\mu{}^A$ under a local gauge transformation are to be determined imposing the condition (12.14), i.e.

$$(D_\mu \varphi)' = (\partial_\mu - i g A'^A_\mu X_A)\,\varphi' = (\partial_\mu - i g A'_\mu{}^A X_A)\,U\varphi =$$

$$= U D_\mu \varphi = U(\partial_\mu - i g A_\mu{}^A X_A)\,\varphi \qquad (12.17)$$

We obtain then the following transformation rule for the gauge potential $A_\mu{}^A$:

$$A'_\mu \equiv A'_\mu{}^A X_A = U A_\mu{}^A X_A U^{-1} - \frac{i}{g}\,(\partial_\mu U)\,U^{-1} \qquad (12.18)$$

The infinitesimal form of this equation is easily obtained putting $U \simeq 1 - i\,\varepsilon^A X_A$, and neglecting terms of order ε^2 and higher. From (12.18) then

$$A'_\mu \simeq (1 - i\varepsilon^B X_B)\,A_\mu\,(1 + i\varepsilon^C X_C) - \frac{i}{g}\,\partial_\mu(-i\varepsilon^A X_A)$$

$$\simeq A_\mu + i A_\mu \varepsilon^C X_C - i\varepsilon^C X_C A_\mu - \frac{1}{g} X_A \partial_\mu \varepsilon^A$$

$$= A_\mu + i\,\varepsilon^C A_\mu{}^B\,(X_B X_C - X_C X_B) - \frac{1}{g} X_A \partial_\mu \varepsilon^A \qquad (12.19)$$

and using the commutation rules (12.3) we have

$$\delta A_\mu = (\delta A_\mu{}^A)\,X_A = A'_\mu - A_\mu = -\frac{1}{g} X_A \partial_\mu \varepsilon^A - f_{BC}{}^A X_A A_\mu{}^B \varepsilon^C \qquad (12.20)$$

from which

$$\delta A_\mu{}^A = -\frac{1}{g}\partial_\mu \varepsilon^A - f_{BC}{}^A A_\mu{}^B \varepsilon^C \qquad (12.21)$$

The new matter Lagrangian

$$\mathcal{L}(\varphi, D\varphi) = (D_\mu \varphi)^+ D^\mu \varphi - m^2 \varphi^+ \varphi \qquad (12.22)$$

obtained from the original one (12.5) by the replacement $\partial_\mu \longrightarrow D_\mu$, is invariant under the following local gauge transformations

$$\delta \varphi = -i\varepsilon^A X_A \varphi$$
$$\delta A_\mu{}^A = -\frac{1}{g}\partial_\mu \varepsilon^A - f_{BC}{}^A A_\mu{}^B \varepsilon^C. \qquad (12.23)$$

In fact, from (12.15) we have

$$\delta \mathcal{L} = (D_\mu \varphi)^+ (-i\varepsilon^A X_A D^\mu \varphi) + i\varepsilon^A X_A (D_\mu \varphi)^+ D^\mu \varphi$$
$$\equiv 0 \qquad (12.24)$$

Therefore, given an action invariant under a group of global continuous transformations, depending on a set of constant parameters ε^A, we can achieve its invariance also under the corresponding local transformations, with $\varepsilon^A = \varepsilon^A(x)$, provided we introduce a compensating vector field $A_\mu{}^A$, and replace the partial derivatives with the gauge covariant ones (12.16). This standard procedure prescribes a coupling between the gauge potential A_μ and the matter field φ which is called "minimal coupling".

Noting that the commutator of two covariant derivatives is generally nonvanishing, we can introduce a tensor field $F_{\mu\nu}{}^A$ (also called "field strength" or curvature) related to the gauge potential as follows:

$$(D_\mu D_\nu - D_\nu D_\mu)\varphi = -ig F_{\mu\nu}{}^A X_A \varphi \qquad (12.25)$$

where

$$F_{\mu\nu}{}^A = \partial_\mu A_\nu{}^A - \partial_\nu A_\mu{}^A + g f_{BC}{}^A A_\mu{}^B A_\nu{}^C = -F_{\nu\mu}{}^A \qquad (12.26)$$

Its transformation properties can be easily obtained noting that, from (12.17), one has

$$D'_\mu = U D_\mu U^{-1} \tag{12.27}$$

Therefore, putting $F_{\mu\nu} = F_{\mu\nu}{}^A X_A$, from (12.25) it follows

$$F'_{\mu\nu} = \frac{i}{g}\,[D'_\mu, D'_\nu] = \frac{i}{g}\,(U D_\mu D_\nu U^{-1} -$$
$$- U D_\nu D_\mu U^{-1}) = U F_{\mu\nu} U^{-1} \tag{12.28}$$

and, for an infinitesimal transformation, $U = 1 - i\varepsilon^A X_A$, we have

$$\delta F_{\mu\nu}{}^A = f_{BC}{}^A \varepsilon^B F_{\mu\nu}{}^C \tag{12.29}$$

Finally, it is important to mention that the tensor $F_{\mu\nu}$ satisfies the following identity

$$\varepsilon^{\mu\nu\alpha\beta} D_\nu F_{\alpha\beta}{}^A = 0 \tag{12.30}$$

which is analogous to the Bianchi identity satisfied by the curvature tensor (see also Chapter XIV, Sect.3).

As a new field has been introduced into the theory, A_μ we must complete the action by adding, to the matter Lagrangian, a kinetic term for the gauge field. The simplest choice, satisfying the requirements of Lorentz invariance and local gauge invariance, is

$$\mathcal{L} = -\frac{1}{4} F_{\mu\nu}{}^A F^{\mu\nu}{}_A \tag{12.31}$$

(note the analogy with the Maxwell Lagrangian). The total Lagrangian is then

$$\mathcal{L} = (D_\mu \varphi)^+ D^\mu \varphi - m^2 \varphi^+ \varphi - \frac{1}{4} F_{\mu\nu}{}^A F^{\mu\nu}{}_A \tag{12.32}$$

and it describes a matter field φ, minimally coupled to a massless vector field A_μ.

It is important to stress that the addition of a mass term for the gauge field, like $M^2 A_\mu A^\mu$, would break explicitly the local gauge invariance, as $\delta(M^2 A_\mu A^\mu) \neq 0$. This point could be regarded as an evidence against the hypothesis that all the interactions are described by gauge fields: in fact we know that the nuclear forces have a very short range, so that they must be mediated by the exchange of massive bosons (like the weak interactions). It would seem, therefore, that these interactions cannot be described by a gauge invariant theory in which the exchanged vector bosons are compelled to be massless.

Fortunately, however, this difficulty can be overcome because a method has been invented to make the gauge field massive, the so-called Higgs mechanism, in the hypothesis of spontaneous breaking of the local gauge symmetry; for more details, the interested reader is referred to same specialistic book on this subject[1].

To conclude this Section, we note that performing the variation of the Lagrangian (12.32) with respect to the vector potential $A_\mu{}^A$, the following field equations are obtained

$$\partial_\nu F^{\mu\nu}{}_A = ig\left(J^\mu{}_A + i f_{AB}{}^C A_\nu{}^B F^{\mu\nu}{}_C\right) \qquad (12.33)$$

where

$$J^\mu{}_A = \varphi^+ X_A D^\mu \varphi - (D^\mu \varphi)^+ X_A \varphi \qquad (12.34)$$

The right-hand side of (12.33), which defines the total conserved current, contains also the contributions of the gauge field itself if the symmetry group is non-abelian, i.e. if $f_{AB}{}^C \neq 0$. In this case the particles corresponding to the gauge fields are themselves sources of interaction, or, in other words, they are "charged", with respect to the charge g of that field.

In the case of the electromagnetism, which corresponds to an abelian local symmetry (as we will see in the following Section), one has $f_{AB}{}^C = 0$, in agreement with the fact that photons are electrically neutral particles, so that they are not sources of electromagnetic fields, and the Maxwell equations are then linear field equations.

2.- Electromagnetism as a gauge theory

The gauge potential of the electromagnetic theory can be interpreted as the compensating field needed to preserve the local gauge invariance under the transformations of the U(1) group, i.e. of the one-parameter unitary transformations, which can be represented as follows

$$\psi' = e^{-i\Lambda}\psi \qquad (12.35)$$

The constant parameter of these global transformations is Λ, and the generator is the identity, so that the structure constants are vanishing and the group is abelian.

Consider, for example, the Dirac Lagrangian in flat space ($\hbar=c=1$)

$$L = \frac{i}{2}(\bar{\psi}\gamma^\mu \partial_\mu \psi - \partial_\mu \bar{\psi}\gamma^\mu \psi) - m\bar{\psi}\psi \qquad (12.36)$$

It is invariant under global U(1) transformations (Λ = const) which, in infinitesimal form, are

$$\delta\psi = -i\Lambda\psi \quad , \qquad \delta\bar{\psi} = i\Lambda\bar{\psi} \qquad (12.37)$$

but not under the local transformations corresponding to $\Lambda = \Lambda(x)$; in fact, from (12.37), we have the following infinitesimal variation of the Lagrangian

$$\delta L = J^\mu \partial_\mu \Lambda \qquad (12.38)$$

where

$$J^\mu = \bar{\psi}\gamma^\mu\psi \qquad (12.39)$$

is the current density for a Dirac spinor.

In order to obtain local invariance, we replace the partial derivatives with the following covariant derivatives

$$\partial_\mu\psi \rightarrow D_\mu\psi = \partial_\mu\psi - ieA_\mu\psi \qquad (12.40)$$

$$\partial_\mu\bar{\psi} \rightarrow D_\mu\bar{\psi} = \partial_\mu\bar{\psi} + ieA_\mu\bar{\psi} \qquad (12.41)$$

where e is a coupling constant (the electric charge in this case) and the compensating potential A_μ is minimally couple to ψ, and satisfies the abelian transformation rules

$$\delta A_\mu = -\frac{1}{e}\partial_\mu\Lambda \qquad (12.42)$$

under a local gauge transformation (eq.12.21 with $f_{AB}{}^C = 0$).

The covariant derivative (12.40) transforms then like the field ψ,

$$\delta(D_\mu\psi) = \partial_\mu(\delta\psi) - ie\,\delta A_\mu\psi - ieA_\mu\,\delta\psi =$$

$$= \partial_\mu(-i\Lambda\psi) + i\psi\partial_\mu\Lambda - eA_\mu\Lambda\psi =$$

$$= -i\Lambda(\partial_\mu - ieA_\mu)\psi = -i\Lambda D_\mu \psi \tag{12.43}$$

and the new matter Lagrangian is invariant

$$\delta\mathcal{L}(\psi, D\psi) = \frac{i}{2}\left[\bar{\psi}\gamma^\mu(-i\Lambda D_\mu\psi) + i\Lambda\bar{\psi}\gamma^\mu D_\mu\psi\right.$$
$$\left. - D_\mu\bar{\psi}\gamma^\mu(-i\Lambda\psi) - i\Lambda D_\mu\bar{\psi}\gamma^\mu\psi\right] \equiv 0 \tag{12.44}$$

By adding the kinetic term for the gauge field

$$\mathcal{L} = -\frac{1}{4}F_{\mu\nu}F^{\mu\nu} \tag{12.45}$$

where

$$F_{\mu\nu} = \partial_\mu A_\nu - \partial_\nu A_\mu \tag{12.46}$$

(following the definition 12.26 for $f_{AB}{}^C = 0$), we get finally the total Lagrangian

$$\mathcal{L} = \frac{i}{2}(\bar{\psi}\gamma^\mu D_\mu\psi - D_\mu\bar{\psi}\gamma^\mu\psi) - m\bar{\psi}\psi - \frac{1}{4}F_{\mu\nu}F^{\mu\nu} \tag{12.47}$$

which is invariant under the following local gauge transformations

$$\delta\psi = -i\Lambda(x)\psi, \qquad \delta\bar{\psi} = i\Lambda(x)\bar{\psi}$$
$$\delta A_\mu = -\frac{1}{e}\partial_\mu\Lambda(x) \tag{12.48}$$

This Lagrangian, using (12.40,41), can also be written as

$$\mathcal{L} = \frac{i}{2}(\bar{\psi}\gamma^\mu\partial_\mu\psi - \partial_\mu\bar{\psi}\gamma^\mu\psi) - m\bar{\psi}\psi - \frac{1}{4}F_{\mu\nu}F^{\mu\nu}$$
$$+ eA_\mu J^\mu \tag{12.49}$$

and then it coincides exactly with the Lagrangian for the electromagnetic field interacting with a charged Dirac field, provided that the gauge coupling constant e is identified with the electric charge of the spinor field.

The field equations, obtained by varying (12.49) with respect to A_μ,

$$\partial_\nu F^{\mu\nu} = eJ^\mu \tag{12.50}$$

toghether with the property (12.30) of the field strength, which in this case is reduced simply to

$$\varepsilon^{\mu\nu\alpha\beta}\partial_\nu F_{\alpha\beta} = 0 \qquad (12.51)$$

coincide with the usual Maxwell equations in the the four-dimensional Minkowski space. They are linear in A_μ because photons are neutral, i.e. the gauge group is abelian and the gauge field does not contribute directly to the current density J^μ of the sources.

3.- Local Lorentz invariance and the gravitational interaction

Consider the Lagrangian density $L(\psi, \partial\psi)$ for a matter field ψ in the flat Minkowski space, and suppose that the corresponding action is invariant under global Lorentz transformations represented by the constant matrices $\Lambda^i{}_k$,

$$x'^i = \Lambda^i{}_k x^k \qquad (12.52)$$

$$\psi'(x') = U(\Lambda)\psi(x) \qquad (12.53)$$

where

$$\Lambda^a{}_i \Lambda_a{}^k = \delta_i^k \qquad (12.54)$$

and the matrices $U(\Lambda)$ represent a Lorentz transformation acting on the matter field ψ (troughout this Section Latin letters denote four-dimensional Lorentz indices).

Considering an infinitesimal transformation near the the identity I, the matrices Λ and U can be expanded as follows (see also Chapter X, Sect.5)

$$\Lambda_{ik} \simeq \eta_{ik} + \omega_{ik} \qquad (12.55)$$

$$U(\Lambda) = I + \frac{1}{2}\omega_{ik} S^{ik} \qquad (12.56)$$

where ω_{ik} is a constant antisymmetric matrix corresponding to six infinitesimal parameters (from (12.54) we have in fact $\omega_{ik} = -\omega_{ki}$), and $S_{ik} = -S_{ki}$ represents the six generators of the Lorentz group, satisfying the following commutation relations

$$[S_{ab}, S_{cd}] = f_{abcd}{}^{ik} S_{ik} \qquad (12.57)$$

where

$$f_{abcd}{}^{ik} = \eta_{bc} S_{ad} + \eta_{ad} S_{bc} - \eta_{bd} S_{ac} - \eta_{ac} S_{bd} \qquad (12.58)$$

The infinitesimal form of the tranformations (12.52,53) is then

$$\delta x^i = x'^i - x^i = \omega^i{}_k x^k \qquad (12.59)$$

$$\delta\psi = \psi' - \psi = \frac{1}{2} \omega^{ik} S_{ik} \psi \qquad (12.60)$$

Note that the parameters ω^{ik} play the role of the parameters ε^A introduced in the first Section of the Chapter. Therefore we have a global Lorentz transformation if ω_{ik} = const, while a local Lorentz transformation will correspond to $\omega_{ik} = \omega_{ik}(x)$. Moreover we can see, from (12.57), that the Lorentz group is non-abelian, as its structure constants $f_{abcd}{}^{ik}$ are nonvanishing. Therefore we can expect that the field equations of a gauge theory with a local Lorentz invariance will be, in general, nonlinear equations.

To give an explicit example, consider the flat space Dirac Lagrangian

$$\mathcal{L} = \frac{i}{2} \bar{\psi} \gamma^k \partial_k \psi - \frac{1}{2} m \bar{\psi} \psi + \text{h.c.} \qquad (12.61)$$

(where h.c. denotes the hermitian conjugate of the first two terms). A Lorentz transformation of the spinor field is represented in flat space by a non-unitary matrix U which satisfies

$$U^{-1} = \gamma^4 U^+ \gamma^4 \qquad (12.62)$$

so that

$$\psi' = U\psi \qquad (12.63)$$

$$\bar{\psi}' = (U\psi)^+ \gamma^4 = \psi^+ U^+ \gamma^4 = \psi^+ \gamma^4 (\gamma^4 U^+ \gamma^4) = \bar{\psi} U^{-1} \qquad (12.64)$$

(since $(\gamma^4)^2 = 1$). Imposing on the Lagrangian (12.61) to be invariant under a Lorentz transformation, i.e. $\mathcal{L} = \mathcal{L}'$,

where

$$\mathcal{L}' = \frac{i}{2} \bar{\psi} U^{-1} \gamma^k \Lambda_k{}^i \partial_i U \psi - \frac{m}{2} \bar{\psi} U^{-1} U \psi + h.c. \tag{12.65}$$

we obtain, in the case of global transformations (U = const),

$$U^{-1} \gamma^k \Lambda_k{}^i U = \gamma^i \tag{12.66}$$

and then, using (12.54),

$$U^{-1} \gamma^k U = \Lambda^k{}_i \gamma^i \tag{12.67}$$

as expected (remember 10.114). This condition is satisfied by the infinitesimal transformation (12.56) if the Lorentz generators, for a Dirac field, are given by

$$S_{ik} = \frac{1}{2} \gamma_{[i} \gamma_{k]} = \frac{1}{4} (\gamma_i \gamma_k - \gamma_k \gamma_i) \tag{12.68}$$

It is easy to verify directly, by using the properties of the gamma matrices, $\gamma^{(i}\gamma^{k)} = \eta^{ik}$, that the Dirac Lagrangian (12.61) is invariant under global Lorentz transformations, i.e. $\delta \mathcal{L} = 0$ for

$$\delta x^i = \omega^i{}_k x^k \tag{12.69}$$

$$\delta \psi = \frac{1}{4} \omega^{ik} \gamma_{[i} \gamma_{k]} \psi \tag{12.70}$$

$$\delta \bar{\psi} = -\frac{1}{4} \omega^{ik} \bar{\psi} \gamma_{[i} \gamma_{k]} \tag{12.71}$$

It should be stressed, however, that global Lorentz invariance is a physically sufficient requirement if the geometry of the manifold we are considering is flat, i.e. if the background world metric can be reduced everywhere to the Minkowski form.

In the more general case of the curved manifold, we can still introduce, at each given point, a flat tangent space in which the physical laws are Lorentz invariant (according to the principle of equivalence). However, the four orthonormal vectors spanning this Minkowski tangent space are defined at a given point of the manifold, and they vary from a point to another. Therefore in general local Lorentz invariance is to be required[2,3].

In particular, as already noted in Chapter X, Sect.5, the transformations properties of the spinor fields can be defined in a flat Minkowski space-time; in the framework of a theory with a curved geometry, the Minkowski tangent space is defined only locally, and then a local Lorentz invariance of the spinor Lagrangian is needed.

To this aim one introduces, as usual, the vierbein fields $e_\mu{}^\kappa$, relating at each point a general curvilinear coordinate system on the manifold, ξ^μ, to the coordinates of the flat Minkowski space, x^κ, (notations: Latin indices are flat, tangent space Lorentz indices; Greek letters denote curved indices in the world manifold).

We have then the usual relations (see Chapter X, Sect.5)

$$e_\mu{}^\kappa = \frac{\partial x^\kappa}{\partial \xi^\mu} \quad , \qquad e^\mu{}_\kappa = \frac{\partial \xi^\mu}{\partial x^\kappa} \tag{12.72}$$

$$g_{\mu\nu} = e_\mu{}^i e_\nu{}^\kappa \eta_{i\kappa} \; , \qquad \eta_{i\kappa} = e^\mu{}_i e^\nu{}_\kappa g_{\mu\nu} \tag{12.73}$$

$$e = \det e_\mu{}^\kappa = \sqrt{-g} = (-\det g_{\mu\nu})^{1/2} \tag{12.74}$$

Remember that each world index behaves like a vector index under general coordinate transformations, but as a scalar under a local Lorentz transformation, and the converse is valid for the flat Lorentz indices. Therefore, contracting world indices with the vierbein field, a world tensor may be transformed into a world scalar. Example: A^μ is a controvariant vector under general coordinate transformations, $A^\kappa = e_\mu{}^\kappa A^\mu$ is a scalar under coordinate, but a vector under local Lorentz transformations.

By using the vierbein field, the Dirac Lagrangian (12.61) in a general curvilinear system of coordinates can be rewritten as

$$\mathcal{L} = \frac{i}{2} e \bar{\psi} \gamma^\kappa e^\mu{}_\kappa \partial_\mu \psi - e \frac{m}{2} \bar{\psi} \psi + h.c. \tag{12.75}$$

This expression is a scalar density, and the corresponding action is invariant under general coordinate transformations. It is only globally Lorentz invariant, however, because if we consider the local infinitesimal Lorentz transformation

$$\begin{aligned} \delta e_\mu{}^i &= \omega^i{}_\kappa(x)\, e_\mu{}^\kappa \\ \delta \psi &= \frac{1}{4} \omega^{i\kappa}(x)\, \gamma_{[i}\gamma_{\kappa]} \psi \\ \delta \bar{\psi} &= -\frac{1}{4} \omega^{i\kappa}(x)\, \bar{\psi} \gamma_{[i}\gamma_{\kappa]} \end{aligned} \tag{12.76}$$

we have

$$\delta \mathcal{L} = - J^{\mu}{}_{ab} \partial_\mu \omega^{ab}(x) \neq 0 \qquad (12.77)$$

where

$$J^{\mu}{}_{ab} = -\frac{i}{4} e e^{\mu\kappa} \bar{\Psi} \gamma_{[\kappa} \gamma_a \gamma_{b]} \Psi \qquad (12.78)$$

is the(totally antisymmetric) spin current of the Dirac field.

The lack of invariance, as usual in the context of a gauge theory, is due to the presence of a partial derivative $\partial_K = e^{\mu}{}_K \partial_\mu$, which in this case is a world scalar, but does not behave like a vector, in the index "k", under a local Lorentz transformation.

In fact, when U = U(x), we have

$$(\partial_\kappa \psi)' = \Lambda_\kappa{}^i \partial_i \psi' = \Lambda_\kappa{}^i \partial_i (U\psi) =$$

$$= \Lambda_\kappa{}^i U \partial_i \psi + \Lambda_\kappa{}^i (\partial_i U) \psi \qquad (12.79)$$

The local invariance can be restored, as seen in the previous Sections of this Chapter, introducing for each generator a compensating vector field, and defining a generalized covariant derivative. In this case we have six generators, as $S_{i\kappa} = - S_{\kappa i}$, and then we introduce a gauge potential $A_\mu{}^{i\kappa} = - A_\mu{}^{\kappa i}$, which is a world vector in the index μ, and an antisymmetric Lorentz tensor, with six components, in the tangent space indices i,k . We are led then to define the following gauge covariant derivative (see 12.16)

$$D_\mu = \partial_\mu + \frac{1}{2} A_\mu{}^{i\kappa} S_{i\kappa} \qquad (12.80)$$

(note that in this case the indices of the internal symmetry group, denoted in eq.(12.16) by capital letters, are now anholonomic Lorentz indices, since now we are concerned with a gauge theory with local Lorentz invariance).

The generalized Lagrangian for the Dirac field becomes then, according to the minimal coupling procedure applied to eq.(12.75)

$$\mathcal{L} = \frac{i}{2} e \bar{\Psi} \gamma^\kappa D_\kappa \psi - \frac{e}{2} m \bar{\Psi} \psi + h.c. \qquad (12.81)$$

where

$$D_K \psi = e^{\mu}{}_K D_\mu \psi \tag{12.82}$$

and is locally Lorentz invariant under the transformation $\psi' = U(x)\psi$, provided that

$$(D_K \psi)' = \Lambda_K{}^i D_i'(U\psi) = \Lambda_K{}^i U D_i \psi \tag{12.83}$$

In fact, if this condition is satisfied, we have

$$\mathcal{L}'(\psi', D'\psi') = \frac{i}{2} e \bar{\psi} U^{-1} \gamma^K \Lambda_K{}^i U D_i \psi - \frac{e}{2} m \bar{\psi} \psi + h.c. =$$

$$= \frac{i}{2} e \bar{\psi} \gamma^i D_i \psi - \frac{e}{2} m \bar{\psi} \psi + h.c. = \mathcal{L}(\psi, D\psi) \tag{12.84}$$

(we have used (12.66)).

From the condition (12.83) we can easily obtain the transformation rules for the gauge potential $A_\mu{}^{iK}$, which compensates the variation of the Lagrangian. Putting $A_\mu = A_\mu{}^{iK} S_{iK}$, eq.(12.83) implies

$$D_\mu' (U\psi) = \left(\partial_\mu + \frac{1}{2} A_\mu'\right) U\psi =$$

$$= U D_\mu \psi = U\left(\partial_\mu + \frac{1}{2} A_\mu\right)\psi \tag{12.85}$$

and gives

$$A_\mu' = U A_\mu U^{-1} - 2(\partial_\mu U) U^{-1} \tag{12.86}$$

(result analogous to eq.(12.18)). The infinitesimal transformation rule can be obtained from this expression, expanding U as in (12.56), and using the commutation rules (12.57). To first order in the parameters ω^{iK} we get

$$\delta A_\mu{}^{iK} = -\partial_\mu \omega^{iK} + \frac{1}{2} f_{abcd}{}^{iK} \omega^{ab} A_\mu{}^{cd} \tag{12.87}$$

As expected from the general theory developed in Sect.1, we have a term due to the derivative of the local parameters $\omega(x)$, and a term arising from the non-abelian structure of the invariance group (compare with eq.(12.21)).

The generalized Dirac Lagrangian (12.81), in which the partial derivatives are replaced by the Lorentz covariant derivatives (12.80), is a scalar density as regards general coordinate transformations, and is invariant under local Lorentz rotations.

Computing the commutator of two covariant derivatives we have

$$[D_\mu, D_\nu]\psi = \frac{1}{2}\left(\partial_\mu A_\nu - \partial_\nu A_\mu + \frac{1}{2}[A_\mu, A_\nu]\right)\psi \tag{12.88}$$

(where $A_\mu = A_\mu{}^{ik} S_{ik}$) and, using the commutation relations (12.37),

$$\frac{1}{2}[A_\mu, A_\nu] = \frac{1}{2} A_\mu{}^{ab} A_\nu{}^{cd} [S_{ab}, S_{cd}] =$$

$$= \frac{1}{2} A_\mu{}^{ab} A_\nu{}^{cd}\, 2\left(-\eta_{b[c} S_{d]a} - \eta_{a[d} S_{c]b}\right) =$$

$$= -A_\mu{}^a{}_c A_\nu{}^{cd} S_{da} - A_\mu{}_d{}^b A_\nu{}^{cd} S_{cb} =$$

$$= \left(A_\mu{}^a{}_c A_\nu{}^{cb} - A_\nu{}^a{}_c A_\mu{}^{cb}\right) S_{ab} \tag{12.89}$$

Therefore we can put

$$D_\mu D_\nu - D_\nu D_\mu = \frac{1}{2} R_{\mu\nu}{}^{ik} S_{ik} \tag{12.90}$$

where the tensor $R_{\mu\nu}{}^{ik}$ is given by

$$R_{\mu\nu}{}^{ik} = \partial_\mu A_\nu{}^{ik} - \partial_\nu A_\mu{}^{ik} + A_\mu{}^i{}_c A_\nu{}^{ck} - A_\nu{}^i{}_c A_\mu{}^{ck} \tag{12.91}$$

and satisfy the following identities

$$R_{\mu\nu}{}^{ik} = -R_{\nu\mu}{}^{ik} = -R_{\mu\nu}{}^{ki} \tag{12.92}$$

$$\varepsilon^{\mu\nu\alpha\beta} D_\nu R_{\alpha\beta}{}^{ik} = 0 \tag{12.93}$$

Remembering Sect.1, it is easy to note that the tensor (12.91) represents the gauge field strength (12.26) for the Lorentz group. The presence of terms quadratic in the potential $A_\mu{}^{ik}$ is due to the non-abelian structure of this group.

In order to obtain the relation between the potential $A_\mu{}^{ik}$ (i.e. the gauge field introduced to restore a local Lorentz symmetry), and the usual gravitational field, consider the gauge covariant derivative (12.80) applied to a Lorentz vector V^a. The Lorentz generators are in this cas

$$(S_{ik})^{ab} = \delta^a_i \, \delta^b_k - \delta_k^a \, \delta_i^b \qquad (12.94)$$

(they can be obtained by applying the infinitesimal transformation (12.60) to a vector field, and comparing with eq.(12.59)). Writing explicitly the vector indices we have then

$$(D_\mu)^a{}_b V^b = \left[\partial_\mu \delta^a{}_b + \frac{1}{2} \dot{A}_\mu{}^{ik} (\delta^a_i \eta_{kb} - \delta^a_k \eta_{ib}) \right] V^b$$

$$= \partial_\mu V^a + A_\mu{}^a{}_b V^b \qquad (12.95)$$

and since

$$V^a = e_\nu{}^a V^\nu \qquad (12.96)$$

then

$$D_\mu V^a = \partial_\mu (e_\nu{}^a V^\nu) + A_\mu{}^a{}_b e_\nu{}^b V^\nu$$

$$= e_\nu{}^a \partial_\mu V^\nu + V^\nu (\partial_\mu e_\nu{}^a + A_\mu{}^a{}_b e_\nu{}^b) \qquad (12.97)$$

On the other hand, the geometrical covariant derivative of the world vector field V^ν can be expressed in terms of the holonomic affine connection $\Gamma_{\mu\nu}{}^\alpha$:

$$\nabla_\mu V^\nu = \partial_\mu V^\nu + \Gamma_{\mu\alpha}{}^\nu V^\alpha \qquad (12.98)$$

Assuming, like in the Einstein-Cartan theory, that the vierbein field is covariantly constant, i.e. $\nabla_\mu e_\nu{}^i = 0$ (remember 10.134) , we obtain then

$$\nabla_\mu V^a = \nabla_\mu (e_\nu{}^a V^\nu) = e_\nu{}^a \nabla_\mu V^\nu =$$

$$= e_\nu{}^a (\partial_\mu V^\nu + \Gamma_{\mu\alpha}{}^\nu V^\alpha) \qquad (12.99)$$

Comparing (12.97,99) we obtain a relation between the world affine connection Γ and the gauge field A :

$$\partial_\mu e_\nu{}^a + A_\mu{}^a{}_b e_\nu{}^b - \Gamma_{\mu\nu}{}^\alpha e_\alpha{}^a = 0 \qquad (12.100)$$

that is

$$A_\mu{}^{ab} = e_\alpha{}^a e^{\nu b} \Gamma_{\mu\nu}{}^\alpha - e^{\nu b} \partial_\mu e_\nu{}^a \qquad (12.101)$$

which, putting $A_\mu{}^{ab} = \omega_\mu{}^{ab}$ coincides with eq.(10.145) relating the world affine connection $\Gamma_{\mu\nu}{}^\alpha$ and the anholonomic connection $\omega_\mu{}^{ab}$.

Therefore the Lorentz gauge covariant derivative (12.8 can be geometrically interpreted and identified with the covariant derivative acting on the flat tangent space indices, provided that the compensating fields for the local Lorentz invariance, $A_\mu{}^{ik}$, are identified with the coefficients of the anholonomic connection, $\omega_\mu{}^{ik}$, introduced in Chapter X to define the covariant derivative of a spinor field.

Note that with this identification, eqs.(12.82) and (10.84) coincide, i.e. in the case of a spinor field we have $D_\mu \psi = \nabla_\mu \psi$, because ψ does not carries holonomic indices; moreover, eq.(12.100) is equivalent to the condition (10.134) requiring a vanishing covariant derivative for the vierbein field.

It is important to note also that the gauge potential, from a geometrical point of view, correspond to a connecti and that this connection is in general nonsymmetric.

In fact, the Lorentz covariant derivative of the flat Minkowski metric gives

$$\nabla_\mu \eta^{ik} = D_\mu \eta^{ik} = \partial_\mu \eta^{ik} + A_\mu{}^i{}_j \eta^{jk} + A_\mu{}^k{}_j \eta^{ij} =$$

$$= A_\mu{}^{ik} + A_\mu{}^{ki} = 0 \qquad (12.102)$$

(because $A_\mu{}^{ik} = A_\mu{}^{[ik]}$). Therefore

$$\nabla_\mu g^{\alpha\beta} = \nabla_\mu (e^\alpha{}_i e^\beta{}_k \eta^{ik}) = e^\alpha{}_i e^\beta{}_k D_\mu \eta^{ik} = 0 \qquad (12.103$$

and solving the equation $\nabla_\mu g^{\alpha\beta} = 0$ for the connection we obtain, as shown in Chapters II and X,

$$\Gamma_{\alpha\beta}{}^\mu = \left\{ {}^{\mu}_{\alpha\beta} \right\} + Q_{\alpha\beta}{}^\mu + Q^\mu{}_{\alpha\beta} - Q_\beta{}^\mu{}_\alpha \qquad (12.10$$

where $Q_{\mu\nu}{}^\alpha = \Gamma_{[\mu\nu]}{}^\alpha$ is the torsion tensor. In the context of a gravitational theory formulated as a gauge theory for the local Lorentz invariance, we are led then naturally to the geometry of a Riemann-Cartan manifold U_4. The standard riemannian geometry of general relativity can be obtained only imposing "ad hoc" the symmetry of the connection, $\Gamma_{\mu\nu}{}^\alpha = \Gamma_{\nu\mu}{}^\alpha$.

Another point worth stressing is that, as the gauge

potential $A_\mu{}^{ik}$ coincides with the anholonomic connection $\omega_\mu{}^{ik}$, the field strength (12.91) coincides then with the Riemann-Cartan curvature tensor of the U_4 theory. In fact, putting

$$R_{\mu\nu}{}^{ik}(\omega) = \partial_\mu \omega_\nu{}^{ik} - \partial_\nu \omega_\mu{}^{ik} + \omega_\mu{}^i{}_j \omega_\nu{}^{jk} - \omega_\nu{}^i{}_j \omega_\mu{}^{jk} \qquad (12.105)$$

remembering the definition (10.27)

$$R_{\mu\nu\alpha}{}^\beta(\Gamma) = \partial_\mu \Gamma_{\nu\alpha}{}^\beta - \partial_\nu \Gamma_{\mu\alpha}{}^\beta + \Gamma_{\mu\rho}{}^\beta \Gamma_{\nu\alpha}{}^\rho - \Gamma_{\nu\rho}{}^\beta \Gamma_{\mu\alpha}{}^\rho \qquad (12.106)$$

and using the relation (12.101) between the Lorentz and world connection one obtains, in general,

$$e_\beta{}^i R_{\mu\nu\alpha}{}^\beta - e_{\alpha k} R_{\mu\nu}{}^{ik} = (\nabla_\mu \nabla_\nu - \nabla_\nu \nabla_\mu) e_\alpha{}^i + (\Gamma_{\mu\nu}{}^\beta - \Gamma_{\nu\mu}{}^\beta) \nabla_\beta e_\alpha{}^i \qquad (12.107)$$

In the hypothesis $\nabla_\mu e_\nu{}^i = 0$ then the anhlonomic curvature tensor can be expressed in terms of the vierbeins $e_\mu{}^k$ and of the Lorentz connection $\omega_\mu{}^{ik}$ as follows:

$$R_{\mu\nu\alpha}{}^\beta(\Gamma) = - e_{\alpha i} e^\beta{}_k R_{\mu\nu}{}^{ik}(\omega) \qquad (12.108)$$

The identity (12.93) is equivalent then to the Bianchi identity (as regards this point see in particular Chapter XIV, Sect.3), and the scalar curvature can be expressed as

$$R(g,\Gamma) = g^{\nu\alpha} \delta^\mu{}_\beta R_{\mu\nu\alpha}{}^\beta(\Gamma) = - g^{\nu\alpha} \delta^\mu{}_\beta e_{\alpha i} e^\beta{}_k R_{\mu\nu}{}^{ik}(\omega) = e^\mu{}_i e^\nu{}_k R_{\mu\nu}{}^{ik}(\omega) \equiv R(e,\omega) \qquad (12.109)$$

Therefore, if we assume the scalar curvature as the Lagrangian for the free gauge field

$$\mathcal{L}_g = - \frac{e}{2\chi} e^\mu{}_i e^\nu{}_k R_{\mu\nu}{}^{ik}(\omega) \qquad (12.110)$$

we obtain a gauge theory of gravity equivalent to the Einstein-Cartan theory, in which torsion, minimally coupled to matter through a covariant derivative, is a nonpropagating field, as the resulting field equations relate algebraically torsion to the spin sources.

However, if the gravitational Lagrangian is chosen in analogy to the standard gauge formalism, we are led to a Lagrangian quadratic in the curvature, $\mathcal{L}_g \propto R_{\mu\nu}{}^{ik} R^{\mu\nu}{}_{ik}$ (see 12.31), which contains a kinetic term for the torsion. In this way torsion becomes a propagating field, however a theory with such Lagrangian is no longer equivalent to general relativity even if the torsion is vanishing.

Summarizing the arguments of this Section, we have seen that the requirements of general covariance and local Lorentz invariance are closely related, and that starting with a globally Lorentz invariant matter Lagrangian in flat space, local invariance can be arranged introducing a gauge covariant derivative which coincides with the geometrical covariant derivative of a Riemann-Cartan manifold. The Einstein-Cartan theory (and, in particular, general relativity) can be interpreted then as a gauge theory with a local Lorentz invariance.

A gravitational theory can be formulated, however, as a gauge theory also starting with the Poincarè group, the De Sitter group, and other, still more general, groups.

What should be, among the possible groups, the most appropriate symmetry needed to describe the gravitational interaction, is still an open question. In any case, the local Lorentz invariance is expected to play a fundamental role in every gauge theory of gravity.

In conclusion, it is perhaps worth stressing that a gauge theory of gravity provides a formal and physical justification, through the definition of gauge covariant derivative, to the principle of minimal gravitational coupling, which otherwise could be considered only an empirical procedure.

In this framework, such a procedure may be applied only when coupling gravity to matter fields, and not to another gauge field. This prevents, for example, a minimal coupling between the electromagnetic field and the torsion, which would destroy the U(1) local gauge invariance.

Consider in fact the Maxwell Lagrangian

$$\mathcal{L} = -\frac{1}{4} F_{\mu\nu} F^{\mu\nu} \quad , \quad F_{\mu\nu} = \partial_\mu A_\nu - \partial_\nu A_\mu \qquad (12.111)$$

The minimal coupling procedure

$$\partial_\mu A_\nu \to \nabla_\mu A_\nu = \partial_\mu A_\nu - \Gamma_{\mu\nu}{}^\alpha A_\alpha \qquad (12.112)$$

implies a coupling with torsion

$$F_{\mu\nu} \to F'_{\mu\nu} = \nabla_\mu A_\nu - \nabla_\nu A_\mu = F_{\mu\nu} - 2\,\Gamma_{[\mu\nu]}{}^\alpha A_\alpha$$

$$= F_{\mu\nu} - 2\,Q_{\mu\nu}{}^\alpha A_\alpha \qquad (12.113)$$

To first order in $Q_{\mu\nu}{}^\alpha$, the field Lagrangian becomes

$$\mathcal{L} \to \mathcal{L}' = -\frac{1}{4} F'_{\mu\nu} F'^{\mu\nu} = \mathcal{L} + Q_{\mu\nu}{}^\alpha A_\alpha F^{\mu\nu} \qquad (12.114)$$

and under a local gauge transformation, $\delta A_\mu = -\partial_\mu \Lambda$, we have

$$\delta \mathcal{L}' = -F^{\mu\nu} Q_{\mu\nu}{}^\alpha \partial_\alpha \Lambda \neq 0 \qquad (12.115)$$

so that the local U(1) invariance in this case would be lost.

Even if the gauge fields are not minimally coupled, classically, among them, they can indirectly interact through the presence of matter fields at a microscopical and quantum level. In the case of torsion, consider in fact the following example. A propagating electromagnetic field, from the point of view of quantum field theory, is represented by photons, which can spontaneously disintegrate, for a small fraction of time, into a virtual electron-positron pair, according to the so-called "vacuum polarization" effect. But electrons and positrons are fermion matter fields, which can couple minimally to torsion. Therefore, if a photon is propagating through a torsion background, the associated virtual pairs of fermions are affected by the presence of torsion, and the propagation of the torsion itself is modified.

At a macroscopical level, this produces a modification of the classical Maxwell equations, which can be represented by adding to the electromagnetic Lagrangian a non-minimal coupling term between torsion and the electromagnetic potential, like that considered in Chapter X, Sect.6 (see 10.206).

REFERENCES

1) See also E.S.Abers and B.W.Lee: Phys.Rep.9,1(1973)

2) R.Utiyama: Phys.Rev.101, 1597 (1956)

3) T.W.Kibble: J.Math.Phys.2, 212 (1961)

CHAPTER XIII

SUPERGRAVITY

1.- Introduction

In Chapter XII the Einstein-Cartan theory has been formulated as a gauge theory with a local Lorentz symmetry.

Supergravity is a gauge theory of gravity with a local supersymmetry.

A supersymmetry is an invariance property transforming boson fields into fermions and viceversa. A Lagrangian is supersymmetric if the corresponding action is invariant when boson and fermion fields are interchanged, according to some prescribed transformation rule. Examples of global supersymmetry can be given also in the flat space-time, but local supersymmetry can be realized only in a curved manifold: this shows directly that local supersymmetry and gravity are closely related.

In the simplest example of supersymmetric theory of gravity, the gravitational interaction is represented by two fields: a tensor field (the metric) corresponding to a massless, spin-two particle (the graviton), and a Rarita-Schwinger field, corresponding to a massless, spin 3/2 particle (the gravitino). The latter field produces corrections to the gravitational force only at a microscopical level (because the exchange of fermions, even if massless, leads to a short range potential), so that the classical gravitational theory is not modified at a macroscopic level. The contributions of the gravitino field, which become important at short distances, i.e. at high energy, improve however the quantum behaviour of the theory, as are expected to cancel infinities and produce a finite quantum theory of gravity.

Therefore supergravity is to be regarded as an important step toward a satisfactory quantization of gravity, and a possible unification of gravity with the other fundamental interactions.

Moreover, even from a formal point of view, supergravity is interesting because, unlike the other gauge theories, bosons and fermions are placed on the same footing, and the sharp distinction between matter and gauge fields is lost. For example, the compensating field needed to restore supersymmetry in the case of local transformations is the gravitino, i.e. a fermion.

Let us introduce briefly this field: it is represented by a "vector-spinor" $\psi_\mu{}^\alpha$, which is a geometrical object carrying a vector indix, μ, and a spinor index, α, and then it behaves like a covariant vector under general coordinate transformations, and like a spinor under local Lorentz rotations in the flat tangent space. As usual, we will write explicitly only the vector index, ψ_μ, and the spinor index is to be understood.

The Lagrangian density for this field in flat space is due to Rarita and Schwinger,

$$\mathcal{L}_{3/2} = -\frac{i}{2}\varepsilon^{\mu\nu\alpha\beta}\,\bar{\psi}_\mu\gamma_5\gamma_\nu\partial_\alpha\psi_\beta \tag{13.1}$$

and it is invariant under the infinitesimal local transformation

$$\delta\psi_\mu = \partial_\mu\lambda(x) \tag{13.2}$$

where the gauge parameter λ is a spinor object.

Performing the variation with respect to $\bar{\psi}$ we obtain the field equation

$$\varepsilon^{\mu\nu\alpha\beta}\gamma_5\gamma_\nu\partial_\alpha\psi_\beta = 0 \tag{13.3}$$

which can also be written as

$$\gamma^\mu(\partial_\mu\psi_\nu - \partial_\nu\psi_\mu) = 0 \tag{13.4}$$

In fact, defining $R^\mu = i\varepsilon^{\mu\nu\alpha\beta}\gamma_5\gamma_\nu\partial_\alpha\psi_\beta$, we have the identities

$$R_\mu - \frac{1}{2}\gamma_\mu(\gamma^\alpha R_\alpha) = -\gamma^\nu\partial_\nu\psi_\mu + \partial_\mu(\gamma^\nu\psi_\nu) \tag{13.5}$$

$$\frac{1}{2}\gamma_\mu R^\mu = \partial^\mu\psi_\mu - \gamma^\alpha\partial_\alpha(\gamma^\nu\psi_\nu) \tag{13.6}$$

(remember eqs.(10.179,180)). Therefore $R^\mu = 0$ is equivalent to (13.4). Moreover, because of the gauge invariance (13.2), we can impose the gauge condition

$$\gamma^\nu\psi_\nu = 0 \tag{13.7}$$

In this gauge, we can see that the field equation $R^{\mu} = 0$, using (13.5,6), reduces simply to the massless Dirac equation

$$\gamma^{\nu} \partial_{\nu} \psi_{\mu} = 0 \qquad (13.8)$$

with the condition

$$\partial^{\mu} \psi_{\mu} = 0 \qquad (13.9)$$

2.- Global supersymmetry

Consider the infinitesimal transformation of a boson field, B(x), into a fermion, F(x), i.e. $B \to B' = B + \delta B$, where

$$\delta B = \varepsilon F \qquad (13.10)$$

(tensor and spinor indices, for simplicity, are suppressed) and ε is the infinitesimal parameter of the transformation (ε = const if the supersymmetry is global).

Eq.(3.10) represents correctly a supersymmetric transformation, provided that the parameter ε satisfies the following requirements.

1) Since B has integer spin, and F half-integer spin, ε must have half-integer spin. The simplest choice is spin 1/2, i.e. a four component spinor.

2) To ensure the correct statistical properties, since boson fields commute, and fermion fields anticommute, the parameter ε and $\bar{\varepsilon} = \varepsilon^{+} \gamma^{4}$ must commute with B and anticommute with F . Moreover

$$\{\varepsilon^{\alpha}, \varepsilon^{\beta}\} = \{\varepsilon^{\alpha}, \bar{\varepsilon}^{\alpha}\} = 0 \qquad (13.11)$$

(note that in this Section the Greek letters α , β , denote spinor indices, when explicitly written).

3) It is convenient, in the supersymmetric theories, to work with real boson fields, $B = B^{*}$, and Majorana fermion fields $F = F^{c}$; it follows that ε must be a Majorana spinor, $\varepsilon = \varepsilon^{c}$, where ε^{c} is the charged conjugate spinor given by

$$\varepsilon^{c} = C \bar{\varepsilon}^{T} \qquad (13.12)$$

and the charge conjugation matrix C satisfies

$$C^T = -C \,, \qquad C^{-1}\gamma^\mu C = -(\gamma^\mu)^T \tag{13.13}$$

(the superscript "T" denotes transposition).

4) Finally we note that the dimensions of the Lagrangian density, in natural units (i.e. $\hbar=c=1$) are $[\mathcal{L}] = M^4$ (because energy has dimension of a mass, M , and lengths have dimensions M^{-1}). Therefore a boson has dimension $[B] = M$ (consider for example the Maxwell Lagrangian) and a fermion $[F] = M^{3/2}$ (consider the Dirac Lagrangian). It follows, from (3.10), that the parameter ε must have dimension $[\varepsilon] = M^{-1/2}$.

Therefore (3.10) represents in general an infinitesimal transformation of global supersymmetry, provided that ε is a four-component, constant Majorana spinor with dimension $M^{-1/2}$.

The inverse transformation is given, in general, by

$$\delta F = \partial B \varepsilon \tag{13.14}$$

where the presence of a derivative is needed, on the grounds of dimensional considerations ($[\partial] = M$).

It is worth stressing that it is just the derivative appearing in eq.(13.14) that leads directly to relate supersymmetry and gravity, because two consecutive supersymmetry transformations produce a spacetime translation.

Consider in fact the following global supersymmetry transformations, with anticommuting parameter ε , relating a scalar field φ and a Majorana spinor ψ :

$$\delta\varphi = i\bar{\varepsilon}\psi \tag{13.15}$$

$$\delta\psi = \gamma^\mu \partial_\mu \varphi\, \varepsilon \tag{13.16}$$

Performing two transformations, with parameters ε_1 and ε_2 , on the boson field we have

$$\delta_1\varphi \equiv \delta(\varepsilon_1)\varphi = i\bar{\varepsilon}_1\psi$$

$$\delta_2(\delta_1\varphi) = i\bar{\varepsilon}_1\gamma^\mu\varepsilon_2\partial_\mu\varphi \tag{13.17}$$

and the commutator of two such infinitesimal transformations gives an infinitesimal translation with parameter $a^\mu = 2\bar{\varepsilon}_2\gamma^\mu\varepsilon_1$.

In fact

$$[\delta(\varepsilon_2), \delta(\varepsilon_1)]\varphi = (\delta_2\delta_1 - \delta_1\delta_2)\varphi =$$

$$= i(\bar{\varepsilon}_1\gamma^\mu\varepsilon_2 - \bar{\varepsilon}_2\gamma^\mu\varepsilon_1)\partial_\mu\varphi \tag{13.18}$$

But since ε is a Majorana spinor, remebering (13.12,13) we have

$$\bar{\varepsilon} = -\varepsilon^T C^{-1} \tag{13.19}$$

and then

$$\bar{\varepsilon}_1\gamma^\mu\varepsilon_2 = -\varepsilon_1^T C^{-1}\gamma^\mu C\bar{\varepsilon}_2^T = -\bar{\varepsilon}_2\gamma^\mu\varepsilon_1 \tag{13.20}$$

Therefore

$$[\delta_2, \delta_1]\varphi = -2i(\bar{\varepsilon}_2\gamma^\mu\varepsilon_1)\partial_\mu\varphi \tag{13.21}$$

After two consecutive transformations (13.15) we obtain again the same field, but translated in space-time of the constant quantity a^μ. We have then a relation between supersymmetry and motion in space-time.

If the supersymmetry is local, $\varepsilon = \varepsilon(x)$, the induced translations differ from point to point, $a^\mu = a^\mu(x)$, depending on the coordinates. It is then evident that, if we make a theory invariant under local supersymmetry transformations, this theory must also be invariant under general coordinate transformations, and this relates local supersymmetry to gravity.

Before concluding this Section, it must be mentioned that the infinitesimalsupersymmetry transformations can be represented introducing fermionic generators of Majorana type, Q_α. The infinitesimal transformation of the boson field (13.15) can be rewritten then as

$$\delta\varphi = (\bar{\varepsilon}^\alpha Q_\alpha)\varphi \tag{13.22}$$

where Q_α is a Majorana spinor (note the analogy of this equation with (12.4): the only difference is that the infinitesimal parameter ε and the generator Q are spinor objects). We have then

$$[\delta_2, \delta_1]\varphi = (\bar{\varepsilon}_2{}^\alpha Q_\alpha\bar{\varepsilon}_1{}^\beta Q_\beta - \bar{\varepsilon}_1{}^\beta Q_\beta\bar{\varepsilon}_2{}^\alpha Q_\alpha)\varphi \tag{12.23}$$

Using the properties of the Majorana spinors, it follows that

$$\bar{\varepsilon} Q = - \varepsilon^T C^{-1} C \bar{Q}^T = \bar{Q} \varepsilon \qquad (13.24)$$

and then the right-hand side of (13.24) can be rewritten as

$$(\bar{\varepsilon}_2{}^\alpha Q_\alpha \bar{Q}_\beta \varepsilon_1{}^\beta - \bar{Q}_\beta \varepsilon_1{}^\beta \bar{\varepsilon}_2{}^\alpha Q_\alpha)\varphi =$$

$$= (\bar{\varepsilon}_2{}^\alpha Q_\alpha \bar{Q}_\beta \varepsilon_1{}^\beta + \bar{\varepsilon}_2{}^\alpha \bar{Q}_\beta Q_\alpha \varepsilon_1{}^\beta)\varphi \qquad (13.25)$$

We obtain then

$$[\delta_2, \delta_1]\varphi = \bar{\varepsilon}_2{}^\alpha \{Q_\alpha, \bar{Q}_\beta\} \varepsilon_1{}^\beta \varphi \qquad (13.26)$$

(where $\{A,B\} = AB + BA$), and, comparing this relation with (13.21), one has

$$\{Q_\alpha, \bar{Q}_\beta\} = -2i(\gamma^\mu)_{\alpha\beta}\partial_\mu = -2(\gamma^\mu P_\mu)_{\alpha\beta} \qquad (13.27)$$

where P_μ is the generator of translations. Therefore the supersymmetry generators Q may be regarded, in this sense, as the square root of the translations.

It is important also to stress that the generators Q_α obey anticommutation rules. When we consider then the supersymmetric extension of a given Lie symmetry, the original Lie algebra, involving only commutators of the bosonic generators (see 12.3), must be generalized to include also fermionic generators and anticommutators (forming the so-called "Z_2-graded" Lie algebra).

As will be shown in the last Chapter of this book, the geometrical structure of general relativity and of the Einstein-Cartan theory is directly related to the algebraic structure of the Poincarè group. The simplest example of supergravity, i.e. of a gravitational theory with a local supersymmetry, can be obtained then as a gauge theory obtained gauging the supersymmetric extension of the Poincarè algebra, which includes, besides the Lorentz and translation generators (M_{ab} and P_a), also a fermion generator Q_α : in addition to the Lorentz connection $\omega_\mu{}^{ab}$ and the vierbein field $e_\mu{}^a$, one must introduce then another gauge field, $\psi_\mu{}^\alpha$ (the gravitino) carrying spinor indices (and of course also a covariant world index, to transform like a compensating vector potential). The reader interested in this approach is referred to the extensive supergravity review of van Nieuwenhuizen[1] .

3.- Supergravity and local supersymmetry.

Consider a pure gravitational field in the absence of matter: according to the Einstein-Cartan theory in this case the torsion tensor is vanishing, the affine connection is symmetric and reduces simply to the Christoffel connection (2.60). The classical gravitational field is then entirely described by the metric tensor, $g_{\mu\nu}$, or, alternatively, by the vierbein field $e_\mu{}^\kappa$. The Einstein Lagrangian can be written (remember 12.109)

$$\mathcal{L}_g = -\frac{1}{2}\sqrt{-g}\, R(g,\Gamma) = -\frac{e}{2}\, e^\mu{}_i\, e^\nu{}_\kappa\, R_{\mu\nu}{}^{i\kappa}(\omega) \qquad (13.28)$$

where the curvature tensor is given by (12.105) and the Lorentz connection is related to the vierbein by (10.141) (in which, obviously, $K_{\mu\nu\alpha} = 0$ in this case). Latin letters denote flat Lorentz indices, Greek letters are curved world indices. Moreover, throughout this Section, the gravitational coupling constant will be set equal to unity, $\chi = 1$, for simplicity.

Consider the following global supersymmetry transformation

$$\delta e_\mu{}^\kappa = i\,\bar{\varepsilon}\,\gamma^\kappa\,\psi_\mu \qquad (13.29)$$

where $\bar{\varepsilon} = \varepsilon^+\gamma^4$, ε is a constant Majorana spinor, $[\varepsilon] = M^{-1/2}$, and ψ_μ is a Majorana spin 3/2 Rarita Schwinger field.

The corresponding infinitesimal transformation of the metric tensor is given by

$$\delta g_{\mu\nu} = \delta(e_\mu{}^i e_\nu{}^\kappa \eta_{i\kappa}) = \eta_{i\kappa}(e_\mu{}^i\,\delta e_\nu{}^\kappa + \delta e_\mu{}^i\, e_\nu{}^\kappa)$$

$$= i\,\bar{\varepsilon}\,(\gamma_\mu\psi_\nu + \gamma_\nu\psi_\mu) \qquad (13.30)$$

where $\gamma_\mu = e_\mu{}^\kappa\gamma_\kappa$. In the weak field approximation, putting $g_{\mu\nu} = \eta_{\mu\nu} + h_{\mu\nu}$, the Einstein Lagrangian (13.28) can be linearized retaining at most quadratic terms in the field variable $h_{\mu\nu}$, and we obtain (remember Chapter VIII, Sect.1) the following kinetic term for a spin two field $h_{ab} = h_{ba}$ in the flat space

$$\mathcal{L}_g \simeq \mathcal{L}_g^{(0)} = -\frac{1}{4}\left[\partial^a h^{bc}\partial_a h_{bc} - 2\partial_a h^{ab}\partial_c h_b{}^c - \right.$$
$$\left. - \partial_a h \partial^a h + 2\partial_a h^{ab}\partial_b h\right] \qquad (13.31)$$

where $h = \eta^{ab} h_{ab}$. It is straightforward to verify that if we consider the linearized Einstein Lagrangian and the Lagrangian for a free Rarita Schwinger field

$$\mathcal{L}_\psi^{(0)} = -\frac{i}{2}\varepsilon^{abcd}\bar{\psi}_a\gamma_5\gamma_b\partial_c\psi_d \qquad (13.32)$$

then the total action

$$S = \int d^4x \left(\mathcal{L}_g^{(0)} + \mathcal{L}_\psi^{(0)}\right) \qquad (13.33)$$

describing a spin-2 and spin-3/2 particle non-interacting in the flat space-time, is invariant under the following global supersymmetry transformations, with constant parameter ε,

$$\delta h_{mn} = i\bar{\varepsilon}\left(\gamma_m\psi_n + \gamma_n\psi_m\right)$$
$$\delta\psi_m = \frac{1}{2}\partial_a\left(h_{mb}\right)\gamma^{[a}\gamma^{b]}\varepsilon \qquad (13.34)$$

If however we assume $\varepsilon = \varepsilon(x)$, i.e. we consider local infinitesimal transformations, then the action (13.33) for the non-interacting graviton-gravitino system is no longer invariant, and the variation of the Lagrangian contains terms which are proportional to the partial derivative of the Majorana spinor, $\partial_\mu\varepsilon^\alpha(x)$.

This variation can be compensated, as stressed in Chapter XII, Sect.1, by introducing the coupling with a gauge field which transforms like the derivative of the gauge parameter, (remember 12.21). As the parameter ε carries a spinor index, α, and its derivative a vector index, μ, it follows that the gauge field needed to compensate the variation of the Lagrangian, restoring the local supersymmetry, must carry both these indices, and then it must be a a Rarita Schwinger field (spin 3/2). A locally supersymmetric theory of gravity can be obtained, therefore, only if the linearized theory described by (13.33) is promoted to a fully interacting theory, that is the gravitational field is coupled to the spin 3/2 field.

To this aim, the total locally supersymmetric action

must contain the graviton ($e_\mu{}^\kappa$) and the gravitino ($\psi_\mu{}^\alpha$); it must be invariant under general coordinate and local Lorentz transformations, to be a generalization of the Einstein Lagrangian which satisfies these requirements. Moreover the affine connection will be in general nonsymmetric, because now a fermion field is present, which is source of torsion according to the Einstein-Cartan theory.

The Lagrangian which satisfies all these conditions is

$$\mathcal{L} = -\frac{1}{2} e R(e,\omega) - \frac{i}{2} \varepsilon^{\mu\nu\alpha\beta} \bar{\psi}_\mu \gamma_5 \gamma_\nu D_\alpha \psi_\beta \qquad (13.35)$$

where $\gamma_\mu = e_\mu{}^\kappa \gamma_\kappa$ and $R(e,\omega)$ is the Einstein-Cartan scalar curvature constructed from an U_4 affine connection. It is convenient to write R in terms of the vierbein field $e_\mu{}^\kappa$ and of the Lorentz anholonomic connection ω (see 12.109,105), because to define the covariant derivative of the spinor field ψ_μ the tangent-space formalism is needed, as shown in Chapter X, Sect.5 . Finally, the covariant derivative of eq. (13.35) is given by

$$D_\mu \psi_\nu = \left(\partial_\mu + \frac{1}{4} \omega_\mu{}^{i\kappa} \gamma_{[i} \gamma_{\kappa]} \right) \psi_\nu \qquad (13.36)$$

It is important to stress that this derivative operates only on the spinor index of the gravitino, and not on its vector index. In this sense, the gravitational coupling of ψ_μ is non-minimal, because it is missing the coupling between the world connection and the vector indices of ψ_μ .

Note that the spinor part of the lagrangian (13.35) is a scalar density of weight w = -1 under general coordinate transformations, without the presence of the factor $e = \sqrt{-g}$, because $\varepsilon^{\mu\nu\alpha\beta}$ is already a tensor density of weight -1 (as shown in Chapter III, Appendix A).

It should be noted also that the compensating field in this case is a fermion (the gravitino), while the analogous of the matter field, which was our starting point, is a boson; the roles are inverted with respect to the procedure of the usual nonsupersymmetric gauge theories.

It is perhaps worth stressing that the Lagrangian (13.35) could be obtained directly, even without any reference to supersymmetry, as the Einstein-Cartan theory for a massless spin 3/2 particle coupled to gravity, starting from the flat space Lagrangian (13.1), introducing the gravitational field and adding the scalar curvature of the U_4 theory.

The use of a nonsymmetric connection follows from the

presence of a spinor field. The absence of interaction between the vector part of the gravitino and the torsion (analogous to the absence of direct interaction between torsion and the electromagnetic vector potential, see Chapter XII, Sect.3), that is the non-minimal character of the coupling (13.36), finds however a convincing motivation oly because it is in that case that the local supersymmetry of the total Lagrangian is achieved.

The total action corresponding to the Lagrangian (13.35) is invariant under the following infinitesimal transformations of local supersymmetry:

$$\delta e_\mu{}^\kappa = i\,\bar{\varepsilon}\,\gamma^\kappa\,\psi_\mu \qquad (13.37)$$

$$\delta\psi_\mu = 2\,D_\mu\,\varepsilon = 2\left(\partial_\mu + \frac{1}{4}\,\omega_\mu{}^{i\kappa}\gamma_{[i}\gamma_{\kappa]}\right)\varepsilon \qquad (13.38)$$

(the Lagrangian (13.35) is invariant only modulo a total divergence, and this shows that supersymmetry behaves not as an internal symmetry, but as a space-time symmetry, like general coordinate transformations).

It should be noted that there is no need of considering explicitly the transformation of the connection, because ω may be expressed entirely in terms of the vierbein and of the gravitino field, as we will see from the field equations.

In fact, performing the variation of the Lagrangian (13.35) with respect to $\omega_\mu{}^{ab}$, we find

$$D_\mu\,e_\nu{}^\kappa - D_\nu\,e_\mu{}^\kappa = \frac{i}{2}\,\bar{\psi}_\mu\,\gamma^\kappa\,\psi_\nu \qquad (13.39)$$

where

$$D_\mu\,e_\nu{}^\kappa = \partial_\mu\,e_\nu{}^\kappa + \omega_\mu{}^a{}_b\,e_\nu{}^b \qquad (13.40)$$

Therefore (13.39) can be written also as

$$2\,\partial_{[\mu}\,e_{\nu]}{}^a + 2\,\omega_{[\mu}{}^a{}_{\nu]} - \frac{i}{2}\,\bar{\psi}_\mu\,\gamma^a\,\psi_\nu = 0 \qquad (13.41)$$

On the other hand, if we assume the tetrad condition $\nabla_\mu e_\nu{}^\kappa = 0$ (remeber 10.134), i.e.

$$\partial_\mu\,e_\nu{}^\kappa + \omega_\mu{}^\kappa{}_i\,e_\nu{}^i - \Gamma_{\mu\nu}{}^\alpha\,e_\alpha{}^\kappa = 0 \qquad (13.42)$$

antisymmetrizing this equation we have

$$\partial_{[\mu} e_{\nu]}{}^{\kappa} + \omega_{[\mu}{}^{\kappa}{}_{\nu]} - \Gamma_{[\mu\nu]}{}^{\kappa} = 0 \qquad (13.43)$$

Comparing eqs.(13.41,43) we obtain immediately that in this theory the gravitino field is algebraically related to the torsion tensor $Q_{\mu\nu}{}^{\alpha} = \Gamma_{[\mu\nu]}{}^{\alpha}$ as follows

$$Q_{\mu\nu}{}^{\alpha} = \frac{i}{4} \bar{\psi}_{\mu} \gamma^{\alpha} \psi_{\nu} \qquad (13.44)$$

Solving (13.41) we can obtain an explicit expression for the anholonomic connection ω (see for example (10.136-141), and we get

$$\omega_{\mu}{}^{i\kappa} \equiv \omega_{\mu}{}^{i\kappa}(e, \psi) = \omega_{\mu}{}^{i\kappa}(e) - K_{\mu}{}^{\kappa i} \qquad (13.45)$$

where $\omega_{\mu}{}^{i\kappa}(e)$ is the usual riemannian part of the connection, given in terms of the Ricci rotation coefficients, depending on the vierbeins and on their first partial derivatives (see 10.142); and $K_{\mu}{}^{i\kappa} = K_{\mu\alpha\beta}\, e^{\alpha i} e^{\beta\kappa}$ is the contorsion tensor due to the gravitino contributions (13.44). From (10.143) we have explicitly

$$K_{\mu\alpha\beta} = \frac{i}{4} \left(- \bar{\psi}_{\mu} \gamma_{\beta} \psi_{\alpha} - \bar{\psi}_{\beta} \gamma_{\alpha} \psi_{\mu} + \bar{\psi}_{\alpha} \gamma_{\mu} \psi_{\beta} \right) \qquad (13.46)$$

The variation of the action with respect to the vierbein field gives

$$G^{\mu\nu} = \theta^{\mu\nu} \qquad (13.47)$$

where $G_{\mu\nu}$ is the usual nonsummetric Einstein tensor of the Einstein-Cartan theory, obtained from the connection (13.45), and $\theta_{\mu\nu}$ is the canonical (nonsymmetric) energy-momentum tensor of the spin 3/2 field

$$\theta^{\mu\nu} = - \frac{i}{2} \eta^{\lambda\nu\alpha\beta} \bar{\psi}_{\lambda} \gamma_5 \gamma^{\mu} D_{\alpha} \psi_{\beta} \qquad (13.48)$$

where

$$\eta^{\mu\nu\alpha\beta} = \frac{1}{e} \varepsilon^{\mu\nu\alpha\beta} \qquad (13.49)$$

is a density of zero weight.

Finally, by varying with respect to $\bar{\psi}_{\mu}$ we obtain the gravitino field equation

$$R^{\mu} = \varepsilon^{\mu\nu\alpha\beta} \gamma_5 \gamma_{\nu} D_{\alpha} \psi_{\beta} = 0 \qquad (13.50)$$

which satisfies the consistency requirements, i.e. $D_\mu R^\mu = 0$, as can be verified using the other field equations.

REFERENCES

1) P.van Nieuwenhuizen: Phys.Rep.68,189(1981) and references therein.

CHAPTER XIV

GRAVITATIONAL THEORY IN THE LANGUAGE OF THE EXTERIOR FORMS

1.- Forms, exterior product and exterior derivative

The aim of this final Chapter is to provide a very short introduction to the notion of differential forms and to the use of the exterior calculus in the context of a theory of gravity. A rigorous and more detailed discussion of these concepts may be found, by the interested reader, in any mathematical book on the theory of differential forms[1,2].

This Chapter may be regarded as an appendix to the whole book, in which the basic geometrical and dynamical structure of general relativity and of the Einstein-Cartan theory, already presented and discussed in the tensorial language, are re-derived using the exterior formalism, mainly to provide the reader with some explicit example of this computation technique which is more and more used in the framework of the modern gravitational and unified theories.

Consider the following composition of differentials, called "exterior product"

$$dx^{\mu} \wedge dx^{\nu} \qquad (14.1)$$

which is antisymmetric and associative

$$dx^{\mu} \wedge dx^{\nu} = - dx^{\nu} \wedge dx^{\mu} \qquad (14.2)$$

$$(dx^{\mu} \wedge dx^{\nu}) \wedge dx^{\alpha} = dx^{\mu} \wedge (dx^{\nu} \wedge dx^{\alpha}) =$$
$$= dx^{[\mu} \wedge dx^{\nu} \wedge dx^{\alpha]} \qquad (14.3)$$

A p-form A is en element of the vector space spanned by the exterior product of p differentials, and can be represented as a polynomial in the dx^{μ} as follows

$$A = A_{\mu_1 \dots \mu_p} dx^{\mu_1} \wedge \dots \wedge dx^{\mu_p} \qquad (14.4)$$

where $A_{\mu_1 \dots \mu_p} = A_{[\mu_1 \dots \mu_p]}$ is a totally antisymmetric covariant tensor of rank p . A scalar field is a zero form; given a covariant vector field A_μ , we can construct the one-form $A = A_\mu dx^\mu$, given an antisymmetric tensor $F_{\mu\nu} = -F_{\nu\mu}$ the corresponding two-form is $F = F_{\mu\nu} dx^\mu \wedge dx^\nu$, and so on.

The exterior product of a p-form A and a q-form B is a (p+q)-form :

$$A \wedge B = A_{[\mu_1 \cdots \mu_p} B_{\mu_{p+1} \cdots \mu_{p+q}]} dx^{\mu_1} \wedge \cdots \wedge dx^{\mu_{p+q}} \tag{14.5}$$

For example, the exterior product of two one-forms, $A = A_\mu dx^\mu$ and $B = B_\mu dx^\mu$, is the two-form $P = A \wedge B$, where

$$P = \frac{1}{2} (A_\mu B_\nu - A_\nu B_\mu) dx^\mu \wedge dx^\nu \tag{14.6}$$

This product is associative

$$A \wedge (B \wedge C) = (A \wedge B) \wedge C \tag{14.7}$$

but in general non-commutative, and if A is a p-form and B is a q-form, we have

$$A \wedge B = (-1)^{pq} B \wedge A \tag{14.8}$$

Finally, the exterior product is bilinear, that is

$$(\alpha A + \beta B) \wedge C = \alpha (A \wedge C) + \beta (B \wedge C) \tag{14.9}$$

where α and β are scalars, and A,B, are forms of the same degree.

Introducing a vierbein field $V_\mu{}^a$, such that (see also Chapter X, Sect.5, and Chapter XII, Sect.3)

$$g_{\mu\nu} = V_\mu{}^a V_\nu{}^b \eta_{ab} \tag{14.10}$$

$$\eta^{ab} = g^{\mu\nu} V_\mu{}^a V_\nu{}^b \tag{14.11}$$

($g_{\mu\nu}$ is the metric tensor of the world manifold, and η_{ab} the Minkowski metric of the flat local tangent space; Greek letters denore holonomic world indices, Latin letters anholonomic tangent space indices), we can define a set of fundamental frames (or basic one-forms) as follows

$$V^a = V_\mu{}^a dx^\mu \tag{14.12}$$

The general p-form of eq.(14.4) then can be expressed also as

$$A = A_{[a_1 \cdots a_p]} V^{a_1} \wedge \cdots \wedge V^{a_p} \qquad (14.13)$$

where holonomic and anholonomic components of the tensor $A_{\mu_1 \cdots \mu_p}$ are related, as usual, by

$$A_{\mu_1 \cdots \mu_p} = A_{a_1 \cdots a_p} V_{\mu_1}^{a_1} \cdots V_{\mu_p}^{a_p} \qquad (14.14)$$

Note that if the geometric objects of a theory are referred to the basic frames (14.12), the formalism become independent of the choice of a local system of coordinate in the world manifold, and the whole theory is generally covariant.

The other fundamental notion to be introduced when dealing with differential forms is the notion of "exterior derivative".

Given a p-form A its exterior derivative is a (p+1)-form defined by

$$dA = \partial_{[\nu} A_{\mu_1 \cdots \mu_p]} dx^{\nu} \wedge dx^{\mu_1} \wedge \cdots dx^{\mu_p} \qquad (14.15)$$

For example, the exterior derivative of a zero-form φ is a one-form $d\varphi$ whose components correspond to the usual definition of gradient

$$d\varphi = (\partial_\mu \varphi) dx^\mu \qquad (14.16)$$

The exterior derivative of a one-form $A_\mu dx^\mu$ is a two-form $F = dA$, whose components correspond to the "curl" of A:

$$F = dA = \frac{1}{2} (\partial_\mu A_\nu - \partial_\nu A_\mu) dx^\mu \wedge dx^\nu \qquad (14.17)$$

and so on.

It is easy to note that the second exterior derivative of a form is vanishing, $d^2 A = 0$: in fact

$$d^2 A = d(dA) = \partial_{[\mu} \partial_\nu A_{\alpha_1 \cdots \alpha_p]} dx^\mu \wedge dx^\nu \wedge dx^{\alpha_1} \cdots dx^{\alpha_p} \equiv 0 \qquad (14.18)$$

because $\partial_{[\mu} \partial_{\nu]} \psi = 0$. A p-form F is called "closed" if $dF = 0$, and "exact" if it can be expressed as the exterior derivative of a (p-1)-form, i.e. $F = dA$. Therefore an exact form is also closed.

Finally, using the commutative and associative properties (14,7,8) , we obtain the following Leibnitz rule for the derivative of an exterior product:

$$d(A \wedge B) = dA \wedge B + (-1)^p A \wedge dB \qquad (14.19)$$

where A is a p-form, and B a q-form.

2.- Gauge theory

The standard formalism of the conventional gauge theories can be conveniently expressed in the language of the exterior forms.

Consider a Lie group G, whose generators X_A satisfy the commutation rules of the Lie algebra of G:

$$[X_A , X_B] = f_{AB}{}^C X_C \qquad (14.20)$$

($f_{AB}{}^C$ are the structure constants of G, and the capital Latin letters denote indices in the group manifold).

In order to formulate a gauge theory with G as local symmetry group, one introduces, corresponding to each generators X_A , the covariant vectors $h_\mu{}^A$ (the compensating fields, remember Chapter XII, Sect.1), and defines the following gauge potential one form h

$$h = h_\mu{}^A X_A \, dx^\mu \qquad (14.21)$$

with values in the Lie algebra of G.

The gauge covariant exterior derivative D is then defined as (remember 12.56)

$$D = d + h \qquad (14.22)$$

and by taking the exterior product of two exterior covariant derivatives we obtain the curvature two-form R (or field-strength, see 12.25 and 12.88). Applying for simplicity the derivative operators to a scalar field φ , we get

$$R\varphi = (D \wedge D)\varphi = d^2\varphi + d(h\varphi) + h \wedge d\varphi + \\ + h \wedge h \,\varphi \qquad (14.23)$$

But $d^2\varphi = 0$, and using (14.19) we have

$$d(h\varphi) = (dh)\varphi - h \wedge d\varphi \qquad (14.24)$$

It follows that

$$R = dh + h \wedge h \qquad (14.25)$$

and putting $R = R^A X_A$, $h = h^A X_A$, we obtain

$$R^A X_A = (dh^A) X_A + \frac{1}{2} h^B \wedge h^C [X_B, X_C] \qquad (14.26)$$

Using the commutation rules (14.20), the components R^A of the curvature two-form can be expressed then in terms of the components of the potential one-form, h^A , as

$$R^A = dh^A + \frac{1}{2} f_{BC}{}^A h^B \wedge h^C \qquad (14.27)$$

Writing explicitly the holonomic indices, we have $R^A = R_{\mu\nu}{}^A dx^\mu \wedge dx^\nu$, where

$$R_{\mu\nu}{}^A = \partial_{[\mu} h_{\nu]}{}^A + \frac{1}{2} f_{BC}{}^A h_\mu{}^B h_\nu{}^C \qquad (14.28)$$

(compare with (12.26)).

The equations $R^A = 0$, with R^A given in (14.27), are called Maurer-Cartan equations, and it can be shown that the set of conditions $R^A = 0$, $dR^A = 0$, are equivalent to the Jacobi identities of the Lie algebra of the group G.

Finally, a gauge invariant action for the gauge field can be constructed using the components of the curvature two-form, R^A , and of the potential one-form, h^A .

3.- Torsion, curvature, and the algebra of the Poincaré group

The simplest example of gauge theory of gravity is obtained considering the Poincaré group. In this case the generators are

$$X_A = \{ P_a , M_{ab} \} \qquad (14.29)$$

where P_a are the four generators of translations, and $M_{ab} = - M_{ba}$ the six generators of Lorentz rotations (the indices A,B run in this case over the set $\{a, (ab)\}$. The corresponding gauge fields are the fundamental frames V^a and The connection one-form $\omega^{ab} = - \omega^{ba}$,

$$h^A = \{ V^a, \tfrac{1}{2}\omega^{ab} \} \qquad (14.30)$$

where

$$V^a = V_\mu{}^a dx^\mu \quad , \qquad \omega^{ab} = \omega_\mu{}^{ab} dx^\mu$$

$V_\mu{}^a$ is the vierbein field (previously denoted with $e_\mu{}^a$) and $\omega_\mu{}^{ab}$ the usual Lorentz anholonomic connection (see ChapterX, Sect.5 and Chapter XII, Sect.3). The factor 1/2 in the definition of the Lorentz gauge potential is required to be consistent with our previous definition of Lorentz covariant derivative, eq.(12.80)).

By using the general expression (14.27), which defines the curvature in terms of the gauge one-forms h and of the structure constants f of the Lie algebra, and putting

$$R \equiv R^A X_A = R^a P_a + \tfrac{1}{2} R^{ab} M_{ab} \qquad (14.31)$$

we can easily calculate the field strength corresponding to translations, R^a, and that relative to the Lorentz rotations, R^{ab} .

The commutation rules of the generators of the Poincaré group are

$$[P_a, P_b] = 0 \qquad (14.32)$$

$$[M_{ab}, P_c] = -\eta_{ac} P_b + \eta_{bc} P_a \qquad (14.33)$$

$$[M_{ab}, M_{cd}] = \eta_{ad} M_{bc} + \eta_{bc} M_{ad} - \eta_{ac} M_{bd} - \eta_{bd} M_{ac} \qquad (14.34)$$

The only contributions to R^a come from the commutators (14.33), which can be rewritten as

$$[M_{ab}, P_c] = f_{ab,c}{}^d P_d \qquad (14.35)$$

where

$$f_{ab\,c}{}^d = -\eta_{ac}\delta^d_b + \eta_{bc}\delta^d_a \qquad (14.36)$$

From (14.27) we have then, for the Poincaré group,

$$R^a = dh^a + f_{bc\,d}{}^a\, h^{bc} \wedge h^d \qquad (14.37)$$

and since (see 14.30) in this case $h^a = V^a$, $h^{ab} = \frac{1}{2}\omega^{ab}$, we obtain

$$R^a = dV^a + \frac{1}{2}\left(\eta_{cd}\,\delta^a_b - \eta_{bd}\,\delta^a_c\right)\omega^{bc}\wedge V^d =$$

$$= dV^a + \omega^a{}_b \wedge V^b \qquad (14.38)$$

In the same way, the structure constants obtained from the commutators (14.34) contribute to the curvature R^{ab}, and following the same line of computation we obtain, from (14.34),

$$f_{ab\,cd}{}^{ik} = \eta_{ad}\,\delta^i_b\,\delta^k_c + \eta_{bc}\,\delta^i_a\,\delta^k_d -$$

$$- \eta_{ac}\,\delta^i_b\,\delta^k_d - \eta_{bd}\,\delta^i_a\,\delta^k_c \qquad (14.39)$$

Therefore, remembering (14.31),

$$\frac{1}{2}R^{ik} = \frac{1}{2}d\omega^{ik} + \frac{1}{8}f_{abcd}{}^{ik}\left(\omega^{ab}\wedge\omega^{cd}\right) =$$

$$= \frac{1}{2}d\omega^{ik} + \frac{1}{8}\left(\omega^{ab}\wedge\omega^{cd}\right)\left(2\eta_{a[d|}\,\delta^i_b\,\delta^k_{|c]} + 2\eta_{b[c|}\,\delta^i_a\,\delta^k_{|d]}\right)$$

which gives finally

$$R^{ik} = d\omega^{ik} + \omega^i{}_c \wedge \omega^{ck} \qquad (14.40)$$

The two equations (14.38,40) are called "structure e- quations", as they describe the geometrical structure of the manifold. In fact, introducing the holonomic components of the curvature two-form, and defining

$$R^a = R_{\mu\nu}{}^a\, dx^\mu \wedge dx^\nu \qquad (14.41)$$

$$R^{ab} = \frac{1}{2}R_{\mu\nu}{}^{ab}\, dx^\mu \wedge dx^\nu \qquad (14.42)$$

(the factor 1/2 is needed here to agree with the previous conventions used to define the curvature tensor in Chapter XII), we have, from the structure equations,

$$R_{\mu\nu}{}^{a} = \partial_{[\mu} V_{\nu]}{}^{a} + \frac{1}{2}\omega_{\mu}{}^{a}{}_{b} V_{\nu}{}^{b} - \frac{1}{2}\omega_{\nu}{}^{a}{}_{b} V_{\mu}{}^{b} \qquad (14.43)$$

$$\frac{1}{2} R_{\mu\nu}{}^{ab} = \partial_{[\mu}\omega_{\nu]}{}^{ab} + \frac{1}{2}\omega_{\mu}{}^{a}{}_{c}\omega_{\nu}{}^{cb} - \frac{1}{2}\omega_{\nu}{}^{a}{}_{c}\omega_{\mu}{}^{cb} \qquad (14.44)$$

Imposing on the vierbein the usual constraint $\nabla_{\mu} V_{\nu}{}^{a} = 0$, i.e.

$$\partial_{\mu} V_{\nu}{}^{a} + \omega_{\mu}{}^{a}{}_{b} V_{\nu}{}^{b} - \Gamma_{\mu\nu}{}^{\alpha} V_{\alpha}{}^{a} = 0 \qquad (14.45)$$

(remember 10.134), by comparing the antisymmetric part of this condition with (14.43) we have immediately

$$R_{\mu\nu}{}^{a} = \Gamma_{[\mu\nu]}{}^{\alpha} = Q_{\mu\nu}{}^{a} \qquad (14.46)$$

that is the components of the curvature two-form R^{a} are to be identified with the torsion tensor.

Moreover, comparing the expression (14.44) with eq. (12.105), we are led to identify $R_{\mu\nu}{}^{ab}$ with the Riemann-Cartan curvature tensor.

Therefore the first structure equation (14.38) defines the connection in terms of the torsion and of the derivatives of the vierbein field (solving this equation one obtains in fact the explicit expression for the coefficients of the Riemann-Cartan anholonomic connection, as shown in Chapter X, eqs.(10.136-143)). The second structure equation (14.40) defines the curvature tensor in terms of the anholonomic connection $\omega_{\mu}{}^{ab}$.

It is interesting to note that, in this framework, torsion is to be interpreted as the field strength relative to the translations, R^{a}, and curvature as the field strength of the Lorentz rotations, R^{ab}.

Moreover it should be expressed that the explicit expression for the torsion and the curvature, given by the two structure equations (14.38,40) in terms of the gauge potentials V^{a}, ω^{ab}, has been obtained as a direct consequence of the commutation relations of the Lie algebra of the Poincaré group. The use of this formalism shows explicitly therefore the close relationship existing between the geometric structure of a manifold, and the algebraic structure of the underlying symmetry group (in this case Poincare invariance).

The theory of general relativity is obtained imposing on the connection to be torsion-free, i.e. $R^{a} = 0$. With this constraint, the connection is no longer an independent field, as it can be entirely expressed in terms of the derivative of the vierbeins (see 10.142), and the simplest action for a gravitational theory, which satisfies the requirements of being a scalar under local Lorentz and general coordinate transformations, can be written in terms of the frames, V^a, and of the curvature R^{ab} as follows:

$$S = \frac{1}{4\chi} \int R^{ab} \wedge V^c \wedge V^d \, \varepsilon_{abcd} \tag{14.47}$$

This integral is nothing but the usual Einstein action, written in the language of the exterior forms. In fact, writing explicitly the holonomic indices, we have

$$S = \frac{1}{4\chi} \int \frac{1}{2} R_{\mu\nu}{}^{ab} V^c_\alpha V^d_\beta \, dx^\mu \wedge dx^\nu \wedge dx^\alpha \wedge dx^\beta \varepsilon_{abcd}$$

$$= \frac{1}{8\chi} \int R_{\mu\nu}{}^{ab} \varepsilon_{ab\alpha\beta} \, dx^\mu \wedge dx^\nu \wedge dx^\alpha \wedge dx^\beta \tag{14.48}$$

where

$$dx^\mu \wedge dx^\nu \wedge dx^\alpha \wedge dx^\beta = \varepsilon^{\mu\nu\alpha\beta} d^4x \tag{14.49}$$

As regards the product $\varepsilon_{ab\alpha\beta}\varepsilon^{\mu\nu\alpha\beta}$, we note that it is a Lorentz tensor (in the indices a,b), but is not a contravariant tensor under general coordinate trandformations (with respect to the indices μ, ν), because $\varepsilon^{\mu\nu\alpha\beta}$ is a density of weight w = -1 (see the Appendix A of Chapter III). The true tensor is $\eta^{\mu\nu\alpha\beta}\varepsilon_{ab\alpha\beta}$, and in that case, from eq.(A.8) of Chapter III, we have

$$\eta^{\mu\nu\alpha\beta} \varepsilon_{ab\alpha\beta} = -2 V_a{}^\rho V_b{}^\sigma \delta^{\mu\nu}_{\rho\sigma}$$

Therefore

$$\varepsilon_{ab\alpha\beta} \, \varepsilon^{\mu\nu\alpha\beta} = -2 V V_{ab}^{\mu\nu} \tag{14.50}$$

where V det (V_μ^a) = $(-\det g_{\mu\nu})^{1/2}$, and

$$V_{ab}^{\mu\nu} = \begin{vmatrix} V^\mu_a & V^\nu_a \\ V^\mu_b & V^\nu_b \end{vmatrix} = V^\mu_a V^\nu_b - V^\nu_a V^\mu_b \tag{14.51}$$

The action becomes then

$$S = -\frac{1}{8\chi}\int R_{\mu\nu}{}^{ab}\, 4V\, V_a{}^{[\mu} V_b^{\nu]}\, d^4x =$$

$$= -\frac{1}{2\chi}\int d^4x \sqrt{-g}\; R \qquad (14.52)$$

where $R(e,\omega) = V_a^\mu V_b^\nu R_{\mu\nu}{}^{ab} = -R_{\mu\nu}{}^{\mu\nu} = R_{\mu\nu}{}^{\nu\mu} = R(g,\Gamma)$ is the usual scalar curvature (remember eqs.(12.108,109)).

The Einstein field equations can be easily obtained, in the first order formalism, performing the variation of the action (14.47), with respect to the frames V :

$$\delta S = \frac{1}{4\chi}\int (R^{ab}\wedge V^c \wedge \delta V^d + R^{ab}\wedge \delta V^c \wedge V^d)\,\varepsilon_{abcd} =$$

$$= \frac{1}{4\chi}\int (R^{ab}\wedge V^c \wedge \delta V^d - R^{ab}\wedge V^d \wedge \delta V^c)\,\varepsilon_{abcd} =$$

$$= \frac{1}{4\chi}\int (R^{ab}\wedge V^c \wedge \delta V^d \varepsilon_{abcd} + R^{ab}\wedge V^d \wedge \delta V^d \varepsilon_{abdc}) =$$

$$= \frac{1}{4\chi}\int 2(R^{ab}\wedge V^c\, \varepsilon_{abcd})\wedge \delta V^d \qquad (14.53)$$

Imposing $\delta S = 0$ we obtain the field equations

$$\frac{1}{2} R^{ab}\wedge V^c\, \varepsilon_{abcd} = 0 \qquad (14.54)$$

which are equivalent to the Einstein equations in vacuum.

In fact, introducing explicitly the world indices, (14 54) becomes

$$\frac{1}{4} R_{[\mu\nu}{}^{ab} V_{\alpha]}{}^c\, \varepsilon_{abcd} = 0 \qquad (14.55)$$

or

$$R_{\mu\nu}{}^{ab} V_\alpha{}^c\, \varepsilon^{\mu\nu\alpha\beta}\, \varepsilon_{abcd} = 0 \qquad (14.56)$$

that is

$$R_{\mu\nu}{}^{ab}\, \varepsilon^{\alpha\mu\nu\beta}\, \varepsilon_{\alpha abc} = 0 \qquad (14.57)$$

Using the properties of the totally antisymmetric symbols we have, from eq.(A.8) of Chapter III ,

$$\varepsilon^{\alpha\mu\nu\beta}\,\varepsilon_{\alpha abc} = -V\,V^{\mu\nu\beta}_{abc} \qquad (14.58)$$

where

$$V^{\mu\nu\beta}_{abc} = \begin{vmatrix} V_a^{\mu} & V_a^{\nu} & V_a^{\beta} \\ V_b^{\mu} & V_b^{\nu} & V_b^{\beta} \\ V_c^{\mu} & V_c^{\nu} & V_c^{\beta} \end{vmatrix} =$$

$$= V_a^{\mu} V_b^{\nu} V_c^{\beta} - V_a^{\mu} V_b^{\beta} V_c^{\nu} - V_a^{\nu} V_b^{\mu} V_c^{\beta} + + V_a^{\nu} V_b^{\beta} V_c^{\mu} + V_a^{\beta} V_b^{\mu} V_c^{\nu} - V_a^{\beta} V_b^{\nu} V_c^{\mu} \qquad (14.59)$$

Remembering (12.108), we have

$$-V R_{\mu\nu}{}^{ab} V^{\mu\nu\beta}_{abc} = V R_{\mu\nu}{}^{\gamma\delta} V_\gamma{}^a V_\delta{}^b V^{\mu\nu\beta}_{abc} =$$

$$= V\left(R_{\mu\nu}{}^{\mu\nu} V_c{}^{\beta} - R_{\mu\nu}{}^{\mu\beta} V_c^{\nu} - R_{\mu\nu}{}^{\nu\mu} V_c{}^{\beta} + + R_{\mu\nu}{}^{\nu\beta} V_c{}^{\mu} - R_{\mu\nu}{}^{\beta\mu} V_c{}^{\nu} - R_{\mu\nu}{}^{\beta\nu} V_c{}^{\mu} \right) \qquad (14.60)$$

Putting

$$R = R_{\mu\nu}{}^{\nu\mu} \qquad (14.61)$$

and

$$R_\nu{}^\beta = R_{\mu\nu}{}^{\beta\mu} = R_{\nu\mu}{}^{\mu\beta} \qquad (14.62)$$

and remebering that $R_{\mu\nu}{}^{\alpha\beta} = -R_{\nu\mu}{}^{\alpha\beta} = -R_{\mu\nu}{}^{\beta\alpha}$, we obtain

$$-R_{\mu\nu}{}^{ab} V^{\mu\nu\beta}_{abc} = -R V_c{}^{\beta} + R_c{}^{\beta} - R V_c{}^{\beta} + R_c{}^{\beta} + R_c{}^{\beta} + R_c{}^{\beta} = -2 R V_c{}^{\beta} + 4 R_c{}^{\beta} = 0 \qquad (14.63)$$

Finally, multiplying by $\frac{1}{4} V_\alpha{}^c$,

$$R_\alpha{}^\beta - \frac{1}{2}\,\delta_\alpha{}^\beta R = 0 \qquad (14.64)$$

which are the well known Einstein field equations in vacuum.

By varying the action (14.47) with respect to the connection ω, in the first order formalism, we have

$$\delta S = \frac{1}{4\chi} \int \delta R^{ab} \wedge V^c \wedge V^d \, \varepsilon_{abcd} \tag{14.65}$$

and

$$\begin{aligned} \delta R^{ab} &= \delta \left(d\omega^{ab} + \omega^a{}_c \wedge \omega^{cb} \right) = \\ &= d\,(\delta\omega^{ab}) + \omega^a{}_c \wedge \delta\omega^{cb} + \delta\omega^a{}_c \wedge \omega^{cb} = \\ &= d\,(\delta\omega^{ab}) + \omega^a{}_c \wedge \delta\omega^{cb} + \omega^b{}_c \wedge \delta\omega^{ac} = \\ &= D\,(\delta\omega^{ab}) \end{aligned} \tag{14.66}$$

where D denotes the Lorentz exterior covariant derivative performed with the anholonomic connection ω^{ab}, that is, for a given field ψ,

$$D\psi = \left(d + \frac{1}{2}\omega^{ab} M_{ab} \right) \psi \tag{14.67}$$

where M_{ab} are the generators appropriate to the transformation properties of ψ under a local Lorentz transformation (for a spinor, for example, $M_{ab} = S_{ab} = \frac{1}{2}\gamma_{[a}\gamma_{b]}$, see (10.149)).

Integrating by parts, and imposing $\delta S = 0$, from (14.65,66) we obtain the field equation

$$\frac{1}{2} DV^c \wedge V^d \, \varepsilon_{abcd} = 0 \tag{14.68}$$

According to the first structure equation (14.38), the Lorentz covariant derivative of the frames defines the torsion two-form: in fact

$$DV^a = dV^a + \omega^a{}_b \wedge V^b \equiv R^a \tag{14.69}$$

The field equation (14.68) can be rewritten then also as

$$\begin{aligned} 0 = R^a \wedge V^b \varepsilon_{abcd} &= R_{\mu\nu}{}^a V_\alpha{}^b \varepsilon^{\mu\nu\alpha\beta} \varepsilon_{abcd} = \\ &= -R_{\mu\nu}{}^a \varepsilon^{\alpha\mu\nu\beta} \varepsilon_{\alpha acd} = \end{aligned}$$

$$= V R_{\mu\nu}{}^{a} V^{\mu\nu\beta}_{a\,c\,d} \qquad (14.70)$$

and using (14.59) it reduces to

$$R_{\mu\nu}{}^{\mu} V_c{}^{\nu} V_d{}^{\beta} - R_{\mu\nu}{}^{\mu} V_c{}^{\beta} V_d{}^{\nu} - R_{\mu\nu}{}^{\nu} V_c{}^{\mu} V_d{}^{\beta} +$$

$$+ R_{\mu\nu}{}^{\nu} V_c{}^{\beta} V_d{}^{\mu} + R_{\mu\nu}{}^{\beta} V_c{}^{\mu} V_d{}^{\nu} - R_{\mu\nu}{}^{\beta} V_c{}^{\nu} V_d{}^{\mu} \qquad (14.71)$$

Putting

$$Q_{\mu\nu}{}^{\beta} = R_{\mu\nu}{}^{\beta} = - R_{\nu\mu}{}^{\beta} \qquad (14.72)$$

and

$$Q_{\nu} = Q_{\nu\mu}{}^{\mu} \qquad (14.73)$$

we obtain

$$R_{\mu\nu}{}^{a} V^{\mu\nu\beta}_{a\,c\,d} = 2\left(Q_{cd}{}^{\beta} - Q_c V_d^{\beta} + Q_d V_c^{\beta}\right) = 0 \qquad (14.74)$$

Finally, multiplying by $V_{\mu}^{c} V_{\nu}^{d}$, we get the Einstein-Cartan field equation (10.52) for the torsion tensor in vacuum

$$Q_{\mu\nu}{}^{\beta} + \delta_{\mu}^{\beta} Q_{\nu} - \delta_{\nu}^{\beta} Q_{\mu} = 0 \qquad (14.75)$$

Contracting with δ_{β}^{ν} we have $Q_{\mu} = 0$, and then (14.75) implies, as expected, the vanishing of the torsion tensor, $Q_{\mu\nu}{}^{\alpha} = 0$, in agreement with the constraint $R^{a} = 0$ (which, in general relativity, is assumed from the beginning, and holds also in the presence of matter).

It is worth stressing that, using this formalism, the Bianchi identities can be easily obtained, by applying the Lorentz exterior covariant derivative to the structure equations.

The derivative of the first structure equation (14.38) gives (remember $d^2 = 0$ and the rules (14.19,8))

$$D R^{a} = d R^{a} + \omega^{a}{}_{c} \wedge R^{c} =$$

$$= d(\omega^{a}{}_{b} \wedge V^{b}) + \omega^{a}{}_{c} \wedge dV^{c} + \omega^{a}{}_{c} \wedge \omega^{c}{}_{b} \wedge V^{b} =$$

$$= d\omega^{a}{}_{b} \wedge V^{b} - \omega^{a}{}_{b} \wedge dV^{b} + \omega^{a}{}_{c} \wedge dV^{c} +$$

$$+ \omega^{a}{}_{c} \wedge \omega^{c}{}_{b} \wedge V^{b} =$$

$$= (d\omega^a{}_b + \omega^a{}_c \wedge \omega^c{}_b) \wedge V^b \qquad (14.76)$$

that is, using (14.60),

$$DR^a = R^a{}_b \wedge V^b \qquad (14.77)$$

The derivative of the second structure equation (14.40) gives

$$DR^{ab} = dR^{ab} + \omega^a{}_i \wedge R^{ib} + \omega^b{}_i \wedge R^{ai} =$$

$$= d(\omega^a{}_c \wedge \omega^{cb}) + \omega^a{}_i \wedge (d\omega^{ib} + \omega^i{}_c \wedge \omega^{cb})$$

$$+ \omega^b{}_i \wedge (d\omega^{ai} + \omega^a{}_c \wedge \omega^{ci}) =$$

$$= d\omega^a{}_c \wedge \omega^{cb} - \omega^a{}_c \wedge d\omega^{cb} + \omega^a{}_i \wedge d\omega^{ib}$$

$$+ \omega^a{}_i \wedge \omega^i{}_c \wedge \omega^{cb} + \omega^b{}_i \wedge d\omega^{ai} +$$

$$+ \omega^b{}_i \wedge \omega^a{}_c \wedge \omega^{ci} \equiv 0 \qquad (14.78)$$

In general relativity, $R^a = 0$, and the first identity (14.77) becomes

$$R^a{}_b \wedge V^b = 0 \qquad (14.79)$$

or, writing explicitly indices,

$$\frac{1}{2} (R_{\mu\nu}{}^a{}_b V^b{}_\alpha) dx^\mu \wedge dx^\nu \wedge dx^\alpha = 0 \qquad (14.80)$$

from which we obtain the well known property of the Riemann curvature tensor (see 3.19)

$$R_{[\mu\nu\alpha]}{}^a = 0 \qquad (14.81)$$

In the case of nonzero torsion, the eq.(14.77) becomes explicitly

$$D_{[\mu} Q_{\nu\alpha]}{}^a = \frac{1}{2} R_{[\mu\nu\alpha]}{}^\beta V_\beta{}^a \qquad (14.82)$$

In fact, according to (12.108) and (14.42), the relation

between the Riemann-Cartan curvature tensor $R_{\mu\nu\alpha}{}^{\beta}$ and the curvature two-form R^{ab} is

$$R^{ab} = \frac{1}{2} R_{\mu\nu}{}^{ab} dx^{\mu} \wedge dx^{\nu} = -\frac{1}{2} R_{\mu\nu}{}^{\alpha\beta} V_{\alpha}{}^{a} V_{\beta}{}^{b} dx^{\mu} \wedge dx^{\nu} \tag{14.83}$$

If we remember that the Lorentz covariant derivative operates only on the flat, tangent space indices (as stressed in Chapter X, Sect.5), then the total covariant derivative of the components $Q_{\nu\alpha}{}^{a}$ of the torsion two-form, which carries both holonomic and anholonomic indices, can be written as

$$\nabla_{\mu} Q_{\nu\alpha}{}^{a} = D_{\mu} Q_{\nu\alpha}{}^{a} - \Gamma_{\mu\nu}{}^{\beta} Q_{\beta\alpha}{}^{a} - \Gamma_{\mu\alpha}{}^{\beta} Q_{\nu\beta}{}^{a} \tag{14.84}$$

so that

$$D_{\mu} Q_{\nu\alpha}{}^{a} = \nabla_{\mu} Q_{\nu\alpha}{}^{a} - \Gamma_{\mu\nu}{}^{\beta} Q_{\alpha\beta}{}^{a} + \Gamma_{\mu\alpha}{}^{\beta} Q_{\nu\beta}{}^{a} \tag{14.85}$$

Eq.(14.84) gives then (as $\Gamma_{[\mu\nu]}{}^{\beta} = Q_{\mu\nu}{}^{\beta}$)

$$\frac{1}{2} R_{[\mu\nu\alpha]}{}^{\rho} V_{\rho}{}^{a} = \nabla_{[\mu} Q_{\nu\alpha]}{}^{a} - 2 Q_{[\mu\nu}{}^{\rho} Q_{\alpha]\rho}{}^{a} \tag{14.86}$$

or, multiplying by V^{β}_{a}, and remembering that $\nabla_{\mu} V^{\beta}_{a} = 0$,

$$R_{[\mu\nu\alpha]}{}^{\beta} = 2\left(\nabla_{[\mu} Q_{\nu\alpha]}{}^{\beta} - 2 Q_{[\mu\nu}{}^{\rho} Q_{\alpha]\rho}{}^{\beta}\right) \tag{14.87}$$

in full agreement with eq.(10.29) in the case of the Riemann-Cartan curvature tensor.

The second Bianchi identity (14.78) can be written explicitly

$$\frac{1}{2} D_{[\mu} R_{\nu\alpha]}{}^{ab} = 0 \tag{14.88}$$

and introducing the total covariant derivative we have

$$\nabla_{\mu} R_{\nu\alpha}{}^{ab} = D_{\mu} R_{\nu\alpha}{}^{ab} - \Gamma_{\mu\nu}{}^{\beta} R_{\beta\alpha}{}^{ab} - \Gamma_{\mu\alpha}{}^{\beta} R_{\nu\beta}{}^{ab} \tag{14.89}$$

Therefore

$$\nabla_{[\mu} R_{\nu\alpha]}{}^{ab} = \Gamma_{[\mu\nu}{}^{\beta} R_{\alpha]\beta}{}^{ab} - \Gamma_{[\mu\alpha}{}^{\beta} R_{\nu]\beta}{}^{ab} = $$

$$= 2\, Q_{[\mu\nu}{}^{\beta} R_{\alpha]\beta}{}^{ab} \tag{14.90}$$

We have obtained then eq.(10.30), which, in the absence of torsion, reduces to the usual Bianchi identity of the riemannian geometry,

$$R^{ab}{}_{[\mu\nu;\alpha]} = 0 \qquad (14.91)$$

To complete this short review of the Einstein-Cartan theory in the language of the exterior forms, consider a phenomenological action S_m describing macroscopic matter sources, such that the independent variation of V^a and ω^{ab} leads to

$$\delta S_m = \int (-\Theta_a \wedge \delta V^a - \Sigma_{ab} \wedge \delta\omega^{ab}) \qquad (14.92)$$

where Θ_a is a vector-valued three-form related to the canonical energy-momentum tensor $\Theta_a{}^b$ by

$$\Theta_a = \frac{1}{6}\, \Theta_a{}^\mu \, \varepsilon_{\mu\nu\alpha\beta}\, dx^\mu \wedge dx^\alpha \wedge dx^\beta \qquad (14.93)$$

and $\Sigma_{ab} = -\Sigma_{ba}$ is the three-form corresponding to the canonical spin density tensor S_{abc}, i.e.

$$\Sigma_{ab} = S_{ab}{}^\mu \, \varepsilon_{\mu\nu\alpha\beta}\, dx^\nu \wedge dx^\alpha \wedge dx^\beta \qquad (14.94)$$

The variation of the total action, given by the sum of the Einstein action (14.47) and of the matter action S_m, with respect to V^a, gives

$$\frac{1}{2} R^{ab} \wedge V^c \varepsilon_{abcd} = \chi\, \Theta_d \qquad (14.95)$$

In the usual tensor language this equation can be rewritten

$$\frac{1}{4} R_{\mu\nu}{}^{ab} V_\alpha{}^c \varepsilon_{abcd}\, \varepsilon^{\mu\nu\alpha\beta} = \frac{1}{6} \chi\, \Theta_d{}^a \varepsilon_{a\mu\nu\alpha}\, \varepsilon^{\mu\nu\beta} \qquad (14.96)$$

or

$$-\frac{V}{4} R_{\mu\nu}{}^{ab} V^{\mu\nu\beta}_{abc} = \frac{3!}{6} \chi\, \Theta_c{}^a V V^\beta_a \qquad (14.97)$$

Therefore, using the result of eq.(14.63), we have

$$R_c{}^\beta - \frac{1}{2} V_c{}^\beta R = \chi\, \Theta_c{}^\beta \qquad (14.98)$$

and multiplying by $V_\alpha{}^c$

$$G_\alpha{}^\beta = \chi\, \Theta_\alpha{}^\beta \qquad (14.99)$$

which coincides with the gravitational field equation (10.54) with matter sources.

By varying the total action with respect to the connection ω^{ab} we find

$$\frac{1}{2} R^{a} \wedge V^{b} \varepsilon_{abcd} = \chi \Sigma_{cd} \tag{14.100}$$

which is equivalent to

$$\frac{1}{2} R_{\mu\nu}{}^{a} V_{\alpha}{}^{b} \varepsilon_{abcd} \varepsilon^{\mu\nu\alpha\beta} = \chi S_{cd}{}^{a} \varepsilon_{a\mu\nu\alpha} \varepsilon^{\mu\nu\alpha\beta} \tag{14.101}$$

that is

$$\frac{V}{2} R_{\mu\nu}{}^{a} V^{\mu\nu\beta}_{acd} = \chi V S_{cd}{}^{a} V_{a}{}^{\beta} \tag{14.102}$$

From (14.74) we have then

$$Q_{cd}{}^{\beta} + V^{\beta}{}_{c} Q_{d} - V^{\beta}_{d} Q_{c} = \chi S_{cd}{}^{\beta} \tag{14.103}$$

or, multiplying by $V_{\mu}{}^{c} V_{\nu}{}^{d}$,

$$T_{\mu\nu}{}^{\beta} = \chi S_{\mu\nu}{}^{\beta} \tag{14.104}$$

which coincides with eq.(10.55).

Therefore the equations (14.95,100) are nothing but the Einstein-Cartan field equations, written in the language of the exterior forms.

Finally, we show that, in this formalism, the contracted Bianchi identity of the U_4 theory (already presented in Chapter X, Sect.2) can be obtained by applying the Lorentz exterior derivative to the left-hand side of the two field equations (14.95,1000).

In fact, using the Bianchi identities (14.77,78) and the structure equations (14.38,40) we have

$$\frac{1}{2} D(R^{ab} \wedge V^{c}) \varepsilon_{abcd} = \frac{1}{2} R^{ab} \wedge R^{c} \varepsilon_{abcd} \tag{14.105}$$

$$\frac{1}{2} D(R^{c} \wedge V^{d}) \varepsilon_{abcd} = \frac{1}{2} R^{c}{}_{k} \wedge V^{k} \wedge V^{d} \varepsilon_{abcd} \tag{14.106}$$

The first of these two equations corresponds to the contracted identity (10.36) of the U_4 theory, the second expresses the antisymmetric part of the Ricci tensor, according to (10.35). Consider for example (14.105). Writing explicitly the holonomic components, we obtain

$$\frac{1}{4} D_\mu (R_{\nu\alpha}{}^{ab} V_\beta{}^c) \varepsilon_{abcd} \varepsilon^{\mu\nu\alpha\beta} = \frac{1}{4} R_{\mu\nu}{}^{ab} R_{\alpha\beta}{}^c \varepsilon_{abcd} \varepsilon^{\mu\nu\alpha\beta} \tag{14.107}$$

and since

$$\nabla_\mu (R_{\nu\alpha}{}^{ab} V_\beta{}^c) = D_\mu (R_{\nu\alpha}{}^{ab} V_\beta{}^c) - \Gamma_{\mu\nu}{}^\rho R_{\rho\alpha}{}^{ab} V_\beta{}^c -$$

$$- \Gamma_{\mu\alpha}{}^\rho R_{\nu\rho}{}^{ab} V_\beta{}^c - \Gamma_{\mu\beta}{}^\rho R_{\nu\alpha}{}^{ab} V_\rho{}^c \tag{14.108}$$

the left-hand side of (14.107) becomes

$$\text{L.H.} = \frac{1}{4} \Big[(\nabla_\mu R_{\nu\alpha}{}^{ab} + \Gamma_{\mu\nu}{}^\rho R_{\rho\alpha}{}^{ab} + \Gamma_{\mu\alpha}{}^\rho R_{\nu\rho}{}^{ab}) V V^{\mu\nu\alpha}_{abd} -$$

$$- V \Gamma_{\mu\beta}{}^c R_{\nu\alpha}{}^{ab} V^{\mu\nu\alpha\beta}_{abcd} \Big] =$$

$$= \frac{V}{4} (\nabla_\mu R_{\nu\alpha}{}^{ab} - 2 Q_{\mu\nu}{}^\rho R_{\alpha\rho}{}^{ab}) V^{\mu\nu\alpha}_{abd} - \frac{V}{4} Q_{\mu\beta}{}^c R_{\nu\alpha}{}^{ab} V^{\mu\nu\alpha\beta}_{abcd} \tag{14.109}$$

where $V^{\mu\nu\alpha}_{abd}$ is given in eq.(14.59), and

$$V^{\mu\nu\alpha\beta}_{abcd} = V^{-1} \varepsilon_{abcd} \varepsilon^{\mu\nu\alpha\beta} =$$

$$= \begin{vmatrix} V_a^\mu & V_a^\nu & V_a^\alpha & V_a^\beta \\ V_b^\mu & V_b^\nu & V_b^\alpha & V_b^\beta \\ V_c^\mu & V_c^\nu & V_c^\alpha & V_c^\beta \\ V_d^\mu & V_d^\nu & V_d^\alpha & V_d^\beta \end{vmatrix} \tag{14.110}$$

Performing explicitly the computations, symplifying and contracting with $V_\rho{}^d$ one obtains

$$\text{L.H.} = - V (\nabla_\mu G_\rho{}^\mu + 2 Q_\beta G_\rho{}^\beta) \tag{14.111}$$

where $G_\rho{}^\mu$ is the Einstein tensor. The right-hand side of (14.107) gives

$$\text{R.H.} = \frac{1}{4} R_{\mu\nu}{}^{ab} R_{\alpha\beta}{}^c \varepsilon_{abcd} \varepsilon^{\mu\nu\alpha\beta} = -\frac{V}{4} R_{\mu\nu}{}^{ab} Q_{\alpha\beta}{}^c V^{\mu\nu\alpha\beta}_{abcd}$$

$$= \frac{V}{4} \Big\{ R_{\mu\nu}{}^{\mu\nu} Q_{\alpha\beta}{}^\alpha V_d{}^\beta + R_{\mu\nu}{}^{\mu\alpha} Q_{\alpha\beta}{}^\beta V_d{}^\nu + R_{\mu\nu}{}^{\mu\beta} Q_{\alpha\beta}{}^\nu V_d{}^\beta -$$

$$-R_{\mu\nu}{}^{\mu\nu} Q_{\alpha\beta}{}^{\beta} V_d{}^{\alpha} - R_{\mu\nu}{}^{\mu\alpha} Q_{\alpha\beta}{}^{\nu} V_d{}^{\beta} - R_{\mu\nu}{}^{\mu\beta} Q_{\alpha\beta}{}^{\alpha} V_d{}^{\nu} -$$

$$- R_{\mu\nu}{}^{\nu\mu} Q_{\alpha\beta}{}^{\alpha} V_d{}^{\beta} - R_{\mu\nu}{}^{\nu\alpha} Q_{\alpha\beta}{}^{\beta} V^{\mu}{}_d - R_{\mu\nu}{}^{\nu\beta} Q_{\alpha\beta}{}^{\mu} V_d{}^{\alpha} +$$

$$+ R_{\mu\nu}{}^{\nu\mu} Q_{\alpha\beta}{}^{\beta} V_d{}^{\alpha} + R_{\mu\nu}{}^{\nu\alpha} Q_{\alpha\beta}{}^{\mu} V_d{}^{\beta} + R_{\mu\nu}{}^{\nu\beta} Q_{\alpha\beta}{}^{\alpha} V_d{}^{\mu} +$$

$$+ R_{\mu\nu}{}^{\alpha\mu} Q_{\alpha\beta}{}^{\nu} V_d{}^{\beta} + R_{\mu\nu}{}^{\alpha\nu} Q_{\alpha\beta}{}^{\beta} V_d{}^{\mu} + R_{\mu\nu}{}^{\alpha\beta} Q_{\alpha\beta}{}^{\mu} V_d{}^{\nu} -$$

$$- R_{\mu\nu}{}^{\alpha\mu} Q_{\alpha\beta}{}^{\beta} V_d{}^{\nu} - R_{\mu\nu}{}^{\alpha\nu} Q_{\alpha\beta}{}^{\mu} V_d{}^{\beta} - R_{\mu\nu}{}^{\alpha\beta} Q_{\alpha\beta}{}^{\nu} V_d{}^{\mu} -$$

$$- R_{\mu\nu}{}^{\beta\mu} Q_{\alpha\beta}{}^{\nu} V_d{}^{\beta} - R_{\mu\nu}{}^{\beta\nu} Q_{\alpha\beta}{}^{\alpha} V_d{}^{\mu} - R_{\mu\nu}{}^{\beta\alpha} Q_{\alpha\beta}{}^{\mu} V_d{}^{\nu} +$$

$$+ R_{\mu\nu}{}^{\beta\mu} Q_{\alpha\beta}{}^{\alpha} V_d{}^{\nu} + R_{\mu\nu}{}^{\beta\nu} Q_{\alpha\beta}{}^{\mu} V_d{}^{\alpha} + R_{\mu\nu}{}^{\beta\alpha} Q_{\alpha\beta}{}^{\nu} V_d{}^{\mu} \Big\} =$$

$$= -V \left\{ -R Q_d + 2 R_d{}^{\alpha} Q_{\alpha} + 2 R_{\nu}{}^{\beta} Q_{d\beta}{}^{\nu} - R_{\mu d}{}^{\alpha\beta} Q_{\alpha\beta}{}^{\mu} \right\} \tag{14.112}$$

This expression, introducing the modified torsion tensor $T_{\alpha\beta\nu}$

$$T_{\alpha\beta}{}^{\nu} = Q_{\alpha\beta}{}^{\nu} + \delta^{\nu}_{\alpha} Q_{\beta} - \delta^{\nu}_{\beta} Q_{\alpha} \tag{14.113}$$

and contracting with $V_{\rho}{}^{d}$, can be rewritten also as

$$R.H. = - V \left\{ -2 Q_{\alpha\rho}{}^{\nu} G_{\nu}{}^{\alpha} + R_{\rho\nu}{}^{\alpha\beta} T_{\alpha\beta}{}^{\nu} \right\} \tag{14.114}$$

Equating (14.111,114) we obtain then

$$\nabla_{\mu} G_{\rho}{}^{\mu} = -2 Q_{\alpha} G_{\mu}{}^{\alpha} - 2 Q_{\alpha\rho}{}^{\nu} G_{\nu}{}^{\alpha} + R_{\rho\nu}{}^{\alpha\beta} T_{\alpha\beta}{}^{\nu} \tag{14.115}$$

that is exactly the same result previously presented in eq.(10.36).

Using the contracted Bianchi identities in the form (14.105,106), and the field equations (14.95,100), it is easy to write the following conservation equations

$$D\Theta_d = \frac{1}{2\chi} R^{ab} \wedge R^c \varepsilon_{abcd} \tag{14.116}$$

for the energy-momentum, and

$$D\Sigma_{ab} = \frac{1}{2\chi} R^c{}_k \wedge V^k \wedge V^d \varepsilon_{abcd} \tag{14.117}$$

for the angular momentum.

This last equation provides informations about the antisymmetric part of the canonical energy-momentum tensor. In fact, the field equation (14.95) can be written in the equivalent form

$$R^{ab} \wedge V^{c} = -\frac{1}{3}\chi\, \varepsilon^{abcd}\, \theta_{d} \qquad (14.118)$$

We have then, from (14.117)

$$D\Sigma_{ab} = \frac{1}{3}\left(\theta_{b} \wedge V_{a} - \theta_{a} \wedge V_{b}\right) \qquad (14.119)$$

and this equation, written explicitly displaying the world indices, shows that the antisymmetric part of the canonical energy-momentum tensor is related to the divergence of the spin density tensor. If the spin is vanishing, the canonical stress tensor is symmetric, and coincides with the dynamical one, as already stressed in Chapter X.

REFERENCES

1) See for example J.Plebanski: "Forms and riemannian geometry", Proceedings of the International School of Cosmology and Gravitation, Erice (May 1972)

2) See also A.Trautman: Bul.Ac.Pol.Sci.Ser.Sci.Math.Astr.Phys.20,185,503 and 895 (1972);
and Ist.Naz.Alta Mat.Symp.Mat.12,139(1973)